# CHAOS AND COARSE GRAINING IN STATISTICAL MECHANICS

While statistical mechanics can describe the equilibrium state of systems with many degrees of freedom, and dynamical systems can explain the irregular evolution of systems with few degrees of freedom, new tools are needed to study the evolution of systems with many degrees of freedom. This book presents the basic aspects of chaotic systems, with emphasis on systems composed of a huge number of particles.

The first part of the book introduces the basic concepts of chaotic dynamics. The book then moves on to explore the role of ergodicity and chaos for the validity of statistical laws. The last part of the book is devoted to the treatment of problems characterized by the presence of more than one significant scale. In addition, the authors also discuss the relevance that many degrees of freedom, coarse-graining procedure, and instability mechanisms, have in justifying a statistical description of macroscopic bodies. The book introduces the tools to characterize the non-asymptotic behaviors of chaotic systems. This text will be of interest to researchers and graduate students in statistical mechanics and chaos.

PATRIZIA CASTIGLIONE is a Researcher at the Institut des Nanosciences de Paris. Her main research topics are dynamical systems theory, quantum and classic chaos, turbulence, and the physics of colors applied to the study of artwork.

MASSIMO FALCIONI is a Researcher at the University of Rome "La Sapienza." His research focuses on elementary particle physics, dynamical systems, and statistical mechanics.

ANNICK LESNE is a CNRS Researcher at the Laboratoire de Physique Théorique de la Matière Condensée (CNRS-Paris 6) and temporarily at the Institut des Hautes Études Scientifiques. Her research lies in renormalization methods for dynamical systems, non-equilibrium statistical physics, and applications of dynamical systems theory and statistical mechanics to biological systems.

ANGELO VULPIANI is a Professor of Theoretical Physics at the University of Rome "La Sapienza," and is a Fellow of the Institute of Physics. His research interests are statistical mechanics, dynamical systems, turbulence, transport, and reaction-diffusion in fluids.

# CHAOS AND COARSE GRAINING IN STATISTICAL MECHANICS

PATRIZIA CASTIGLIONE

*Institut des Nanosciences de Paris*

MASSIMO FALCIONI

*University of Rome "La Sapienza"*

ANNICK LESNE

*Laboratoire de Physique Théorique de la Matière Condensée, Paris*

ANGELO VULPIANI

*University of Rome "La Sapienza"*

CAMBRIDGE UNIVERSITY PRESS
Cambridge, New York, Melbourne, Madrid, Cape Town, Singapore, São Paulo, Delhi

Cambridge University Press
The Edinburgh Building, Cambridge CB2 8RU, UK

Published in the United States of America by Cambridge University Press, New York

www.cambridge.org
Information on this title: www.cambridge.org/9780521895934



First published 2008

Printed in the United Kingdom at the University Press, Cambridge

*A catalog record for this publication is available from the British Library*

*Library of Congress Cataloging in Publication data*
Chaos and coarse graining in statistical mathematics / Patrizia Castigilione . . . [et al.].
p. cm.
Includes bibliographical references and index.
ISBN 978-0-521-89593-4 (hardback)
1. Chaotic behavior in systems. 2. Dynamics.
3. Statistical mechanics. 4. Mathematical statistics.
I. Castiglione, Patrizia, 1968– II. Title.
QC174.8.C475 2008
003'.857–dc22

2008013430

ISBN 978-0-521-89593-4 hardback

# Contents

# Preface

Statistical Mechanics has been founded during the XIX-th century by the seminal work of Maxwell, Boltzmann and Gibbs, with the main aim to explain the properties of macroscopic systems from the atomistic point of view. Accordingly, from the very beginning, starting from the Boltzmann's ergodic hypothesis, a basic question was the connection between the dynamics and the statistical properties. This is a rather difficult task and, in spite of the mathematical progress, by Birkhoff and von Neumann, basically ergodic theory had a marginal relevance in the development of the statistical mechanics (at least in the physics community). Partially this was due to a misinterpretation of a result of Fermi[1] and a widely spreaded opinion (based also on the belief of influential scientists as Landau) on the key role of the many degrees of freedom and the practical irrelevance of ergodicity. This point of view found a mathematical support on some results by Khinchin who was able to show that, in systems with a huge number of particles, statistical mechanics works (independently of the ergodicity) just because, on the constant energy surface, the most meaningful physical observables are nearly constant, apart from regions of very small measure,

On the other hand the discovery of the deterministic chaos (from the anticipating work of Poincaré to the contributions, in the second half of the XX-th century, by Chirikov, Hénon, Lorenz and Ruelle, to cite just the most famous) beyond its undoubted relevance for many natural phenomena, showed how the typical statistical features observed in systems with many degrees of freedom, can be generated also by the presence of deterministic chaos in simple systems. For example low dimensional models can emulate spatially extended dynamics modelling transport and conduction processes.

[1] A theorem about non integrable Hamiltonian systems with $N$ degrees of freedom assures the non existence of smooth invariant surfaces of dimension $2N - 2$; from this result Fermi (erroneously) concluded that generic Hamiltonian systems are ergodic.

Surely the rediscovery of deterministic chaos has revitalized investigations on the foundation of Statistical Mechanics forcing the scientists to reconsider the connection between statistical properties and dynamics. However, even after many years, there is not a consensus on the basic conditions which should ensure the validity of the statistical mechanics. Roughly speaking the two extreme positions are the "traditional" one, for which the main ingredient is the presence of many degrees of freedom and the "innovative" one which considers chaos a crucial requirement to develop a statistical approach.

It is unnecessary to stress the role of simplified models and numerical simulation. Because of technical difficulties in the treatment of any realistic system, the numerical study of simple models is essential. One of the first numerical experiments was the celebrated paper *Studies of non-linear problems* by Fermi, Pasta and Ulam, that showed that the ergodic problem was still far from being solved; and pointed out the necessity of using numerical simulation as a research tool complementary to analytical studies.

The main aim of this book is to show how, for understanding the conceptual aspects of the statistical mechanics, one has to combine concepts and techniques developed in the context of the dynamical systems with statistical approaches able to describe systems with many degrees of freedom. We discuss with particular emphasis the relevance of non asymptotic quantities, *e.g* $\epsilon$-entropy, and the role of pseudochaotic systems, *i.e.* non chaotic systems with a non trivial behaviour.

We do not pretend to write a treatise on dynamical systems or statistical mechanics, however we tried to make the book as self-contained as possible.

The book is divided into three parts:

*Part I* : **Deterministic chaos and complexity** (Chapters 1, 2 and 3)
*Part II* : **Foundation of equilibrium and non equilibrium statistical mechanics** (Chapters 4, 5 and 6)
*Part III* : **Effective equations, multiscale and renormalization group** (Chapters 7 and 8)

In the first part we start introducing the basic concepts and ideas on chaotic dynamics. There exist well established ways to define the complexity of a temporal evolution, in terms of either Lyapunov exponents (LE) or Kolmogorov-Sinai (KS) entropy. This approach has been rather successful in deterministic low dimensional systems. On the other hand in high dimensional systems, as well as in low dimensional cases without a unique characteristic time some interesting features cannot be captured by LE or KS entropy. The basic reason of this weakness is that these quantities are properly defined only in specific asymptotic limits, that are: very long times and arbitrary accuracy. On the contrary in realistic situations one has to deal with finite accuracy and finite time, so it is important to take into account these limitations. For instance relaxing the limit of arbitrary high accuracy, one can introduce suitable tools, such as the Finite Size Lyapunov Exponent (FSLE) and the $\epsilon$-entropy.

An analysis in terms of *FSLE* and $\epsilon$-entropy allows for the characterization of non trivial systems in situations far from asymptotic (*i.e.* finite time and finite observational resolution). In particular we discuss the utility of $\epsilon$-entropy and FSLE for a pragmatic classification of signals, and the use of chaotic systems in the generation of sequences of (pseudo) random numbers.

The second part discusses the role of ergodicity and chaos for the validity of statistical laws. Detailed numerical studies show in a clear way that for high dimensional Hamiltonian systems chaos is not a fundamental ingredient for the validity of the equilibrium statistical mechanics. Therefore the point of view that good statistical properties need chaos is unnecessarily demanding: even in the absence of chaos, one can have (according to Khichin ideas) a good agreement between the time averages and the predictions by the equilibrium statistical mechanics.

About the problem of the irreversibility of macroscopic processes it seems to us that Boltzmann was basically able to understand the essence of mechanism of the Second Law. The possible presence of chaos plays a minor role while the relevant aspects are the large number of degrees of freedom, and the selection of "good" initial conditions in such a way that the molecular chaos hypothesis is satisfied. With such assumptions one can eliminate the fluctuations in the time behaviour of $\mathcal{H}(t)$ vs $t$ and therefore the classical objections by Loschmidt and Zermelo are overcome. Exact mathematical results have shown that the original intuitions of Boltzmann were correct.

Usually one deals with the behaviour of single macroscopic systems, and indeed thermodynamics, as a physical theory, has been developed to describe the properties of single systems, made of many microscopic, interacting parts. Thus it seems to us that it is quite fair to conclude that statistical ensembles are just useful mathematical tools. The study of a system made of many weakly coupled subsystems evidences the objective nature of the growing in time of the Boltzmann entropy, *i.e.* its independence from the coarse graining resolution, as far as it is small enough.

There is a rather strong evidence that chaos (in the technical sense of the existence of a positive Lyapunov exponent) is not a necessary ingredient for the validity of the statistical mechanics laws as diffusion and conduction. Numerical results show that the basic elements are: an instability mechanism, able to induce a particle dispersion at small scales, and the suppression of periodic orbits, to allow for a diffusion at large scale. In chaotic systems the instability mechanism is nothing but the sensitivity to the initial condition; however also in systems with zero maximal Lyapunov exponent finite-size instability mechanisms can exist.

The last part is devoted to the treatment of problems characterized by the presence of more than one significant scale, *i.e.* with a variety of degrees of freedom with different time scales. For this class of systems it is necessary, both practically and conceptually, to treat the "slow dynamics" in terms of effective equations. These equations are able to catch some general features and to evidence dominant ingredients which can remain hidden in the detailed description.

We discuss some general aspects of the multiple-scale method and its connection with other important issues as the renormalization group. We see at work, in some simple cases,the basic tools necessary for the study of phenomena as diffusion and mesoscopic description of non-equilibrium statistical mechanics. In multiscale analysis one replaces the original evolution equation with an effective one which is valid at very large time (or at large spatial distance). As an example we can mention the asymptotic behaviour of the transport problem as described by a Fick's equation containing the eddy diffusion coefficients to take into account the inhomogeneity due to the advection field in the original problem.

This book is not an updated text of the most recent progresses in all the fields of statistical physics (in particular those regarding non equilibrium stationary states). Since we want to limit the treatment to some basic aspects, we do not discuss those results like fluctuations theorems which, for the technical aspects would almost deserve another and different book. Of course the selection of issues in this book reflects our scientific interest during the last years. We would like to express our thanks for inspiration, collaboration and correspondence to E. Aurell, L. Biferale, G. Boffetta, F. Cecconi, A. Celani, M. Cencini, E. Charpentier, P. Collet, A. Crisanti, D. del-Castillo-Negrete, P. Grassberger, C. Gruber, S. Isola, M.H. Jensen, K. Kaneko, H. Kantz, G. Lacorata, M. Laguës, R. Livi, V. Loreto, G. Mantica, U. Marini Bettolo Marconi, A. Mazzino, P. Muratore Ginanneschi, E. Olbrich, G. Parisi, L. Palatella, S. Pigolotti, A. Politi, A. Puglisi, L. Rondoni, S. Ruffo, M. Serva, A.Torcini, M. Vergassola and D. Vergni.

A special thank to G. Benettin for having provided us with figures 4.1, 4.2 and 4.3.

Finally we thank F. Cecconi, M. Cencini, R. Livi, P. Muratore Ginanneschi, A. Ponno, A. Puglisi and L. Rondoni for valuable comments on some parts of the manuscript.

# 1

# Basic concepts of dynamical systems theory

Everything should be made as simple as possible, but not simpler.

*Albert Einstein*

## 1.1 Deterministic systems

Since the Pythagorean attempts to explain the tangible world by means of numerical quantities related to integer numbers, western culture has been characterized by the idea that Nature can be described by mathematics. This idea comes from the explicit or hidden assumption that the world obeys some precise rules. It may appear obvious today, but the systematic application of mathematics to the study of natural phenomena dates from the seventeenth century when Galileo inaugurated modern physics with the publication of his major work *Discorsi e Dimostrazioni Matematiche Intorno a Due Nuove Scienze* (Discourses and Mathematical Demonstrations Concerning Two New Sciences) in 1638. The fundamental step toward the mathematical formalization of reality was taken by Newton and his mechanics, explained in *Philosophiae Naturalis Principia Mathematica* (The Mathematical Principles of Natural Philosophy), often referred to as the *Principia*, published in 1687. This was a very important date not only for the philosophy of physics but also for all the other sciences; this great work can be considered to represent the high point of the scientific revolution, in which science as we know it today was born. From the publication of the *Principia* to the twentieth century, for a large community of scientists the main goal of physics has been the reduction of natural phenomena to mechanical laws. A natural phenomenon was considered really understood only when it was explained in terms of mechanical movements.

The idea of determinism was established in a rather vivid way by Pierre Simon de Laplace (1814), in his book *Essai Philosophique sur les Probabilités* (Philosophical Essay on Probability):

We must consider the present state of Universe as the effect of its past state and the cause of its future state. An intelligence that would know all forces of nature and the respective situation of all its elements, if furthermore it was large enough to be able to analyze all these data, would embrace in the same expression the motions of the largest bodies of Universe as well as those of the slightest atom: nothing would be uncertain for this intelligence, all future and all past would be as known as present.

This statement has been a point of reference for scientific thought: a good scientific theory has to describe a natural phenomenon using mathematical methods. Once the temporal evolution equations of the phenomenon are written and the initial conditions are determined, the state of the system can be known at each future time by solving the equations. However, we would like to emphasize that Laplace was not naive at all about the true relevance of determinism (see later), as has sometimes been asserted by some writers of popular science.

### *1.1.1 Dynamical systems*

Let us now introduce the notion of dynamical system. A deterministic dynamical system is essentially described by:

(a) the phase space $\Omega$, containing the vectors $\mathbf{x}$ that determine, in a quantitative way, all the possible states of the system;
(b) an evolution law $U(t, t_0)$, i.e. a rule that allows us to determine the state $\mathbf{x}(t)$ of the system at time $t$, given the state $\mathbf{x}(t_0)$ at time $t_0$. Formally we can write

$$\mathbf{x}(t) = U(t, t_0)\mathbf{x}(t_0) = U(t - t_0)\mathbf{x}(t_0) \equiv U^{t-t_0}\mathbf{x}(t_0),$$

where, in the second equality, the stationarity of the evolution rule has been assumed, i.e. the system undergoes the same evolution from a given state $\mathbf{x}_0$, independently from the time it is found in $\mathbf{x}_0$. Moreover, $U^t$ is a semigroup, that is $U^{r+s} = U^r U^s$ $(r, s > 0)$ and $U^0 = I$, i.e. $\mathbf{x}(t_0) = U^0\mathbf{x}(t_0)$.

The state of the system is typically specified by a $d$-dimensional vector $\mathbf{x}$, whose $d$ components $x_1, x_2, \ldots, x_d$, are called the degrees of freedom of the system. An elementary example is given by the pendulum, whose state is determined by the angle $\theta$ to the vertical and the angular velocity $\omega = \mathrm{d}\theta/\mathrm{d}t$; therefore the phase space is a cylindrical surface defined by $\theta \in [0, 2\pi]$ and $\omega \in [-\infty, +\infty]$: all the states of the pendulum are represented by points on this surface.

The most common deterministic evolution laws are maps and differential equations. In the first case the time is a discrete variable and the evolution law reads

$$\mathbf{x}(t + 1) = \mathbf{g}[\mathbf{x}(t)] \tag{1.1}$$

corresponding to the following system of $d$ equations

$$
\begin{aligned}
x_1(t+1) &= g_1[x_1(t), x_2(t), \ldots, x_d(t)] \\
\cdots \quad & \cdots \quad \cdots \\
x_d(t+1) &= g_d[x_1(t), x_2(t), \ldots, x_d(t)].
\end{aligned}
\tag{1.2}
$$

In the case of differential equations the time is a continuous variable and the evolution law is prescribed as

$$
\frac{\mathrm{d}\mathbf{x}(t)}{\mathrm{d}t} = \mathbf{f}\,[\mathbf{x}(t)] \tag{1.3}
$$

which corresponds to the system of equations

$$
\begin{aligned}
\frac{\mathrm{d}}{\mathrm{d}t}x_1(t) &= f_1[x_1(t), x_2(t), \ldots, x_d(t)] \\
\cdots \quad & \cdots \quad \cdots \\
\frac{\mathrm{d}}{\mathrm{d}t}x_d(t) &= f_d[x_1(t), x_2(t), \ldots, x_d(t)].
\end{aligned}
\tag{1.4}
$$

The functions $\mathbf{g}$ and $\mathbf{f}$ in (1.1) and (1.3) do not contain an explicit time dependence, as a consequence of the stationarity assumption on the evolution. This assumption is not a severe limitation. A system can be made formally time independent by increasing by one unit the number of degrees of freedom.

The deterministic nature of the maps (1.1) is evident: given the initial state $\mathbf{x}(t_0)$, the state $\mathbf{x}(t)$ at time $t > t_0 = t - n$ is given by

$$
\mathbf{x}(t) = \mathbf{g}\,[\mathbf{x}(t-1)] = \mathbf{g}\,[\mathbf{g}\,[\mathbf{x}(t-2)]] = \cdots = \mathbf{g}^{(n)}[\mathbf{x}(t_0)], \tag{1.5}
$$

where $\mathbf{g}^{(2)}(\mathbf{x}) = \mathbf{g}\,[\mathbf{g}\,[\mathbf{x}]]\,, \ldots, \mathbf{g}^{(n)}(\mathbf{x}) = \mathbf{g}\left[\mathbf{g}^{(n-1)}\,[\mathbf{x}]\right].$

The deterministic nature of the differential equations (1.3) is assured, under quite general conditions, by the existence and unicity theorem of the solution to a system of ordinary differential equations (Arnold 1974).

In particular, if $f_1(x_1, \ldots, x_d), \ldots, f_d(x_1, \ldots, x_d)$ are linear functions of the variables $x_1, \ldots, x_d$,

$$
\begin{aligned}
f_1(x_1, \ldots, x_d) &= a_{11}x_1 + a_{12}x_2 + \cdots + a_{1d}x_d \\
\cdots \quad & \cdots \quad \cdots \\
f_d(x_1, \ldots, x_d) &= a_{d1}x_1 + a_{d2}x_2 + \cdots + a_{dd}x_d,
\end{aligned}
\tag{1.6}
$$

and if the $a_{ij}$ coefficients are constant, the solution of the system can be easily written in an explicit form (Arnold 1974):

$$
\mathbf{x}(t) = \mathrm{e}^{\mathbf{A}t}\,\mathbf{x}(0), \tag{1.7}
$$

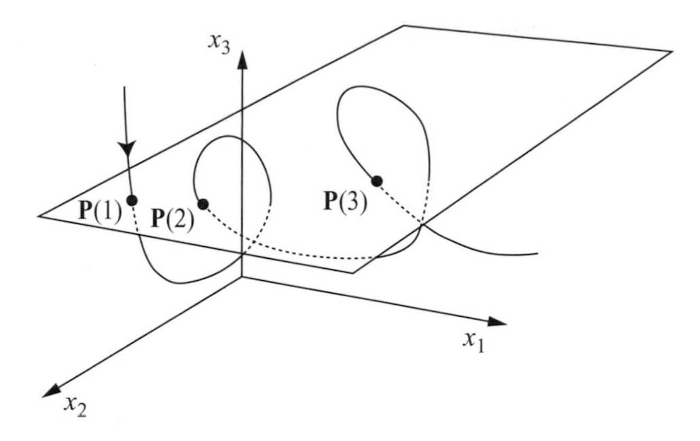


Figure 1.1 The generation of the Poincaré map by means of the Poincaré surface of section method in a three-dimensional flow.

where **A** is the matrix whose elements are $\{a_{ij}\}$. An analogous result holds for linear map systems.

It is not difficult to understand that the maps and the differential equation systems are not completely disconnected representations of dynamical systems. For example, we can consider the simplest algorithm for the numerical integration of (1.4), i.e. the Euler scheme, to compute $\mathbf{x}(t+\tau)$ from $\mathbf{x}(t)$ with $\tau$ small enough: applying the definition of derivative, and neglecting terms of order $\tau^2$, one obtains the map

$$\begin{aligned} x_1(t+\tau) &= x_1(t) + f_1[x_1(t), x_2(t), \ldots, x_d(t)]\tau \\ \cdots &\quad \cdots \quad \cdots \\ x_d(t+\tau) &= x_d(t) + f_d[x_1(t), x_2(t), \ldots, x_d(t)]\tau. \end{aligned} \tag{1.8}$$

Of course the Euler scheme is not very accurate. Nevertheless, more precise algorithms, for example the popular Runge–Kutta method, are nothing but maps which determine $\mathbf{x}(t+\tau)$ from $\mathbf{x}(t)$. Another way to reduce a continuous time dynamical system (or "flow") to a discrete time map is through the Poincaré surface of section method. If we consider the $d$-dimensional flow (1.4), the *Poincaré map* gives its reduction to a $(d-1)$-dimensional map. For illustrative purposes, consider the three-dimensional case. The trajectory $\mathbf{x}(t)$ crosses the plane $x_3 = h$ with $\mathrm{d}x_3/\mathrm{d}t < 0$, the Poincaré surface, in the points $\mathbf{P}(0), \mathbf{P}(1), \ldots, \mathbf{P}(n)$ at times $t_0, t_1, \ldots, t_n$ (see Figure 1.1). Since the point $\mathbf{x}(t_{n+1}) = (x_1(t_{n+1}), x_2(t_{n+1}), h)$ is determined uniquely by the point $\mathbf{x}(t_n)$, one has a deterministic rule connecting $\mathbf{P}(n)$ with $\mathbf{P}(n+1)$, i.e.

the Poincaré map which describes the evolution of the system on the plane:

$$\mathbf{P}(n+1) = \mathbf{g}[\mathbf{P}(n)]. \tag{1.9}$$

In general the explicit form of the Poincaré map associated with a given ordinary differential equation is not known, however its existence is useful for characterizing the behavior of the flow. For example, if the continuous time dynamical system is periodic, there will be only a finite number of isolated points on the Poincaré section. If the trajectory is quasi-periodic,[1] then there will be a regular closed figure, while if the trajectory is very irregular, there will be a non-structured set of points.

### 1.1.2 Attractors

The dynamical systems can be divided into two large classes: the *conservative* and the *dissipative* systems. A conservative dynamical system preserves the volume of the phase space. That is, given a region $\mathcal{A}_0$, whose volume is $V_0$, the points evolved from $\mathcal{A}_0$ define a region $\mathcal{A}_t$ whose volume is $V_t = V_0$. This property is translated in differential terms as

$$\left|\det\left[\frac{\partial}{\partial x_i} g_j(\mathbf{x})\right]\right| = 1 \qquad \text{for maps,} \tag{1.10}$$

and

$$\nabla \cdot \mathbf{f} = \sum_{i=1}^{d} \frac{\partial}{\partial x_i} f_i(\mathbf{x}) = 0 \qquad \text{for flows.} \tag{1.11}$$

An important example of a conservative system is given by Hamilton's equations for the motion of particles without friction. In contrast, a dissipative dynamical system does not preserve the volume of the phase space, i.e. $V_t < V_0$. The mathematical formulation of the contraction of the phase space in differential form is

$$\left|\det\left[\frac{\partial}{\partial x_i} g_j(\mathbf{x})\right]\right| < 1 \qquad \text{for maps,} \tag{1.12}$$

and

$$\nabla \cdot \mathbf{f} = \sum_{i=1}^{d} \frac{\partial}{\partial x_i} f_i(\mathbf{x}) < 0 \qquad \text{for flows.} \tag{1.13}$$

[1] An $N$-frequency quasi-periodic motion can be represented by $N$ independent variables, $f_1(t), f_2(t), \ldots, f_N(t)$, such that each $f_k$ is periodic with period $T_k$ and the $N$ frequencies $\Omega_i = 2\pi/T_i$ are incommensurate, that is, the relation $m_1\Omega_1 + m_2\Omega_2 + \cdots + m_N\Omega_N = 0$ does not hold for any set of integers, $m_1, m_2, \ldots, m_N$, except for the trivial solution $m_1 = m_2 = \cdots = m_N = 0$. A two-frequency quasi-periodic motion lies on a two-dimensional torus.

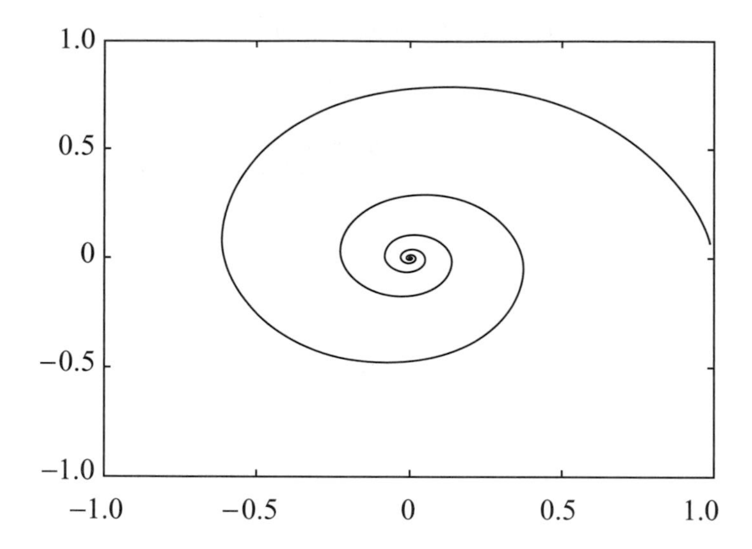


Figure 1.2 Example of a simple attractor: a stable fixed point.

A simple example of a dissipative system is the one-dimensional damped harmonic oscillator

$$\frac{\mathrm{d}^2 x}{\mathrm{d}t^2} + \nu \frac{\mathrm{d}x}{\mathrm{d}t} + \omega^2 x = 0.$$

Because of the friction term $\nu \mathrm{d}x/\mathrm{d}t$, the system is dissipative and, as time goes on, the oscillation amplitude $x$ and the velocity $\dot{x}$ of the oscillator decrease and approach the asymptotic values $x = 0$, $\dot{x} = 0$. A trajectory in the phase space is shown in Figure 1.2 where the orbit spirals to the origin for any initial condition. In this case, the point (0, 0) is an attracting point of the dynamical system.

Another example of a dissipative system is the pendulum clock, where the energy lost due to friction is reintegrated by a non-linear mechanism so that the oscillation amplitude is stabilized, as in the system described by the Van der Pol equation:

$$\frac{\mathrm{d}^2 x}{\mathrm{d}t^2} + (x^2 - \nu)\frac{\mathrm{d}x}{\mathrm{d}t} + \omega^2 x = 0\,.$$

Figure 1.3 shows two typical trajectories of this kind of system: in both cases, the orbit, with time, spirals (inwards or outwards) to approach the closed curve on which it circulates in periodic motion in the $t \to \infty$ limit. The closed curve is a *limit cycle*.

As the above examples show, a very important property of dissipative systems is the presence of attracting sets or *attractors* in the phase space. These are bounded subsets of $\Omega$ to which regions of initial conditions of non-zero phase space volume

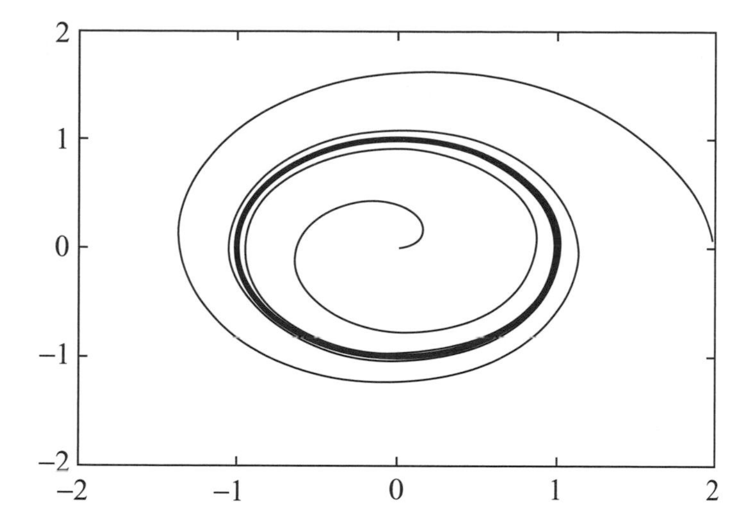


Figure 1.3 Example of a simple attractor: a stable limit cycle.

asymptote as time increases. From the property of volume preservation, it is easy to see that conservative dynamical systems do not possess attractors.

The attractors of the two continuous time systems considered above have a regular geometrical structure (a point, a closed curve) but this is not the case for all dissipative systems. An example of a non-trivial geometrical structure is the attractor of the two-dimensional Hénon map

$$\begin{cases} x(t+1) = 1 - ax^2(t) + y(t) \\ y(t+1) = bx(t). \end{cases} \tag{1.14}$$

Figure 1.4 shows the attractor of the Hénon map, for $a = 1.4$ and $b = 0.3$. The blow-up of the boxed region in Figure 1.4 (see Figure 1.5) reveals a small-scale pattern consisting of almost parallel lines. A further zoom in of a portion of Figure 1.5, shown in Figure 1.6, reveals that the part has the same structure as the whole. On continuing this zooming in procedure we would find a similar structure on arbitrarily small scales. This property of self-similarity qualifies the attractor as a *fractal*; see, e.g., Ott (1993). When the motions on the attractor, as in the case of the Hénon map, are also chaotic (see Section 1.3) the attractor is called a *strange attractor*.

## 1.2 Unpredictability: systems with many degrees of freedom

After Newton's foundation of the dynamical laws, the deterministic approach became a powerful and successful method for the understanding of natural phenomena especially in astronomy. As remarkable examples one can mention the derivation

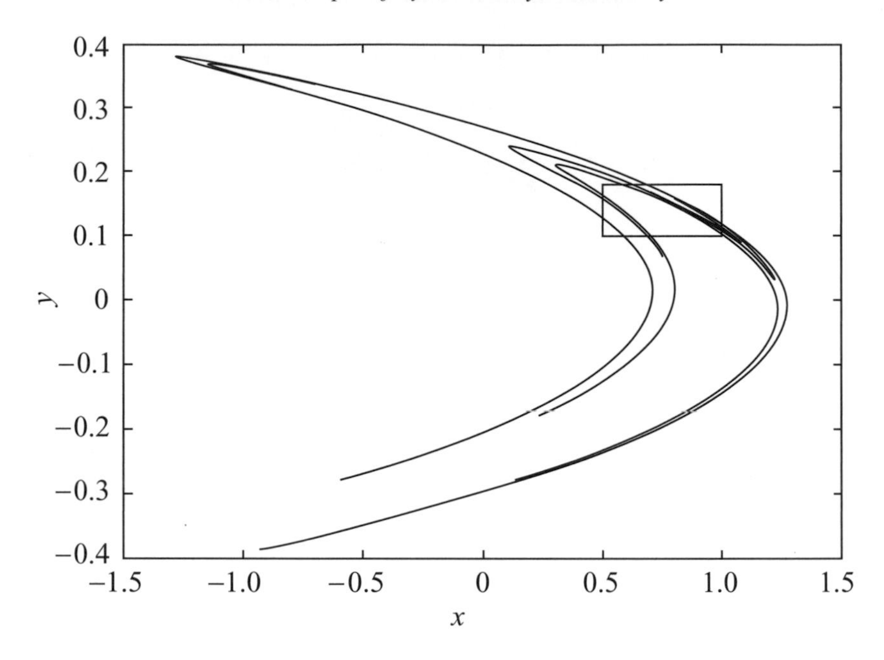


Figure 1.4 The attractor of the Hénon map, obtained using Eq. (1.14) with $a = 1.4$ and $b = 0.3$.

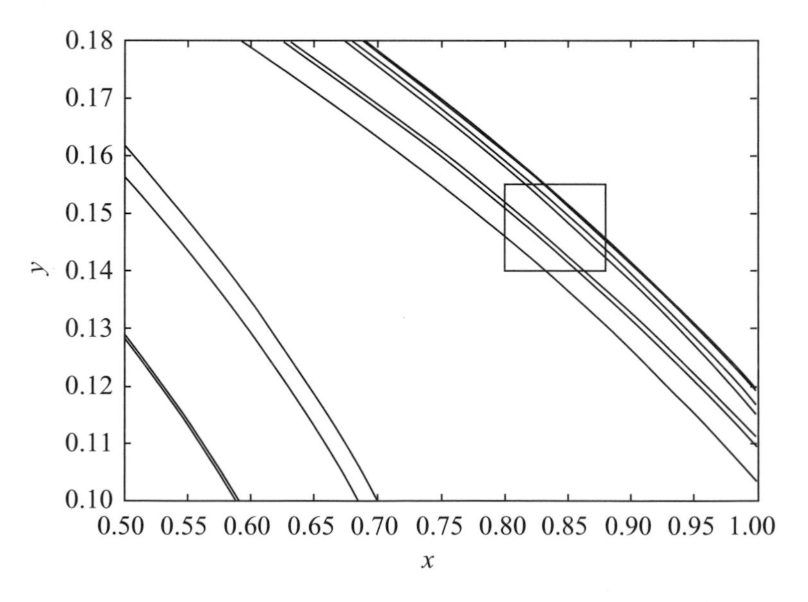


Figure 1.5 Enlargement of the boxed region in Figure 1.4.

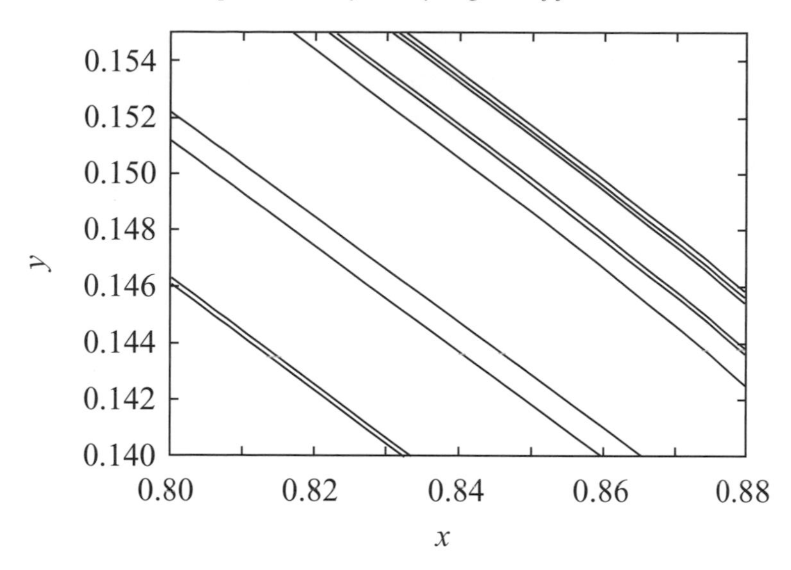


Figure 1.6 Enlargement of the boxed region in Figure 1.5.

of Kepler's laws from the Newtonian dynamical equations, and the gravitational force. Another paradigmatic success of Newtonian mechanics was the discovery of the planet Neptune, whose existence was predicted theoretically by Le Verrier and Adams. Today, the positions of many celestial bodies and artificial satellites can be calculated quickly with good accuracy by the powerful computers of astronomical study centers.

Nevertheless, everyday life is characterized by a lot of phenomena which exhibit unpredictable behaviors like the evolution of the weather or the fall of a leaf. How do we reconcile the deterministic Laplacian assumption with the "irregularity" and "unpredictability" of many natural phenomena? Laplace answered this question, again in his book *Essai Philosophique sur les Probabilités* (Philosophical Essay on Probability), by identifying the origin of the irregularity in our ignorance on the system:

> The curve described by a simple molecule of air or vapor is regulated in a manner just as certain as the planetary orbits; the only difference between them is that which comes from our ignorance. Probability is relative, in part to this ignorance, in part to our knowledge.

Thus, according to the previous point of view, the observed irregularity is more apparent than real: it is due to a large number of simple reasons, for example, a large number of simple mechanical equations that rule the evolution of the system. This interpretation is at the basis of Langevin's approach to Brownian motion.

### 1.2.1 Brownian motion

In 1827 the Scottish botanist Robert Brown noticed that pollen grains suspended in water jiggled about under the lens of the microscope, following a zig-zag path. Initially, he believed that such activity was peculiar to the male sexual cells of plants, but then he observed that pollen of plants dead for over a century showed the same movement. Further study revealed that the same motion could be observed not only with particles of other organic substances but even with chips of glass or granite or particles of smoke.

In 1889 Gouy found that Brownian motion was more rapid for smaller particles, lower viscosity of the surrounding fluid and higher temperatures. These facts suggest that the basic cause of Brownian motion lies in the "thermal molecular motion in the liquid environment." Therefore it is natural (at least today!), following the atomistic point of view, to suppose that a suspended particle is constantly and randomly bombarded from all sides by the molecules of the liquid.

After important and independent works by Einstein (1905) and Smoluchowski (1906), Langevin (1908) proposed an approach in terms of a stochastic differential equation (to use modern terminology) for the particle movement, taking into account the effect of the molecular hits by means of an average force, as given by the fluid friction, and a random fluctuating term.

The basic physical assumptions in both Einstein's and Langevin's approaches are

(a) Stokes's law for the friction of a body moving in a liquid;
(b) equipartition of the kinetic energy among the various degrees of freedom of the system, i.e. between the particles of the fluid and the grain performing Brownian motion.

A colloidal particle suspended in a liquid at temperature $T$ is somehow assimilated to a particle of the liquid, so that it possesses an average kinetic energy $RT/(2N_A)$, in each spatial direction, where $R$ is the perfect gas constant and $N_A$ is the Avogadro number (the number of molecules in one mole); therefore one has:

$$\frac{1}{2}m\langle v_x^2\rangle = \frac{RT}{2N_A}. \tag{1.15}$$

According to Stokes's law, a spherical particle of radius $a$, moving in a liquid with the speed $v_x$ in the $x$ direction experiences a viscous resistance:

$$F_{\text{Stokes}} = -\alpha v_x = -6\pi\eta a v_x,$$

where $\eta$ is the viscosity. The above law holds if $a$ is much larger than the average distance between the liquid molecules, and Stokes's force represents the average macroscopic effect of the large number of irregular impacts of the molecules of the fluid.

Therefore, isolating the average force, the dynamical equation of the particle in the direction $x$ can be written as

$$m\frac{\mathrm{d}v_x}{\mathrm{d}t} = F_{\mathrm{Stokes}} + F_x(t) = -\alpha v_x + F_x(t) \tag{1.16}$$

where $F_x(t)$ is a random fluctuating force mimicking the effects of the molecules. By construction $\langle F_x(t)\rangle = 0$, and, since the characteristic time of $F_x$ is much smaller than $\tau = m/\alpha$ (the characteristic time of the deterministic Stokes equation), one can assume $\langle F_x(t)F_x(t')\rangle = c\delta(t-t')$, where $c$ can be determined by the energy equipartition.

By multiplying Eq. (1.16) by $\Delta x(t) = x(t) - x(0)$, and taking the average over a large number of identical particles one has

$$m\left\langle \frac{\mathrm{d}^2\Delta x}{\mathrm{d}t^2}\Delta x\right\rangle = -\alpha\left\langle \frac{\mathrm{d}\Delta x}{\mathrm{d}t}\Delta x\right\rangle + \langle F_x(t)\Delta x\rangle. \tag{1.17}$$

Using Eq. (1.15) and the fact that $F_x(t)$ and $x(t)$ are not correlated, we obtain

$$\frac{1}{2}\frac{\mathrm{d}}{\mathrm{d}t}\langle \Delta x^2\rangle = \frac{RT}{\alpha N_{\mathrm{A}}}\left(1 - \mathrm{e}^{-(\alpha/m)t}\right), \tag{1.18}$$

the solution of which is

$$\langle (x(t) - x(0)^2)\rangle = \frac{2RT}{\alpha N_{\mathrm{A}}}\left[t - \frac{m}{\alpha}\left(1 - \mathrm{e}^{-(\alpha/m)t}\right)\right]. \tag{1.19}$$

For a grain of size $O(1\ \mu\mathrm{m})$ in a standard liquid (such as water) at room temperature, the characteristic time $\tau = m/\alpha = m/6\pi\eta a$ is $O(10^{-7}\mathrm{s})$, in the limit $t \ll \tau$, we get

$$\langle (x(t) - x(0)^2)\rangle \simeq \frac{2RT}{\alpha N_{\mathrm{A}}}\frac{\alpha}{2m}t^2 = \frac{RT}{N_{\mathrm{A}}m}t^2 \tag{1.20}$$

while for $t \gg \tau$

$$\langle (x(t) - x(0)^2)\rangle \simeq \frac{2RT}{\alpha N_{\mathrm{A}}}t = \frac{RT}{3N_{\mathrm{A}}\pi\eta a}t. \tag{1.21}$$

Now we have the celebrated Einstein relation which gives the diffusion coefficient $D$ in terms of macroscopic variables and the Avogadro number:

$$D = \lim_{t\to\infty}\frac{\langle (x(t) - x(0)^2)\rangle}{2t} = \frac{RT}{6N_{\mathrm{A}}\pi\eta a}. \tag{1.22}$$

Let us stress that the previous equation gives an unambiguous link between the microscopic and macroscopic levels, since it allows the determination of the Avogadro number (i.e. a quantity related to the microscopic level of description) from experimentally accessible macroscopic quantities.

Brownian motion had a central role in the development of physics and mathematics. The theoretical work by Einstein and the experiments by Perrin gave clear and conclusive evidence of the relationship between the diffusion coefficient and Avogadro's number (Perrin 1913). This result could be considered as the definitive "proof" of the existence of atoms: after that even two champions of the energetic point of view, Helm and Ostwald, accepted atomism as a physical fact and not merely a useful hypothesis. In a lecture in Paris in 1911, Arrhenius, summarizing the work of Einstein and Perrin, declared that "after this, it does not seem possible to doubt that the molecular theory entertained by the philosophers of antiquity, Leucippus and Democritos, has attained the truth at least in essentials" (Mehra 2001).

In addition, Langevin's approach to Brownian motion was the first example of a stochastic differential equation, and inspired the development of the mathematical theory of continuous time stochastic processes.

Although Langevin's (as well Einstein's) approach is, from a mathematical point of view, rather simple, there is a very subtle conceptual point at the basis of the theory of Brownian motion. The ingenious idea is the assumption of the validity of Stokes's law (which is macroscopic in nature), and at the same time the assumption that the Brownian particle is in statistical equilibrium with the molecules in the liquid. In other words, in spite of the fact that the mass of the colloidal particle is much larger than the mass of the molecules, energy equipartition is assumed to hold.

## 1.3 Unpredictability: deterministic chaos

According to the view of Laplace, prediction of the system state at any future time is possible whenever exact knowledge of the initial state is available. However, typically our knowledge of the state of a system is not perfect: there is always an uncertainty due at least to the accuracy of the measurement instruments.

Therefore the proper question to ask about the prediction of a deterministic system is: what is the temporal evolution of the difference between two states which are initially close? Henri Poincaré was one of the first to understand that even a deterministic system can have "irregular" behavior. As he writes in his book *Science et Méthode* (1908):

> A very small cause which escapes our notice determines a considerable effect that we cannot fail to see, and then we say that the effect is due to chance. If we knew exactly the laws of nature and the situation of the universe at the initial moment, we could predict exactly the situation of the same universe at a succeeding moment. But even if it were the case that the natural laws had no longer any secret for us, we could still know the situation approximately. If that enabled us to predict the succeeding situation with the same approximation, that is all

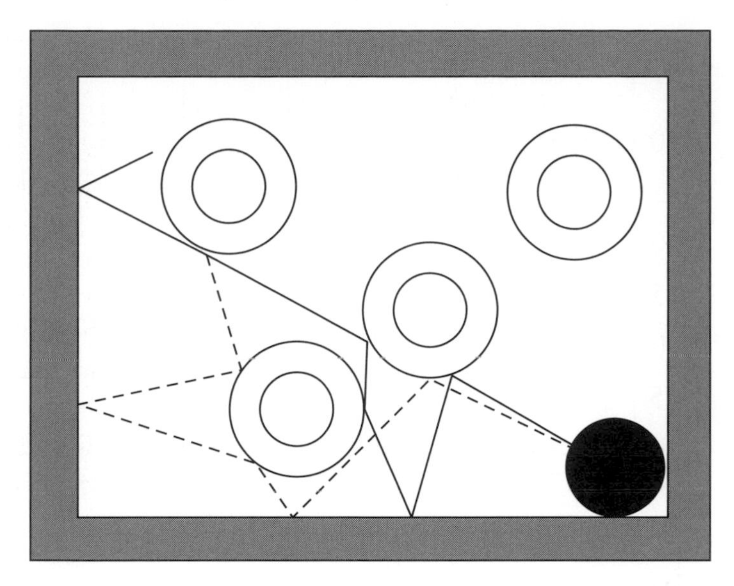

Figure 1.7 Billiard table with convex obstacles. The center of the bead follows the full line; another bead, whose initial direction is slightly different, follows the dashed line. After a few collisions the two trajectories are completely different.

we require, and we should say that the phenomenon had been predicted, that it is governed by the laws. But it is not always so; it may happen that small differences in the initial conditions produce very great ones in the final phenomena. A small error in the former will produce an enormous error in the latter. Prediction becomes impossible and we have the fortuitous phenomenon.

An example of a fast temporal amplification of an initially small difference between two states is found in the system of a bead rolling on a rectangular billiard table with fixed circular obstacles, and undergoing elastic collisions with the obstacles and the borders. If $r$ is the radius of the bead, $R$ the radius of the obstacles, $L_1 \times L_2$ the size of the billiard table, the system can be represented as a point (the center of the bead) moving on a billiard table of size $(L_1 - 2r) \times (L_1 - 2r)$ having fixed circular obstacles of radius $R + r$. It is easy to realize that if two beads have initially slightly different directions, the collisions with the circular obstacles will amplify the angle between the two directions in such a way that after a few collisions the trajectories of the two beads will be very different (see Figure 1.7).

In this case we say that one has *deterministic chaos*, i.e. a deterministic system with *sensitive dependence on initial conditions*.

One of the most celebrated systems with this property is the Lorenz model introduced in 1963 as a drastic simplification of atmospheric convection. Such a

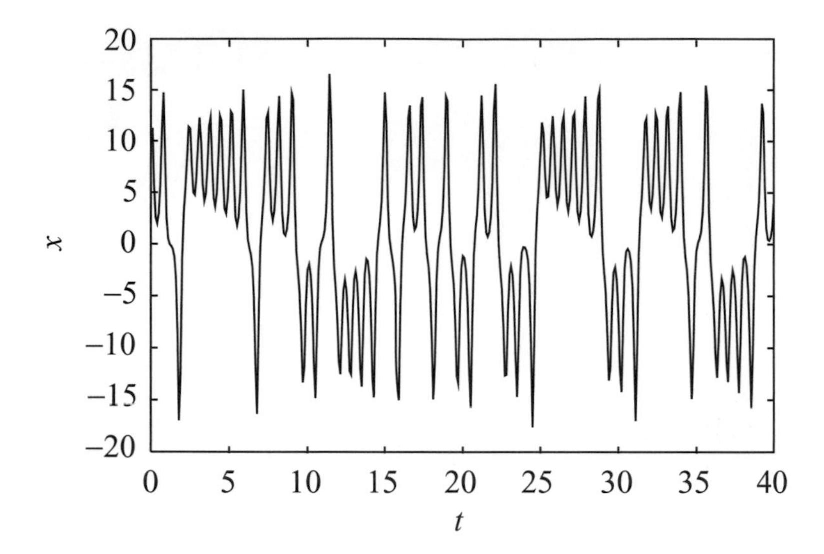


Figure 1.8 The variable $x$ versus $t$ of the Lorenz system, for $\sigma = 10$, $b = 8/3$ and $r = 28$.

system is given by the ordinary differential equations:

$$\begin{cases} \mathrm{d}x/\mathrm{d}t = \sigma(y - x) \\ \mathrm{d}y/\mathrm{d}t = x(r - z) - y \\ \mathrm{d}z/\mathrm{d}t = xy - bz, \end{cases} \tag{1.23}$$

where $x$, $y$ and $z$ are the coefficients of the Fourier expansion of the velocity and the temperature field; $\sigma$, $r$ and $b$ are positive parameters, $r$ being proportional to the Rayleigh number, one of the basic parameters for atmospheric convection (Lorenz 1963).

The Lorenz model has become one of the paradigmatic systems displaying *deterministic chaos*: its generic solutions are aperiodic with a sensitive dependence on the initial conditions. For example, numerical experiments with $\sigma = 10$, $r = 28$ and $b = 8/3$, show that the trajectories achieve aperiodic motion (see Figure 1.8), and after a transient settle on a set of zero volume in the phase space, a strange attractor. A measure of the sensitive dependence on the initial conditions is shown in Figure 1.9.

This example shows in a rather convincing way, how even a deterministic system with a few degrees of freedom can exhibit unpredictable behavior.

The chaotic character of the Lorenz model is evident if one reduces the system to a discrete time map. Let $m(n)$ be the value of the $n$th maximum of $z(t)$. By plotting $m(n+1)$ versus $m(n)$, the successive pairs $(m(n), m(n+1))$ for $n = 1, 2, \ldots$ fall

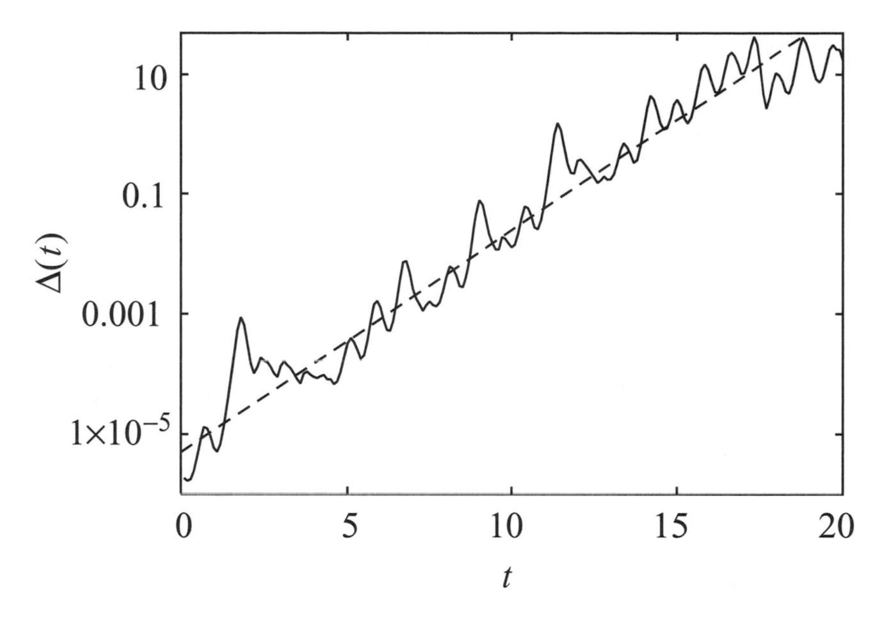


Figure 1.9 The increase in the distance, $\Delta(t)$, between two trajectories initially close, in the Lorenz system, for $\sigma = 10$, $b = 8/3$ and $r = 28$. The dashed line represents exponential growth.

on a graph close to a tent shape:

$$x(n+1) = \begin{cases} 2\,x(n) & \text{if } x(n) < 1/2 \\ 2\,[1 - x(n)] & \text{if } x(n) \geq 1/2, \end{cases} \tag{1.24}$$

where $x$ is now a generic variable, not to be confused with the variable $x$ of Eq. (1.23).

It is fairly easy to study the above map and recover the sensitive dependence on initial conditions as follows. Let the initial condition $x(0) \in [0, 1]$ be represented as a binary number

$$x(0) = \sum_{j=1}^{\infty} 2^{-j} a_j$$

where each of the digits $a_j$ is either 0 or 1.

The sequence $\{a_j\}$ $(j = 1, 2, \ldots)$ is the binary representation of $x(0)$, and to simplify the notation we write $x(0) = [a_1, a_2, \ldots]$. Denoting by $\mathcal{N}$ the negation operator, defined so that $\mathcal{N}0 = 1$, $\mathcal{N}1 = 0$ and $\mathcal{N}^0 = I$ (the identity operator), we can write

$$x(1) = \begin{cases} [a_2, a_3, a_4, \ldots] & \text{if } x(0) < 1/2 \\ [\mathcal{N}a_2, \mathcal{N}a_3, \mathcal{N}a_4, \ldots] & \text{if } x(0) \geq 1/2. \end{cases} \tag{1.25}$$

Using the properties of the negation operator $\mathcal{N}$ one has

$$x(1) = [\mathcal{N}^{a_1} a_2, \mathcal{N}^{a_1} a_3, \ldots]. \tag{1.26}$$

Iterating the above argument one has:

$$x(n) = [\mathcal{N}^{a_1+a_2+\cdots+a_n} a_{n+1}, \mathcal{N}^{a_1+a_2+\cdots+a_n} a_{n+2}, \ldots]. \tag{1.27}$$

Now consider two initial conditions $x_1(0)$ and $x_2(0)$ such that

$$|x_1(0) - x_2(0)| \leq 2^{-M}$$

for some arbitrary (large) integer number $M$; this means that $x_1(0)$ and $x_2(0)$ have the first $M$ binary digits identical, and they may differ only afterwards. However, the expression (1.27) shows that the distance between the points increases rapidly and that as soon as $n > M$ one can only conclude that

$$|x_1(n) - x_2(n)| \leq 1.$$

Therefore, even an arbitrarily small error in the initial conditions eventually dominates the solution of the system, making long-term prediction impossible.

## 1.4 Probabilistic aspects of dynamical systems

Since the time evolution of a deterministic system is determined perfectly by its initial condition, apparently there is no room for chance, so at first glance the use of probabilistic concepts and methods can appear as a paradox or at least as a sort of "spurious" trick. On the other hand, as we saw in the previous sections, even in a deterministic world the use of probability may become necessary.

If for any reason there is not perfect control of the state $\mathbf{x}$ of the system, it is rather natural to introduce the probability density of $\mathbf{x}$ and then to wonder about its time evolution.

The different origins for this necessity can be grouped in two large classes:

(a) the system has a very large number of degrees of freedom, but only a small number of them are accessible or interesting;
(b) the deterministic system is chaotic, and therefore small uncertainties are amplified exponentially.

From a historical point of view the first example of the necessity of using statistical concepts in physics was statistical mechanics, i.e. case (a), where one wants to study the small set of collective variables describing the thermodynamic properties of a macroscopic system. The Brownian motion, which was at the origin of the modern theory of stochastic processes, is another example of situation (a), where one is interested in the small number of degrees of freedom of the Brownian particle.

We start with the probabilistic description of a discrete time system (1.1). Suppose a probability density $\rho_t(\mathbf{x})$ is given such that the probability of finding the state of the system in a region $b$ of its phase space is: $\text{Prob}[\mathbf{x}(t) \in b] = \int_b \rho_t(\mathbf{x})\mathrm{d}\mathbf{x}$. The time evolution of $\rho_t(\mathbf{x})$ is ruled by

$$\rho_{t+1}(\mathbf{x}) = \mathcal{L}_{\text{PF}}\rho_t(\mathbf{x}), \tag{1.28}$$

where $\mathcal{L}_{\text{PF}}$ is referred to as the Perron–Frobenius operator (Lasota and Mackey 1985, Ott 1993):

$$\mathcal{L}_{\text{PF}}\rho_t(\mathbf{x}) = \int \rho_t(\mathbf{y})\delta(\mathbf{x} - \mathbf{g}[\mathbf{y}])\mathrm{d}\mathbf{y}; \tag{1.29}$$

for the one-dimensional case one can write

$$\mathcal{L}_{\text{PF}}\rho_t(x) = \sum_k \frac{\rho_t(y_k)}{|g'(y_k)|} \tag{1.30}$$

where $y_k$ are the pre-images of $x$, i.e. the points such that $g(y_k) = x$, and $g'$ indicates the derivative.

Equation (1.29), or (1.30) in the one-dimensional case, says that

$$\text{Prob}[\mathbf{x}(t) \in b] = \text{Prob}[\mathbf{x}(t-1) \in \mathbf{g}^{(-1)}(b)]$$

where $\mathbf{g}^{(-1)}(b)$ is the pre-image of $b$.

In an analogous way, when a dynamical system is described by a set of ordinary differential equations (1.3), for the evolution of a density $\rho(\mathbf{x}, t)$ one has

$$\frac{\partial \rho}{\partial t} = \mathbf{L}\rho = -\nabla \cdot (\mathbf{f}\rho) \tag{1.31}$$

where $\mathbf{L}$ is known as the Liouville operator.

In spite of the different specific reasons for a probabilistic treatment, deterministic systems and systems with an explicit random component in the evolution law share many practical (and conceptual) aspects. One of the most common (and important) stochastic processes is described by the stochastic differential equations (often in physics the term Langevin equation is used):

$$\frac{\mathrm{d}x_j}{\mathrm{d}t} = f_j(\mathbf{x}) + R_j(t), \tag{1.32}$$

where $j = 1, \dots, N$ and $\{R_j(t)\}$ is a multi-dimensional white noise, i.e. a Gaussian stochastic process with

$$\langle R_j(t)\rangle = 0, \qquad \langle R_j(t)R_i(t')\rangle = Q_{ij}\delta(t - t'), \tag{1.33}$$

where the brackets indicate averaging over the various realizations of $R_j(t)$ and the covariance matrix $\{Q_{ij}\}$ is positive definite (Chandrasekhar 1943). One can think of the random variables $R_j(t)$ as emulating the effects of a fast internal dynamics (this happens in Brownian motion, see Chapter 7) or the action of external random disturbances, as in noisy electric circuits. The presence of the noise terms $\{R_j\}$ changes the evolution equation for the probability density: Eq. (1.31) is replaced by the Fokker–Planck equation (Gardiner 1990):

$$\frac{\partial \rho}{\partial t} = -\nabla \cdot (\mathbf{f}\rho) + \frac{1}{2} \sum_{ij} Q_{ij} \frac{\partial^2 \rho}{\partial x_j \partial x_j}. \tag{1.34}$$

The evolution laws (1.28), or (1.31), have a general validity, they hold for generic deterministic systems. On the other hand, the behavior of $\rho_t(\mathbf{x})$, or $\rho(\mathbf{x}, t)$, depends on the deterministic dynamics (1.1), or (1.3), in particular whether the system is chaotic or not.

Before a formal treatment we discuss a simple example, the logistic map

$$x(t+1) = rx(t)(1 - x(t)), \tag{1.35}$$

where $r$ is a control parameter varying in the interval $[0, 4]$. We do not repeat here the analysis of the stability for different values of $r$, the reader can find this in any texbook, e.g. Ott (1993). For $r < 3$ one has a unique attracting fixed point $x^*$, that is, $x^* = 0$ for $r < 1$ and $x^* = 1 - 1/r$ for $1 < r < 3$. It is easy to understand that for $r < 3$, because of the presence of an attracting fixed point, at large time

$$\rho_t(x) \to \delta(x - x^*), \tag{1.36}$$

independently of the initial probability density $\rho_0(x)$.

For $3 < r < r_1 = 3.448$, one has an attracting periodic trajectory taking values $x^{(1)}(r)$ and $x^{(2)}(r)$. In this case one has that, after a transient,

$$\rho_t(x) = c_1(t)\delta(x - x^{(1)}) + c_2(t)\delta(x - x^{(2)}), \tag{1.37}$$

where $c_1(t)$ and $c_2(t)$ evolve periodically in time, i.e. $c_1(t+1) = c_2(t)$, $c_2(t+1) = c_1(t)$, and they depend on the initial probability density $\rho_0(x)$.

In an analogous way, in the range $r_{n-1} < r < r_n$ one has an attracting periodic trajectory, of period $2^n$, taking values $x^{(1)}, x^{(2)}, \ldots, x^{(2^n)}$. After a transient,

$$\rho_t(x) = \sum_{k=1}^{2^n} c_k(t)\delta(x - x^{(k)}), \tag{1.38}$$

where the coefficients $c_1(t), c_2(t), \ldots, c_{2^n}(t)$ evolve periodically in time (in a cyclic way, i.e. $c_1(t+1) = c_{2^n}(t)$; $c_2(t+1) = c_1(t)$; $c_3(t+1) = c_2(t)$; $\ldots$) and depend on the initial probability density $\rho_0(x)$.

Let us now consider a chaotic case, e.g. $r = 4$. In such a case, as $t \to \infty$

$$\rho_t(x) \to \rho^{\text{inv}}(x) = \frac{1}{\pi\sqrt{x(1-x)}} \tag{1.39}$$

independently of the initial probability density $\rho_0(x)$. By definition, an invariant density $\rho^{\text{inv}}(\mathbf{x})$ satisfies the equation

$$\rho^{\text{inv}}(\mathbf{x}) = \mathcal{L}_{\text{PF}}\rho^{\text{inv}}(\mathbf{x}). \tag{1.40}$$

In the continuous time case an invariant density of probability obeys the equation

$$\mathbf{L}\rho^{\text{inv}}(\mathbf{x}) = 0. \tag{1.41}$$

We can summarize the above results: if the dynamical behavior is regular, i.e. periodic, $\rho_t(x)$ is not able to forget the initial density $\rho_0(x)$ which, in general, does not relax to an invariant density. In contrast, if the system is chaotic then $\rho_t(x)$ relaxes to a well-defined $\rho^{\text{inv}}(x)$ which does not depend on the initial $\rho_0(x)$.

### 1.4.1 Ergodicity

Consider now the abstract definition of dynamical system which is specified by a deterministic evolution law in the phase space $\Omega$,

$$\mathbf{x}(0) \to \mathbf{x}(t) = U^t\mathbf{x}(0), \tag{1.42}$$

and a measure $\mathrm{d}\mu(\mathbf{x})$ invariant under the evolution $U^t$, i.e. $\mathrm{d}\mu(\mathbf{x}) = \mathrm{d}\mu(U^{-t}\mathbf{x})$. Since in dissipative systems the invariant measure is singular with respect to the Lebesgue measure, we use the symbol $\mathrm{d}\mu(\mathbf{x})$ instead of the more familiar $\rho^{\text{inv}}(\mathbf{x})\mathrm{d}\mathbf{x}$ (which can be used if a density exists).

The dynamical system $(\Omega, U^t, \mathrm{d}\mu(\mathbf{x}))$ is called ergodic, with respect to the measure $\mathrm{d}\mu(\mathbf{x})$, if, for every integrable function $A(\mathbf{x})$, one has

$$\overline{A} \equiv \lim_{\mathcal{T}\to\infty} \frac{1}{\mathcal{T}} \int_{t_0}^{t_0+\mathcal{T}} A(\mathbf{x}(t))\mathrm{d}t = \int A(\mathbf{x})\mathrm{d}\mu(\mathbf{x}) \equiv \langle A \rangle, \tag{1.43}$$

where $\mathbf{x}(t) = U^{t-t_0}\mathbf{x}(t_0)$ (Arnold and Avez 1968). Of course in the case of discrete time the integral must be replaced by a sum.

We can say that if a system is ergodic, a very long trajectory gives the same statistical information as the measure $\mu(\mathbf{x})$.

For statistical physics, Hamiltonian systems are the most important ones: in the case of $N$ particles, $\Omega$ is the surface of constant energy, $\mathbf{x} = (\mathbf{q}^{(N)}, \mathbf{p}^{(N)})$, $\mathbf{q}$ and $\mathbf{p}$ being the vectors of position and momentum respectively, $U$ is given by the Hamilton equation, and $\mathrm{d}\mu$ is the microcanonical distribution. Indeed this is the context that gave rise to the *ergodic problem*, that is, whether, in an isolated system Eq. (1.43) is satisfied by the microcanonical measure (see Chapter 4).

It is easy to understand that the ergodicity property can hold even in systems with very regular dynamics. As an example of an ergodic dynamical system whose time evolution does not show any irregular behavior we can mention rotation on a torus:

$$\begin{cases} x(t+1) = x(t) + u \mod 1 \\ y(t+1) = y(t) + v \mod 1. \end{cases} \tag{1.44}$$

A simple computation shows that the Lebesgue measure $d\mu(\mathbf{x}) = dxdy$ is invariant under time evolution. If $u/v$ is rational the evolution (1.44) is periodic and therefore non-ergodic, with respect to $d\mu(\mathbf{x})$; while for irrational $u/v$ the motion is quasi-periodic and ergodic, with respect to $d\mu(\mathbf{x})$.

Note that, in such an ergodic system, one cannot have a relaxation to an invariant density. If one starts with a given $\rho_0(x, y)$ localized around $(\hat{x}, \hat{y})$ then $\rho_t(x, y)$ is the translation of the initially localized function, now around $(U^t(\hat{x}), U^t(\hat{y}))$.

We do not enter into detail here on how to decide whether a generic dynamical system is ergodic or not. In Chapter 4 we will discuss some aspects of ergodicity which are relevant for equilibrium statistical mechanics.

### 1.4.2 Mixing

The dynamical system $(\Omega, U^t, d\mu(\mathbf{x}))$ is called mixing if, for all the sets $a, b \subset \Omega$, one has

$$\lim_{t\to\infty} \mu(a \cap U^t b) = \mu(a)\mu(b). \tag{1.45}$$

The above condition for mixing has a simple interpretation. The points $\mathbf{x}$ belonging to $a \cap U^t b$ are those such that $\mathbf{x} \in a$ and $U^{-t}\mathbf{x} \in b$. Therefore from (1.45) one has that the fraction of points starting from $b$ and ending up in $a$ after a (large) time $t$ is nothing but the product of the measures of $a$ and $b$ and it is independent of the positions of $a$ and $b$ in $\Omega$.

As an example of a mixing system we discuss briefly the two-dimensional area-preserving cat map:

$$\begin{cases} x(t+1) = x(t) + y(t) \mod 1 \\ y(t+1) = x(t) + 2y(t) \mod 1. \end{cases} \tag{1.46}$$

Consider a set of points, $b$, represented as a black region in Figure 1.10. Many iterations "mix" the striations of $U^t b$ more and more finely within the square. We can see that the cat map is mixing with essentially the same meaning that we use

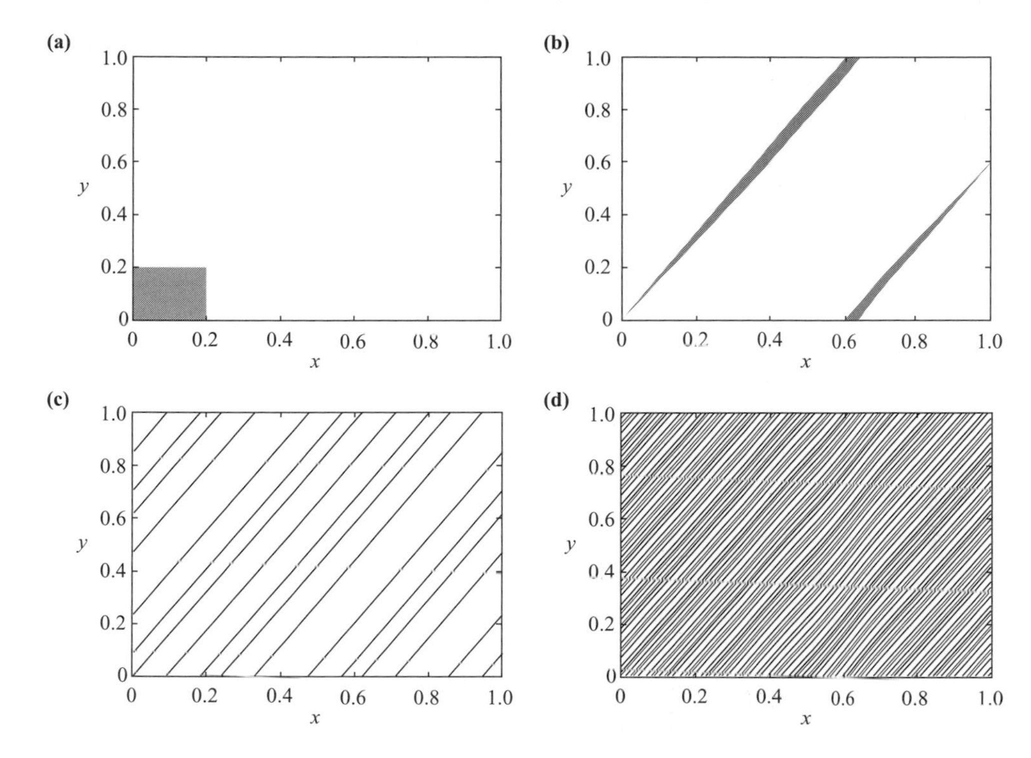


Figure 1.10 Evolution of the cat map. Going from left to right and from top to bottom, the evolutions are plotted with 40 000 points, at times $t = 0, 2, 4, 6$.

when we describe the mixing of cream as a cup of coffee is stirred, if the usual physical space is interpreted as a phase space.

It is easy to see that the mixing condition is stronger than ergodicity, i.e. a mixing system must be ergodic. The opposite in general is not true, as the example given by (1.44) shows.

For the sake of simplicity of notation, consider the discrete time case and assume that a probability density exists, i.e. $d\mu(\mathbf{x}) = \rho^{\text{inv}}(\mathbf{x})d\mathbf{x}$. If a system is mixing then one has a relaxation to the invariant measure, i.e. for large $t$ one has

$$\rho_t(\mathbf{x}) \to \rho^{\text{inv}}(\mathbf{x}). \tag{1.47}$$

The characteristic time $\tau_c$ of the relaxation process toward the invariant measure, i.e. such that

$$\rho_t(\mathbf{x}) = \rho^{\text{inv}}(\mathbf{x}) + O(e^{-t/\tau_c}), \tag{1.48}$$

is also relevant for the decay of the correlation between generic functions $G$ and $F$:

$$\langle G(\mathbf{x}(t))F(\mathbf{x}(0))\rangle = \langle G(\mathbf{x})\rangle\langle F(\mathbf{x})\rangle + O(e^{-t/\tau_c}) \tag{1.49}$$

where the average is on the invariant measure. One can see that $\tau_c$ is related to the spectral properties of the Perron–Frobenius operator $\mathcal{L}_{PF}$ (Lasota and Mackey 1985, Ruelle 1989, Ott 1993).

It is not difficult to repeat the above considerations for systems described by ordinary differential equations, simply replacing the Perron–Frobenius operator with the Liouville operator $\mathbf{L}$. For the stochastic differential equations (1.32) one has just to replace $\mathbf{L}$ with $\mathbf{L} + 1/2 \sum Q_{ij} \partial^2/\partial x_j \partial x_j$.

### *1.4.3 Natural measure*

The mathematical definition of dynamical system as $(\Omega, U^t, \mathrm{d}\mu(\mathbf{x}))$ can sound a bit pedantic, and sometimes in the physical literature one does not specify the invariant measure which is assumed to be the one "selected by the dynamics." This is an important and delicate point which deserves a short discussion.

In a generic dynamical system, there are a lot of invariant ergodic measures so one needs a criterion to select the "right one." In order to clarify this point, let us consider the logistic map (1.35) for $r = 4$. Noting that there exist unstable periodic trajectories $(x^{(1)}, x^{(2)}, \ldots, x^{(2^n)})$ of period $2^n$, with $n = 1, 2, \ldots$, one understands that, in addition to the $\rho^{\mathrm{inv}}(x)$ in (1.39), there are an infinite number of invariant measures:

$$\rho^{(n)}(x) = \sum_{k=1}^{2^n} 2^{-n} \delta(x - x^{(k)}). \tag{1.50}$$

Is there a reason why the $\rho^{\mathrm{inv}}(x)$ in (1.39) must be preferred to the $\rho^{(n)}(x)$? First let us note that the measures $\rho^{(n)}(x)$ in (1.50) are surely invariant but they are unstable, so if one adds to the map (1.35) a small noise term of strength $\epsilon$, the $\rho^{(n)}(x)$ have no role at all for any $\epsilon \neq 0$.

It is reasonable to accept the view that the system under investigation is inherently noisy (e.g., owing to the influence of the external world, not accounted for in the system). Therefore one can assume that the "correct" measure is that measure obtained by adding a (small) noise term (of strength $\epsilon$) to the dynamical system and then performing the limit $\epsilon \to 0$. The measure selected using this procedure is called a "natural" (or physical) measure and it is, by construction, a "dynamically robust" quantity. According to Eckmann and Ruelle (1985) this idea dates back to Kolmogorov.

In any numerical simulation both computers and algorithms are not "perfect," i.e. there are unavoidable "errors" due to truncations, roundoff and so on. In a similar way in laboratory experiments it is not possible to eliminate all the noisy interactions with the environment. So it is evident (at least from a physical point of view) that with numerical simulations and experiments one obtains an approximation of the natural measure.

Let us conclude by noting that the concepts of chaos, ergodicity and mixing, introduced in this chapter, are completely general, i.e. they have mathematical validity for generic (finite-dimensional) dynamical systems. On the other hand we will see in Chapters 4, 5 and 6 that in the cases with a very large number of degrees of freedom (which are those relevant for statistical mechanics) one can see the emergence of concepts and behaviors which are absent in low-dimensional systems.

## References

Arnold, V. I. (1974). *Équations Différentielles Ordinaires*, Moscow: Mir.
Arnold, V. I. and Avez, A. (1968). *Ergodic Problems of Classical Mechanics*, New York: Benjamin.
Chandrasekhar, S. (1943). Stochastic problems in physics and astronomy, *Rev. Mod. Phys.* **15**, 1.
Eckmann, J.-P. and Ruelle, D. (1985). Ergodic theory of chaos and strange attractors, *Rev. Mod. Phys.* **57**, 617.
Einstein, A. (1905). *Ann. Phys.* **17**, 549. English translation in: *Investigations on the Theory of Brownian Movement*, New York: Dover, 1956. This volume contains Einstein's papers on Brownian motion.
Gardiner, C. W. (1990). *Handbook of Stochastic Methods for Physics, Chemistry and the Natural Sciences*, Berlin: Springer.
Langevin, P. (1908). Sur la theorie du mouvement brownien, *C. R. Acad. Sci. (Paris)* **146**, 530. English translation in *Am. J. Phys.* **65**, 1079 (1997).
Laplace, P. S. (1814). *Essai Philosophique sur les Probabilités*, Paris: Seure Courciēre.
Lasota, A. and Mackey, M. C. (1985). *Probabilistic Properties of Deterministic Systems*, Cambridge: Cambridge University Press.
Lorenz, E. N. (1963). Deterministic nonperiodic flow, *J. Atmos. Sci.* **20**, 130.
Mehra, J. (2001). *The Golden Age of Theoretical Physics*, Singapore: World Scientific.
Ott, E. (1993). *Chaos in Dynamical Systems*, Cambridge: Cambridge University Press.
Perrin, J. (1913). *Les Atomes*, Paris: Alcan.
Poincaré, H. J. (1908). *Science et Méthode*, Paris: Flammarion.
Ruelle, D. (1989). *Chaotic Evolution and Strange Attractors*, Cambridge: Cambridge University Press.
Smoluchowski, M. (1906). Zur Kinetischen Theorie der Brownschen Molekularbewegung und der suspensionen, *Ann. Phys.* **21**, 756.

# 2

# Dynamical indicators for chaotic systems: Lyapunov exponents, entropies and beyond

At any time there is only a thin layer separating what is trivial from what is impossibly difficult. It is in that layer that discoveries are made . . .

*Andrei N. Kolmogorov*

An important aspect of the theory of dynamical systems is the formalization and quantitative characterization of the sensitivity to initial conditions. The Lyapunov exponents $\{\lambda_i\}$ are the indicators used to measure the average rate of exponential error growth in a system.

Starting from the idea of Kolmogorov of characterizing dynamical systems by means of entropy-like quantities, following the work by Shannon in information theory, another approach to dynamical systems has been developed in the context of information theory, data compression and algorithmic complexity theory. In particular, the Kolmogorov–Sinai entropy, $h_{\mathrm{KS}}$, can be defined and interpreted as a measure of the rate of information production of a system. Since the ability to produce information is tightly linked to the exponential diversification of trajectories, it is not a surprise that a relation exists between $h_{\mathrm{KS}}$ and $\{\lambda_i\}$, the Pesin relation.

One has to note that quantities such as $\{\lambda_i\}$ and $h_{\mathrm{KS}}$ are properly defined only in specific asymptotic limits, that is, very long times and arbitrary accuracy. Since in realistic situations one has to deal with finite accuracy and finite time – as Keynes said, in the long run we shall all be dead – it is important to take into account these limitations. Relaxing the requirement of infinite time, one can investigate the relevance of finite time fluctuations of the "effective" Lyapunov exponent. In addition, relaxing the limit of arbitrary high accuracy, one can introduce suitable tools, such as the finite size Lyapunov exponent (FSLE) and the $\epsilon$-entropy.

## 2.1 Dynamical systems approach

### 2.1.1 Characteristic Lyapunov exponents

The characteristic Lyapunov exponents (LE) are somehow an extension of linear stability analysis to the case of aperiodic motions. Roughly speaking, they measure the typical rate of exponential divergence of nearby trajectories. In this sense they give information on the rate of growth of a very small error in the initial state of a system.

Consider a dynamical system with an evolution law given, in the case of continuous time, by the differential equation

$$\frac{\mathrm{d}\mathbf{x}}{\mathrm{d}t} = \mathbf{F}(\mathbf{x}), \tag{2.1}$$

or, in the case of discrete time, by the map

$$\mathbf{x}(t+1) = \mathbf{G}(\mathbf{x}(t)). \tag{2.2}$$

The vector $\mathbf{x} \in \mathbb{R}^d$ specifies uniquely one state of the system. We assume that $\mathbf{F}$ and $\mathbf{G}$ are differentiable functions, that the evolution is well defined for time intervals of arbitrary extension, and that the motion takes place in a bounded region of the phase space. We intend to study the separation between two trajectories, $\mathbf{x}(t)$ and $\mathbf{x}'(t)$, starting from two close initial conditions, $\mathbf{x}(0)$ and $\mathbf{x}'(0) = \mathbf{x}(0) + \delta\mathbf{x}(0)$, respectively.

As long as the difference between the trajectories, $\delta\mathbf{x}(t) = \mathbf{x}'(t) - \mathbf{x}(t)$, remains small (infinitesimal, strictly speaking), it can be regarded as a vector, $\mathbf{z}(t)$, in the tangent space. The time evolution of $\mathbf{z}(t)$ is given by the linearized differential equations,

$$\frac{\mathrm{d}z_i(t)}{\mathrm{d}t} = \sum_{j=1}^{d} \left.\frac{\partial F_i}{\partial x_j}\right|_{\mathbf{x}(t)} z_j(t), \tag{2.3}$$

or, in the case of discrete time by the linear maps,

$$z_i(t+1) = \sum_{j=1}^{d} \left.\frac{\partial G_i}{\partial x_j}\right|_{\mathbf{x}(t)} z_j(t). \tag{2.4}$$

Under a rather general hypothesis, Oseledec (1968) proved that for almost all initial conditions $\mathbf{x}(0)$ in the tangent space, an orthonormal basis $\{\mathbf{e}_i\}$ exists such that, for large times,

$$\mathbf{z}(t) = \sum_{i=1}^{d} c_i \mathbf{e}_i \, \mathrm{e}^{\lambda_i t}, \tag{2.5}$$

where the basis $\{\mathbf{e}_i\}$ depends on $\mathbf{x}(0)$ and $t$ and the coefficients $\{c_i\}$ depend on $\mathbf{z}(0)$ (Goldhirsch *et al.* 1987). The quantities $\lambda_1 \geq \lambda_2 \geq \cdots \geq \lambda_d$ are called *characteristic Lyapunov exponents*. If the dynamical system has an ergodic invariant measure, the spectrum of Lyapunov exponents $\{\lambda_i\}$ does not depend on the initial condition, except for a set of measure zero with respect to the invariant measure.

Loosely speaking, (2.5) tells us that in the phase space, where the motion evolves, a $d$-dimensional sphere of small radius $\epsilon$ centered in $\mathbf{x}(0)$ is deformed with time into an ellipsoid of semi-axes $\epsilon_i(t) = \epsilon \exp(\lambda_i t)$, directed along the $\mathbf{e}_i$ vectors. Furthermore, for a generic small perturbation $\delta\mathbf{x}(0)$, the distance between a trajectory and a perturbed trajectory behaves as

$$|\delta\mathbf{x}(t)| \sim |\delta\mathbf{x}(0)|\, \mathrm{e}^{\lambda_1 t} \left[1 + O\left(\mathrm{e}^{-(\lambda_1 - \lambda_2)t}\right)\right]. \tag{2.6}$$

If $\lambda_1 > 0$ we have a rapid (exponential) amplification of an error on the initial condition. In such a case, the system is chaotic and, de facto, unpredictable over long times. Indeed, if we put $\delta_0 = |\delta\mathbf{x}(0)|$ for the initial error, and we want to predict the states of the system with a certain tolerance $\Delta$ (not too large), then the prediction is possible just up to a *predictability time* given by

$$T_{\mathrm{p}} \sim \frac{1}{\lambda_1} \ln\left(\frac{\Delta}{\delta_0}\right). \tag{2.7}$$

Therefore $T_{\mathrm{p}}$ is basically determined by the largest Lyapunov exponent, since its dependence on $\delta_0$ and $\Delta$ is very weak. Because of its pre-eminent role, very often one simply refers to $\lambda_1$ as "the Lyapunov exponent," and one indicates it by $\lambda$.

Equation (2.6) suggests how to compute $\lambda_1$ numerically. We introduce the response, after a time $t$, to an infinitesimal perturbation on $\mathbf{x}(\tau)$, defined as follows:

$$R_\tau(t) \equiv \frac{|\mathbf{z}(\tau + t)|}{|\mathbf{z}(\tau)|} = \frac{|\delta\mathbf{x}(\tau + t)|}{|\delta\mathbf{x}(\tau)|}. \tag{2.8}$$

In ergodic systems the Lyapunov exponent $\lambda_1$ is obtained by averaging the logarithm of the response along a typical trajectory:

$$\lambda_1 = \lim_{t\to\infty} \frac{1}{t} \langle \ln R_\tau(t) \rangle, \tag{2.9}$$

where $\langle\cdot\rangle$ denotes the time average $\lim_{T\to\infty}(1/T)\int_{\tau_0}^{\tau_0+T}(\cdot)\mathrm{d}\tau$. Oseledec's theorem implies that $(1/t)\ln R_\tau(t)$, for large $t$, is a non-random quantity, i.e. for almost all the initial conditions its value does not depend on the specific initial condition. Therefore, for large times, the average in (2.9) can be neglected.

As the typical growth rate of a generic small segment in phase space is driven by the largest Lyapunov exponent, the sum of the first $n$ ($\leq d$) Lyapunov exponents controls the variations of small $n$-dimensional volumes in phase space. This gives us a way to compute the sub-leading Lyapunov exponents. After the selection of $n \leq d$ non-parallel (i.e. independent) tangent vectors $[\mathbf{z}^{(1)}(0), \ldots, \mathbf{z}^{(n)}(0)]$, one introduces the $n$-order response $R_\tau^{(n)}(t)$ (Benettin *et al.* 1980)

$$R_\tau^{(n)}(t) \equiv \frac{|\mathbf{z}_1(t+\tau) \times \mathbf{z}_2(t+\tau) \times \cdots \times \mathbf{z}_n(t+\tau)|}{|\mathbf{z}_1(\tau) \times \mathbf{z}_2(\tau) \times \cdots \times \mathbf{z}_n(\tau)|}. \tag{2.10}$$

Analogously to the Lyapunov exponent, it can be shown that

$$\sum_{i=1}^{n} \lambda_i = \lim_{t \to \infty} \frac{1}{t} \langle \ln R_\tau^{(n)}(t) \rangle. \tag{2.11}$$

Let us stress that the Lyapunov exponents give information on the typical behaviors along a generic trajectory, followed for infinite time and keeping the perturbation infinitesimally small. In this respect, they are global quantities characterizing fine-grained properties of a system.

### 2.1.2 Lyapunov exponents and dimensions

As we discussed briefly in Chapter 1, the motions of dissipative chaotic systems can evolve on strange attractors, that are created by the competitive effect of stretching and folding. These structures typically have a non-integer fractal dimension.[1] It is quite natural to wonder whether a relation exists between this dimension and the Lyapunov exponents, since the positive ones give rise to stretching along the corresponding eigendirections, while the negative ones are responsible for contraction, connected to the folding.

Kaplan and Yorke (1979) conjectured that the fractal dimension of the set where the natural measure is concentrated on the attractor is estimated by the Lyapunov dimension

$$D_{\mathrm{KY}} = J + \frac{\sum_{i=1}^{J} \lambda_i}{|\lambda_{J+1}|} \tag{2.12}$$

where $J$ is the maximum integer such that $\sum_{i=1}^{J} \lambda_i \geq 0$. It can be shown that in some cases (e.g. hyperbolic maps of the plane), $D_{\mathrm{KY}}$ coincides with the information dimension $D_{\mathrm{I}}$ (Young 1982). Let us briefly sketch the argument that leads to the Kaplan and Yorke formula. By the definition of $J$, $J$-dimensional hypervolumes should increase or at least remain constant (since $\sum_{i=1}^{J} \lambda_i \geq 0$), while the $J+1$-dimensional hypervolumes should be contracted to zero (since $\sum_{i=1}^{J+1} \lambda_i < 0$). The dimension of the attractor is thus larger than $J$ and smaller than

[1] A chaotic attractor is characterized by its invariant measure, which is singular with respect to the Lebesgue measure. The scaling properties of the invariant measure are given by Renyi dimensions. For the sake of simplicity we do not discuss the difference between the several dimensions; for an analysis in terms of multifractal objects see Paladin and Vulpiani (1987).

$J + 1$. Finally, a linear interpolation allows one to obtain the fractional part of the dimension.

The Kaplan–Yorke formula has great practical relevance. A direct computation of the dimensions, for example using the celebrated Grassberger and Procaccia (1983) method (or a suitable generalization (Paladin and Vulpiani 1987)), has rather severe limitations. This is due to the obvious fact that the number of points necessary to estimate $D_{\mathrm{I}}$ increases exponentially with $D_{\mathrm{I}}$, so it is practically impossible to measure a value of $D_{\mathrm{I}}$ larger than 5 or 6 (Eckmann and Ruelle 1992). In contrast, the Lyapunov exponents can be easily numerically computed even in high-dimensional systems. Therefore the Kaplan–Yorke formula is the unique way to compute numerically the dimension of a high-dimensional attractor.

## 2.2 Information theory approach

In experimental investigations of physical processes, we typically have access to the system only through a measuring device which produces a time record of a certain observable, i.e. a sequence of data. In this regard a system, whether or not chaotic, generates messages and may be regarded as a source of information. This remark opens the possibility of studying dynamical systems from a very interesting point of view.

Information has been given a rigorous and quantitative definition, in the framework of the theory of communication, to cope with the practical problem of transmitting a message in the cheapest way without losing information. The characterization of the information contained in a sequence can be approached from two very different points of view. The first one, that of information theory (Shannon 1948), is a statistical approach, i.e. it does not consider the transmission of a specific message (sequence) but refers to the statistical properties of all the messages emitted by the source. The information theory approach characterizes the source of information, so that it gives us a powerful method to characterize chaotic systems.

The second point of view considers the problem of characterizing a single sequence. This has led to the theory of algorithmic complexity and algorithmic information theory (Solomonoff 1964, Kolmogorov 1965, Chaitin 1966).

### *2.2.1 Shannon entropy*

At the end of the 1940s Shannon (1948) introduced powerful concepts and techniques for a systematic study of sources emitting sequences of discrete symbols (e.g. binary digit sequences). Originally information theory was introduced in the practical context of electric communications, nevertheless in a few years it became an important branch of both pure and applied probability theory with strong relations with other fields such as computer science, cryptography, biology and physics (Zurek 1990).

For the sake of self-consistency we recall briefly the basic concepts and ideas of Shannon entropy. Consider a source that can output $m$ different symbols; denote by $s(t)$ the symbol emitted by the source at time $t$ and by $P(W_N)$ the probability that a given word $W_N = (s(1), s(2), \ldots, s(N))$, of length $N$, is emitted:

$$P(W_N) = P(s(1), s(2), \ldots, s(N)). \tag{2.13}$$

We assume that the source is statistically stationary, so that the following statistical time translation invariance holds: $P(s(1), \ldots, s(N)) = P(s(t+1), \ldots, s(t+N))$.

We recall that, given a set of events whose probabilities of occurrence are $(p_1, p_2, \ldots, p_n)$, under rather natural assumptions and consistency requirements, one can prove (Khinchin 1957) that

$$H \propto -\sum_{i=1}^{n} p_i \log p_i$$

is the unique quantity (called the entropy of the probabilistic scheme) which measures the average uncertainty about one outcome, or the average information that is supplied by one occurrence. In the definition of $H$ the proportionality sign is due to freedom in the choice of the base of the logarithm.

Now we introduce the $N$-block (or $N$-word) entropy,

$$H_N = -\sum_{\{W_N\}} P(W_N) \ln P(W_N), \tag{2.14}$$

which measures the information content of the $N$-word ensemble, and the "differential" entropy

$$h_N = H_{N+1} - H_N = -\sum_{\{W_N\}} P(W_N) \sum_{s} P(s|W_N) \ln P(s|W_N), \tag{2.15}$$

where $P(s|W_N)$ is the probability of the symbol $s$ conditioned by the word $W_N$. So $h_N$ is the average information supplied by observing the $(N+1)$th symbol, provided the $N$ previous ones are known. One can also say that $h_N$ is the average uncertainty about the $(N+1)$th symbol, provided the $N$ previous ones are given. For a stationary source the limits in the following equations exist, are equal and define the Shannon entropy $h_{\mathrm{Sh}}$:

$$h_{\mathrm{Sh}} = \lim_{N\to\infty} \frac{H_N}{N} = \lim_{N\to\infty} h_N. \tag{2.16}$$

The $h_N$ are non-increasing quantities, $h_{N+1} \leq h_N$, that is, knowledge of a longer past history cannot increase the uncertainty on the next outcome.

In the case of a periodic source, with period $T$, $h_{\rm Sh} = 0$ since $h_N = 0$ as soon as $N$ is larger than $O(\ln T)$. In the case of a non-correlated source (i.e. emitting independent identically distributed symbols) one has $h_N = h_1 = H_1 = h_{\rm Sh}$. In the case of a $k$th order Markov process $h_N = h_{\rm Sh}$ for all $N \geq k$. This is because a $k$th order Markov process has the property that the conditional probability of having a given symbol depends only on the results of the last $k$ times, i.e.

$$P(s(t)|s(t-1), s(t-2), \ldots) = P(s(t)|s(t-1), s(t-2), \ldots, s(t-k)) \,. \tag{2.17}$$

The Shannon entropy is a measure of the "surprise" the source emitting the sequences can give us, since it quantifies the asymptotic uncertainty about the further emission of a symbol in a very long sequence. However, it can also be viewed as a measure of the richness (or "complexity") of the source. This can be expressed precisely by the first theorem of Shannon–McMillan (Khinchin 1957) which applies to stationary ergodic sources. In this context ergodicity can be thought of as implying that the sequences emitted by the source share the same statistical properties; so the frequency of appearance of an $N$-word in a (infinitely) long sequence approaches a definite limit (its probability) independent of the particular sequence.

**Theorem** *If $N$ is large enough, the ensemble of $N$-long sequences can be separated in two classes, $\Omega_1(N)$ and $\Omega_0(N)$, such that all the words $W_N \in \Omega_1(N)$ have (roughly) the same probability $P(W_N) \sim \exp(-N h_{\rm Sh})$ and*

$$\sum_{W_N \in \Omega_1(N)} P(W_N) \to 1 \qquad \text{for } N \to \infty, \tag{2.18}$$

*while*

$$\sum_{C_N \in \Omega_0(N)} P(C_N) \to 0 \qquad \text{for } N \to \infty. \tag{2.19}$$

The meaning of this theorem is the following. An $m$-state process admits in principle $m^N$ possible sequences of length $N$, but the number of sequences effectively observable, $N_{\rm eff}(N)$, (those in $\Omega_1(N)$, also called *typical* sequences) is

$$N_{\rm eff}(N) \sim \exp(N h_{\rm Sh}). \tag{2.20}$$

Note that $N_{\rm eff} \ll m^N$ if $h_{\rm Sh} < \ln m$. The entropy per symbol, $h_{\rm Sh}$, is a property of the source. However, because of the ergodicity it can be obtained by analyzing just one single sequence in the ensemble of typical sequences, and it can also be viewed as a property of each one of the typical sequences. Therefore, as in the following, one may speak about the Shannon entropy of a sequence, in this context.

The above theorem in the case of processes without memory is just the law of large numbers. Let us observe that (2.20) is somehow the equivalent in information theory of the Boltzmann law in statistical thermodynamics: $S \propto \ln W$, $W$ being the number of possible microscopic configurations and $S$ the thermodynamic entropy. This justifies the name "entropy" for $h_{\rm Sh}$.

Let us now mention another important result about the Shannon entropy. It is not difficult to recognize that the quantity $h_{\rm Sh}$ sets the maximum compression rate of a sequence $\{s(1), s(2), s(3), \ldots\}$. Indeed a theorem of Shannon states that, if the length $T$ of a sequence is large enough, one cannot construct another sequence (always using $m$ symbols), from which it is possible to reconstruct the original one, whose length is smaller than $(h_{\rm Sh}/\ln m)T$ (Shannon 1948). In other words $h_{\rm Sh}/\ln m$ is the maximum allowed compression rate.

The relation between Shannon entropy and the data compression problem is well highlighted by considering the Shannon–Fano code to map $\mathcal{N}$ objects (e.g. the $N$-words $W_N$) into sequences of binary digits (0, 1) (Welsh 1989). The main goal in building a code (with the intention to use it in communications and digital storage) is to define the most efficient coding procedure, i.e. that which generates the shortest possible (coded) sequence. The Shannon–Fano code is as follows. First one orders the $\mathcal{N}$ objects according to their probabilities, in a decreasing way, $p_1 \geq p_2 \geq \cdots \geq p_{\mathcal{N}}$. Then the passage from the $\mathcal{N}$ objects to the symbols (0, 1) is obtained by defining the coding $E(r)$, of binary length $\ell(E(r))$, of the $r$th object with the requirement that the expected total length, $\sum_r p_r \ell_r$, be the minimal one. This can be realized with the following choice:

$$-\ln_2 p_r \leq \ell(E(r)) < 1 - \ln_2 p_r. \tag{2.21}$$

In this way highly probable objects are mapped into short code words while the low probability objects are mapped into longer code words. So, in the case of $N$-words, the average code length of a word is bounded by

$$\frac{H_N}{\ln 2} \leq \sum_r p_r \ell(E(r)) \equiv \langle \ell_N \rangle \leq \frac{H_N + 1}{\ln 2}, \tag{2.22}$$

and in the limit $N \to \infty$ one has

$$\lim_{N\to\infty} \frac{\langle \ell_N \rangle}{N} = \frac{h_{\rm Sh}}{\ln 2}. \tag{2.23}$$

This means that in a good binary coding, the mean length of an $N$-word is equal to $N$ times the Shannon entropy, apart from a multiplicative factor, due to the fact that in the definition (2.16) of $h_{\rm Sh}$ we used the natural logarithm (which happens to be used more often in a dynamical system context) and here we want to work with a two-symbol code and measure information in bits.

*The transient behavior of the block entropies*

In addition to the asymptotic values of $h_N$, the transient behavior of $H_N$ may also be important. This has been underlined by Grassberger (1986) who proposed to consider another quantity as well as the Shannon entropy: the "effective measure of complexity," namely

$$C = \sum_{N=1}^{\infty} N(h_{N-1} - h_N) = \sum_{N=0}^{\infty} N(h_N - h_{\mathrm{Sh}}). \tag{2.24}$$

From the above definition, it follows that for large $N$, the block entropies grow as:

$$H_N \simeq C + N h_{\mathrm{Sh}}. \tag{2.25}$$

For trivial processes, for example for Bernoulli schemes or for a Markov chain of order 1, $C = 0$ and $h_{\mathrm{Sh}} > 0$, while in a periodic sequence (of period $\mathcal{T}$) $h_{\mathrm{Sh}} = 0$ and $C \sim \ln(\mathcal{T})$.

The latter example shows the interest of the notion, since it allows discrimination of behaviors associated with the same Shannon entropy. First of all, within all stochastic processes with the same $H_k$, with $k \leq N$, $C$ is minimal for the Markov processes of order $N - 1$ compatible with the block entropies of order $k \leq N$. The decrement $\delta h_N = h_{N-1} - h_N$ is the average amount of information (per symbol) by which the uncertainty of $s(N+1)$ decreases when $s(N), s(N-1), \ldots, s(1)$ are known. Thus $C$ is the average usable part of the information about the past which has to be remembered at any time if one wants to be able to reconstruct the sequence fully from its past.

In Chapter 3 we will see how even systems with $h_{\mathrm{Sh}} = 0$ but large $C$ can have interesting behavior and can be useful, for example for the generation of a sequence of (pseudo) random numbers with deterministic algorithms.

### 2.2.2 *The Kolmogorov–Sinai entropy*

After introducing the Shannon entropy we give a definition of the Kolmogorov–Sinai (or *metric*) entropy (Kolmogorov 1958, Sinai 1959). Consider a trajectory, $\mathbf{x}(t)$, generated by a deterministic system, sampled at the times $t_j = j\,\tau$, with $j = 1, 2, 3, \ldots$. Perform a finite partition $\mathcal{A}$ of the phase space. Since we are considering motions that evolve in a bounded region, all the trajectories visit a finite number of different cells (also known as the atoms of the partition), each one identified by a symbol. With the finite number of symbols $\{s\}_{\mathcal{A}}$ enumerating the cells of the partition, the time-discretized trajectory $\mathbf{x}(t_j)$ determines a sequence $\{s(1), s(2), s(3), \ldots\}$, whose meaning is clear: at the time $t_j$ the trajectory is in the cell labeled $s(j)$. To each subsequence of length $N\tau$ one can associate a word of length $N$: $W_j^N(\mathcal{A}) = (s(j), s(j+1), \ldots, s(j+(N-1)))$. If the system is ergodic, as we suppose, from the observed frequencies of the words one obtains the probabilities by which one calculates the block entropies $H_N(\mathcal{A})$:

$$H_N(\mathcal{A}) = -\sum_{\{W^N(\mathcal{A})\}} P(W^N(\mathcal{A})) \ln P(W^N(\mathcal{A})). \tag{2.26}$$

It is important to note that the probabilities $P(W^N(\mathcal{A}))$, computed by the frequencies of $W^N(\mathcal{A})$ along a trajectory, are essentially dependent on the stationary measure selected by the trajectory. This implies a dependence on this measure of all the quantities defined below, $h_{\mathrm{KS}}$ included. We shall always consider the natural invariant measure and do not indicate this kind of dependence. The entropy per unit time of the trajectory with respect to the partition $\mathcal{A}$, $h(\mathcal{A})$, is defined as follows:

$$h_N(\mathcal{A}) = \frac{1}{\tau}[H_{N+1}(\mathcal{A}) - H_N(\mathcal{A})], \tag{2.27}$$

$$h(\mathcal{A}) = \lim_{N\to\infty} h_N(\mathcal{A}) = \frac{1}{\tau} \lim_{N\to\infty} \frac{1}{N} H_N(\mathcal{A}). \tag{2.28}$$

Notice that, for the deterministic systems we are considering, the entropy per unit time does not depend on the sampling time $\tau$ (Billingsley 1965) (by contrast with the case of stochastic motions, see Gaspard and Wang (1993)). The Kolmogorov–Sinai entropy, by definition, is the supremum of $h(\mathcal{A})$ over all possible partitions (Billingsley 1965, Eckmann and Ruelle 1985)

$$h_{\mathrm{KS}} = \sup_{\mathcal{A}} h(\mathcal{A}). \tag{2.29}$$

It is not simple at all to determine $h_{\mathrm{KS}}$ according to this definition. A useful tool in this respect would be the Kolmogorov–Sinai theorem, by means of which one is granted that $h_{\mathrm{KS}} = h(\mathcal{G})$ if $\mathcal{G}$ is a generating partition. A partition is said to be generating if every infinite sequence $\{s(j)\}_{j=1,\ldots,\infty}$ individuates a single initial point. We have to note that, given a generating partition $\mathcal{G}$, one can consider the set $F^{-\tau}\mathcal{G}$, of the pre-images of the atoms of $\mathcal{G}$, and then the intersections of all the atoms with all the pre-images. This generates the atoms of a new partition, called a dynamical refinement of $\mathcal{G}$, and that is itself a generating partition. Since one can consider pre-images referring to an arbitrary number of times, one obtains an infinity of generating partitions.

However, the difficulty now is that, with the exception of very simple cases, we do not know how to construct a generating partition. We only know that, for an automorphism (i.e. invertible transformation), according to a theorem proved by Krieger (1970), there exists a generating partition with $k$ elements such that $e^{h_{\mathrm{KS}}} \leq k \leq e^{h_{\mathrm{KS}}} + 1$. Then, a more tractable way to define $h_{\mathrm{KS}}$ is based upon considering the partition $\mathcal{A}_\epsilon$ made up by a grid of cubic cells of edge $\epsilon$, from which one has

$$h_{\mathrm{KS}} = \lim_{\epsilon \to 0} h(\mathcal{A}_\epsilon). \tag{2.30}$$

We expect $h(\mathcal{A}_\epsilon)$ to become independent of $\epsilon$ when $\mathcal{A}_\epsilon$ is fine enough to be "contained" in a generating partition.

For discrete time maps the above is still valid, with $\tau = 1$ (however, Krieger's theorem only applies to invertible maps).

The important point to note is that, while the entropy of a deterministic system is finite, for a truly stochastic (i.e. non-deterministic) system, with continuous states, $h(\mathcal{A}_\epsilon)$ is not bounded and $h_{\mathrm{KS}} = \infty$.

### *About the Kolmogorov–Sinai entropy*

In this subsection, to simplify the discussion, we consider the case of discrete time evolutions. All the results can be adapted to the continuous time case.

We begin by underlining that, even though we defined $h_{\mathrm{KS}}$ for a (continuous-valued) deterministic system, in what follows we were faced with Shannon's analysis of finite-alphabet sequences, originating from the partitioning of the phase space. It is clear that the theoretical tools developed by Shannon, to study the statistical properties of the ensembles of sequences, are independent of the way the latter are generated, i.e. either by a deterministic rule or by a probabilistic rule. So the same is true for the Kolmogorov–Sinai entropy, which is a quantity pertaining to both deterministic and probabilistic processes. Bringing our attention to ensembles of evolutions, one has a unifying scheme to study all kinds of dynamical processes. An important fact about the Kolmogorov–Sinai entropy is that it is a (non-negative) real number that characterizes a dynamical system in an invariant way. This means that two systems have the same Kolmogorov–Sinai entropy if a one-to-one transformation between them exists, with correspondence between subsets of equal probability and commuting with the two evolutions, such that the systems are isomorphic and may be seen as different realizations of the same abstract dynamical system. This invariant numerical quantity allowed Kolmogorov to split the class of measure-preserving transformations according to the value of their entropies. In particular, this showed that stationary sequences of independent random variables (Bernoulli shifts) with different entropy values are not isomorphic (this was an open problem at the time). The converse assertion, i.e. two Bernoulli shifts with the same entropy are isomorphic, was proved by Ornstein (1974).

An important conceptual point to stress is the following. When the Kolmogorov–Sinai entropy was introduced a general feeling was that it could also set a quantitative boundary separating deterministic from probabilistic evolutions. Indeed, since $h_{\mathrm{KS}}$ quantifies the uncertainty remaining on the next state of the system, if an infinitely long past is known, the Kolmogorov–Sinai entropy of a deterministic system is expected to be zero, at variance with the probabilistic case. The finding of deterministic evolutions with positive entropy came as a big surprise. However, the theoretical tools underlying the definition of the Kolmogorov–Sinai entropy provide us with a clear quantitative characterization of the notion of *deterministic chaos* or *deterministic randomness*. In fact we know that for a deterministic invertible transformation with $h_{\mathrm{KS}} > 0$ a generating partition exists with a finite number, $k$, of elements such that $\mathrm{e}^{h_{\mathrm{KS}}} \leq k \leq \mathrm{e}^{h_{\mathrm{KS}}} + 1$. This implies that the trajectories of the system are in isomorphic correspondence with those of a $k$-state (discrete time) random process, whose asymptotic average uncertainty about the state is given by $h_{\mathrm{KS}}$. We have a *mathematical equivalence* between a deterministic evolution and a probabilistic process with finite state space. Moreover, this also means that to obtain a faithful description of the original system, when the complete trajectories are involved, we do not need too many details in the single observations.

Since the generating property of a partition permits a faithful representation of a dynamical system as a finite-state process, by means of a kind of "natural" coarse graining, some effort has been devoted to the construction of generating partitions. It is worth recalling that if a generating partition exists, then there is an infinity of them, given by its dynamical refinements. Clearly, the goal is to find partitions with a small (minimum) number of states. In a few cases a solution is known. For the one-dimensional maps a generating partition is obtained by considering the coordinates where maxima, minima and vertical asymptotes (the critical points) are located (Collet and Eckmann 1980). The effect of such a partitioning of the interval is that the pre-images of a single point coming from different monotonic branches of the map have different symbols, i.e. different histories. A solution is also available for some two-dimensional maps. For instance, for the following baker-like map, defined on the unit square,

$$\begin{cases} x(t+1) = \dfrac{1}{p}\, x(t) \\ \\ y(t+1) = p\; y(t) \end{cases} \qquad \text{if} \quad 0 \leq x < p,$$

$$\begin{cases} x(t+1) = \dfrac{1}{q}\, x(t) - \dfrac{p}{q} \\ \\ y(t+1) = q\; y(t) + p \end{cases} \qquad \text{if} \quad p \leq x < 1,$$

with $p + q = 1$, a generating partition is obtained by dividing the square into two rectangles with height 1, whose bases are intervals of length $p$ and $q$. This map turns out to be equivalent to a Bernoulli shift: its sequences, recorded on the described partition, are indistinguishable from those produced by independent extractions of two symbols with probability $p$ and $q$. It is an example of a general class of systems usually referred to as *hyperbolic* systems (Eckmann and Ruelle 1985). For these systems one can introduce the notion of *Markov partition*, that is, a generating partition enjoying additional properties, very useful when computing, for example, invariant measures and entropies (Badii and Politi 1997, Dorfman 1999). However, it is difficult to construct Markov partitions, even when they exist; moreover a generic system is not hyperbolic. Thus the construction of generating partitions in general relies upon reasonable recipes. For instance, Grassberger and Kantz (1985) consider the two-dimensional Hénon map which is one of the simplest of the chaotic dissipative systems. By analogy with the one-dimensional maps, where the generating partitions are determined by the critical points, they adopt a guiding principle that the dividing lines of the partition must connect points where the stable and unstable directions are tangential, the so-called *homoclinic tangencies* (see, for instance, Badii and Politi (1997)). Of course the construction of such a partition is not a simple task. For an explanation of this idea, its extension to conservative systems and discussion of the difficulties, see Grassberger *et al.* (1989), Giovannini and Politi (1991), Christiansen and Politi (1995, 1996), Jaeger and Kantz (1997), and Lai *et al.* (1999). A different strategy for finding generating partitions has been proposed by Plumecoq and Lefranc (2000a, 2000b) and Davidchack *et al.* (2000), for systems endowed with unstable periodic orbits that are dense in the attractor. Some algorithms to estimate generating partitions from observed data are proposed by Kennel and Buhl (2003) and Hirata *et al.* (2004).

### *Kolmogorov–Sinai entropy and Lyapunov exponents*

We can give an interpretation of the Kolmogorov–Sinai entropy based on the Shannon–McMillan theorem. We construct a partition of the phase space with cells of volume $\epsilon^d$, then we consider the trajectories starting from a region of volume $V_0 = \epsilon^d$ at $t_0$ and sampled at discrete times $t_j = j\,\tau$ $(j = 1, 2, 3, \ldots, t)$; in the case of a map one can put $\tau = 1$. In this way a unique sequence of symbols $\{s(0), s(1), s(2), \ldots\}$ is associated with one trajectory. In a chaotic system a great number of different symbolic sequences originates from the same initial cell, because of the divergence of nearby trajectories. The total number of admissible symbolic sequences, $\widetilde{N}(\epsilon, t)$, increases exponentially with a rate given by the topological entropy

$$h_{\mathrm{T}} = \lim_{\epsilon \to 0} \lim_{t \to \infty} \frac{1}{t} \ln \widetilde{N}(\epsilon, t). \tag{2.31}$$

However, if we discard the sequences of very small probability and we consider only the number of remaining sequences $N_{\mathrm{eff}}(\epsilon, t) \leq \widetilde{N}(\epsilon, t)$, whose total probability in the long time limit is as near to 1 as one wants, we arrive at a more physical quantity which is the Kolmogorov–Sinai entropy:

$$h_{\mathrm{KS}} = \lim_{\epsilon \to 0} \lim_{t \to \infty} \frac{1}{t} \ln N_{\mathrm{eff}}(\epsilon, t) \leq h_{\mathrm{T}}. \tag{2.32}$$

Thus $h_{\mathrm{KS}}$ quantifies the long time exponential rate of growth of the number of "effective" trajectories of a system, those that can be detected numerically or experimentally and that are associated with the natural measure.

By means of a heuristic reasoning, we can obtain a relation between $h_{\mathrm{KS}}$ and the Lyapunov exponents. We may wonder what is the number of cells where, at a time $t > t_0$, the points that evolved from $V_0$ can be found, i.e. we wish to know how big is the coarse-grained volume $V(\epsilon, t)$, occupied by the states evolved from $V_0$, if the minimum volume we can observe is $\epsilon^d$. As stated at the end of Section 2.1.1, we have $V(t) \sim V_0 \exp(t \sum_{i=1}^{d} \lambda_i)$. However, this is true only in the limit $\epsilon \to 0$. In this (unrealistic) limit, $V(t) = V_0$ for a conservative system (where $\sum_{i=1}^{d} \lambda_i = 0$) and $V(t) < V_0$ for a dissipative system (where $\sum_{i=1}^{d} \lambda_i < 0$). As a consequence of limited resolution power, in the evolution of the volume $V_0 = \epsilon^d$ the effect of the contracting directions (associated with the negative Lyapunov exponents) is completely lost. We can experience only the effect of the expanding directions, associated with the positive Lyapunov exponents. As a consequence, in the typical case, the coarse-grained volume behaves as

$$V(\epsilon, t) \sim V_0 \, \mathrm{e}^{(\sum_{\lambda_i > 0} \lambda_i) t}, \tag{2.33}$$

when $V_0$ is small enough. Since $N_{\mathrm{eff}}(\epsilon, t) \propto V(\epsilon, t)/V_0$, one has

$$h_{\mathrm{KS}} = \sum_{\lambda_i > 0} \lambda_i \, . \tag{2.34}$$

This argument can be made more rigorous (Pesin 1976) to derive the Pesin relation:

$$h_{\mathrm{KS}} \leq \sum_{\lambda_i > 0} \lambda_i . \tag{2.35}$$

Because of its relation with the Lyapunov exponents – or by the definition (2.32) – it is clear that $h_{\mathrm{KS}}$ is also a fine-grained and global characterization of a dynamical system.

*Extended systems*

Let us discuss briefly *extended* dynamical systems, whose degrees of freedom depend on space and time, and which can display unpredictable behaviors in both time and space evolution, i.e. *spatiotemporal chaos*. Following Hohenberg and Shraiman (1988) we can give a more precise meaning to the terms *spatiotemporal* chaos and *extended* systems. For a generic system of size $L$, three characteristic lengths can be defined: $\ell_D$, $\ell_E$, $\xi$ respectively associated with the scales of *dissipation* (e.g. the scale at which dissipation becomes effective, smaller scales can be considered as inactive), *excitation* (e.g. the scale at which energy is injected in the system) and *correlation* (that we assume can be suitably defined). Note that $\ell_E$ is an external parameter, to be assumed as a fixed quantity, for example one can consider $\ell_E \sim O(L)$. Now one has two limiting situations.

When the characteristic lengths (generated by the dynamics) are of the same order ($\ell_D, \xi \sim O(L)$) distant regions of the system are strongly correlated. Because of the coherence, the spatial nature is not very important and one speaks of *temporal* chaos, i.e. the system is basically low dimensional. More interesting is the case $L \gg \xi \gg \ell_D$ where distant parts of the system are weakly correlated so that the number of (active) degrees of freedom is an extensive quantity, i.e. it increases with the system size, and is asymptotically proportional to $L^d$, where $d$ is the spatial dimension. As a consequence, the number of positive Lyapunov exponents, the Kolmogorov–Sinai entropy and the attractor dimension, $D_{KY}$, are extensive quantities. The above picture is just an approximate scenario (see Hohenberg and Shraiman (1988) for further details) but sufficiently broad to include systems ranging from fluid dynamics to biological and chemical reactions.

A great simplification in the study of extended systems can be achieved by considering discrete time and space models, and introducing the *coupled map lattices* (CML) (Kaneko 1993), i.e. maps defined on a discrete lattice. A typical one-dimensional CML (the extension to $d$-dimensions is straightforward) can be written in the following way:

$$\mathbf{x}_i(t+1) = (1-\varepsilon)\mathbf{f}_a[\mathbf{x}_i(t)] + \frac{1}{2}\varepsilon\,(\mathbf{f}_a[\mathbf{x}_{i+1}(t)] + \mathbf{f}_a[\mathbf{x}_{i-1}(t)]). \qquad (2.36)$$

Here $i = -L/2, \ldots, L/2$, where $L$ is the lattice size, $\mathbf{x} \in \mathbb{R}^n$ is the state variable which depends on the site and time, and $\mathbf{f}_a$ is a map, which drives the local dynamics and depends on a control parameter $a$. Usually, periodic boundary conditions $\mathbf{x}_{i+L} = \mathbf{x}_i$ are assumed and, for scalar variables ($n = 1$), one studies coupled logistic maps, $f_a(x) = ax(1-x)$ or tent maps, $f_a(x) = a|1/2 - |x - 1/2||$.

Lyapunov exponents, attractor dimensions and the Kolmogorov–Sinai entropy are also well-defined quantities for extended systems with a finite number of degrees of freedom. In order to construct a statistical description of spatiotemporal chaos, as Ruelle (1982) pointed out, one requires the existence of a good thermodynamic limit for the Lyapunov spectrum $\{\lambda_i(L)\}_{i=1,L}$. This means the existence of the limit

$$\lim_{L\to\infty} \lambda_i(L) = \Lambda(x), \qquad (2.37)$$

where $x = i/L$, in the limit $L \to \infty$, becomes a continuous index in [0, 1], and $\Lambda(x)$ is a non-increasing function. The function $\Lambda(x)$ can be viewed as a *density* of Lyapunov exponents. If such a limit did not exist, it would be impossible to construct a statistical description of spatiotemporal chaos, i.e. the phenomenology of these systems would depend on $L$.

Once the existence of a Lyapunov density is proved, one can use the Kaplan–Yorke formula and the Pesin relation, and introduce an entropy for the degree of freedom $\mathcal{H}_{\mathrm{KS}}$ and a *dimension density* $\mathcal{D}_{\mathrm{KY}}$ $\mathcal{D}_{\mathrm{KY}}$, that is to say a density of active degrees of freedom. Indeed one can write

$$h_{\mathrm{KS}} = \sum_{\lambda_i > 0} \lambda_i \longrightarrow \sum_{\Lambda > 0} \Lambda\left(\frac{i}{L}\right) = L \int_0^1 \Lambda(x)\theta(\Lambda(x))\mathrm{d}x, \tag{2.38}$$

where $\theta(x)$ is the step function, so that

$$\mathcal{H}_{\mathrm{KS}} = \lim_{L\to\infty} \frac{h_{\mathrm{KS}}}{L} = \int_0^1 \mathrm{d}x\, \Lambda(x)\theta(\Lambda(x)). \tag{2.39}$$

Using the Kaplan–Yorke conjecture, the *dimension density*

$$\mathcal{D}_{\mathrm{KY}} = \lim_{L\to\infty} \frac{D_{\mathrm{KY}}}{L} \tag{2.40}$$

is given implicitly by:

$$\int_0^{\mathcal{D}_{\mathrm{KY}}} \mathrm{d}x\, \Lambda(x) = 0\,. \tag{2.41}$$

The existence of a good thermodynamic limit is supported by numerical simulations (Kaneko 1986, Livi *et al.* 1986) and some exact results (Bunimovich and Sinai 1993).

### 2.2.3 Algorithmic complexity

We saw that the Shannon entropy puts a limit on how efficiently the ensemble of the messages emitted by a source can be coded. We may wonder about the compressibility properties of a single sequence. This problem can be addressed using the notion of algorithmic complexity, which is concerned with the difficulty of reproducing a given string of symbols.

Everybody agrees that the sequence of binary digits

$$0111010001011001011010\ldots \tag{2.42}$$

is, in some sense, more random than

$$1010101010101010101010\ldots\ ; \tag{2.43}$$

the notion of algorithmic complexity, introduced independently by Kolmogorov (1965), Chaitin (1966) and Solomonoff (1964), is a way to formalize the intuitive idea of randomness of a sequence.

Consider a binary digit sequence (this does not constitute a limitation) of length $N$ $(i_1, i_2, \ldots, i_N)$, generated by a certain computer code on some machine $\mathcal{M}$. One defines the algorithmic complexity, or algorithmic information content, $K_{\mathcal{M}}(N)$, of the sequence as the bit length of the shortest computer program able to produce the $N$-sequence and to stop afterward. Of course, such a length depends not only on the sequence but also on the machine. However, a result of Kolmogorov (1965) proves the existence of a universal computer, $\mathcal{U}$, that is able to perform the same computation a program $p$ makes on $\mathcal{M}$ with a modification of $p$ that depends only on $\mathcal{M}$ (and not on $p$). This implies that for all finite strings:

$$K_{\mathcal{U}}(N) \leq K_{\mathcal{M}}(N) + C_{\mathcal{M}}, \tag{2.44}$$

where $K_{\mathcal{U}}(N)$ is the complexity with respect to the universal computer and $C_{\mathcal{M}}$ depends only on the machine $\mathcal{M}$ (and not on $N$). At this point we can consider the algorithmic complexity with respect to a universal computer – and we can drop the machine dependence in the symbol for the algorithmic complexity, $K(N)$. The reason for this is that we are interested in the limit of very long sequences, $N \to \infty$, for which one defines the algorithmic complexity per unit symbol:

$$\mathcal{C} = \lim_{N\to\infty} \frac{K(N)}{N}, \tag{2.45}$$

which, because of (2.44), is an intrinsic quantity, i.e. independent of the machine.

Now coming back to the two $N$-sequences (2.42) and (2.43), it is obvious that the second one can be obtained with a small-length minimal program, i.e.

$$\text{“PRINT 10 } \frac{N}{2} \text{ TIMES.”}$$

The bit length of the above program is $O(\ln N)$ (the number $N$ requires $\ln N$ bits to be written) and therefore when taking the limit $N \to \infty$ in (2.45), one obtains $\mathcal{C} = 0$. Of course $K(N)$ cannot exceed $N$, since the sequence can always be obtained with a trivial program (of bit length $N$)

$$\text{“PRINT } i_1, i_2, \ldots, i_N.\text{”}$$

Therefore, in the case of a very irregular sequence, e.g. (2.42), one expects $K(N) \propto N$, i.e. $\mathcal{C} \neq 0$. In such a case one calls the sequence complex (i.e. of non-zero algorithmic complexity) or random.

Algorithmic complexity cannot be computed. Indeed, since the algorithm which computes $K(N)$ cannot have less than $K(N)$ binary digits and since in the case of random sequences $K(N)$ is not bounded in the limit $N \to \infty$, then it cannot be computed in the most interesting cases. The un-computability of $K(N)$ may be understood in terms of Gödel's incompleteness theorem (Chaitin 1990). In addition to the problem of whether or not $K(N)$ is computable in a specific case, the concept of algorithmic complexity allows us to clarify the vague and intuitive notion of randomness. For a systematic treatment of algorithmic complexity, information theory and data compression see Li and Vitanyi (1997).

There exists a relation between the Shannon entropy, $h_{\text{Sh}}$, and the algorithmic complexity $\mathcal{C}$. It is possible to show that

$$\lim_{N\to\infty} \frac{\langle K(N)\rangle}{H_N} = \frac{1}{\ln 2}, \tag{2.46}$$

where $\langle K(N)\rangle = \sum_{C_N} P(C_N) K_{C_N}(N)$, and $K_{C_N}(N)$ is the algorithmic complexity of the $N$-words. Therefore the expected complexity per symbol $\langle K(N)/N\rangle$ is asymptotically equal to the Shannon entropy (apart from the $\ln 2$ factor).

Equation (2.46) stems from the results of the Shannon–McMillan theorem about the two classes of sequences (i.e. $\Omega_1(N)$ and $\Omega_0(N)$). Indeed, in the limit of very large $N$, the probability of observing a sequence in $\Omega_1(N)$ goes to 1, and the entropy of such a sequence as well as its algorithmic complexity equals the Shannon entropy. Apart from the numerical coincidence of the values of $\mathcal{C}$ and $h_{\text{Sh}}/\ln 2$ there is a conceptual difference between information theory and algorithmic complexity theory. The Shannon entropy essentially refers to the information content in a statistical sense, i.e. it refers to an ensemble of sequences generated by a certain source. On the other hand, the algorithmic complexity defines the information content of an individual sequence. Of course, as noted above, the fact that it is possible to use probabilistic arguments on an individual sequence is a consequence of the ergodicity of the system, which allows the assumption of good and uniform statistical properties of arbitrarily long $N$-words.

For a dynamical system one can define the notion of algorithmic complexity of the trajectory starting from the point $\mathbf{x}$, $\mathcal{C}(\mathbf{x})$. This requires the introduction of finite open coverings of the phase space, the consideration of symbolic sequences thus generated, for the given trajectory, sampled at constant time intervals, and a search for the supremum of the algorithmic complexity per symbol on varying the coverings (Alekseev and Yakobson 1981). Then it can be shown (Brudno 1983, White 1993) that for almost all $\mathbf{x}$ (we always mean with respect to the natural invariant measure) one has:

$$\mathcal{C}(\mathbf{x}) = \frac{h_{\text{KS}}}{\ln 2}, \tag{2.47}$$

where, as before, the factor $\ln 2$ is a conversion factor between natural logarithms and bits.

This result says that the Kolmogorov–Sinai entropy quantifies not only the richness, or surprise, of a dynamical system but also the difficulty of describing (almost) any of its typical sequences.

*Algorithmic complexity and Lyapunov exponents*

Summing up, the theorem of Pesin together with those of Brudno and White shows that a chaotic dynamical system can be seen as a source of messages that cannot be described in a concise way, i.e. they are complex. We present here two examples that may help in understanding the previous conclusion and the relation between the Lyapunov exponent, the Kolmogorov–Sinai entropy and the algorithmic complexity.

Following Ford (1983), let us consider the shift map

$$x(t+1) = 2x(t) \mod 1, \tag{2.48}$$

whose natural measure is the Lebesgue measure, and $\lambda = \ln 2$. From Eqs. (2.47) and (2.34) one expects that in this case $\mathcal{C}(\mathbf{x}) = 1$, i.e. almost all the trajectories generate binary sequences of maximum complexity. If one writes an initial condition in binary representation, i.e., $x(0) = \sum_{j=1}^{\infty} a_j\, 2^{-j}$, such that $a_j = 0$ or 1, it is clearly seen that the action of the map (2.48) on $x(0)$ is just a shift of the binary coordinates:

$$x(1) = \sum_{j=1}^{\infty} a_{j+1}\, 2^{-j} \quad \cdots \quad x(t) = \sum_{j=1}^{\infty} a_{j+t}\, 2^{-j}. \tag{2.49}$$

With this observation it is possible to verify that $K(N) \simeq N$ for almost all the solutions of (2.48). Let us consider $x(t)$ with accuracy $2^{-k}$ and $x(0)$ with accuracy $2^{-l}$, of course $l = t + k$. This means that, in order to obtain the $k$ binary digits of the output solution of (2.48), we must use a program of length no less than $l = t + k$. Basically one has to specify $a_1, a_2, \ldots, a_l$. Therefore we are faced with the problem of the algorithmic complexity of the binary sequence $(a_1, a_2, \ldots, a_\infty)$ which determines the initial condition $x(0)$. Martin-Löf (1966) proved a remarkable theorem stating that, with respect to the Lebesgue measure, almost all the binary sequences $(a_1, a_2, \ldots, a_\infty)$, which represent a real number in $[0, 1]$, have maximum complexity, i.e. $K(N) \simeq N$. In practice no human being will ever be able to distinguish the typical sequence $(a_1, a_2, \ldots, a_\infty)$ from the output of a fair coin toss.

As a second example we consider a one-dimensional chaotic map

$$x(t+1) = f(x(t)). \tag{2.50}$$

If one wants to transmit to a friend on Mars the sequence $\{x(t),\ t = 1, 2, \ldots, T\}$ accepting only errors smaller than a tolerance $\Delta$, one can apply the following strategy (Paladin *et al.* 1995).

(1) Transmit the rule (2.50): for this task one has to use a number of bits independent of the length of the sequence $T$.
(2) Specify to the friend the initial condition $x(0)$ with a fixed precision $\delta_0 < \Delta$ using a finite number of bits which is independent of $T$.
(3) Let the system evolve until the first time $\tau_1$ such that the distance between two trajectories, that was initially $\delta x(0) = \delta_0$, equals $\Delta$ and then specify again the new initial condition $x(\tau_1)$ with precision $\delta_0$.
(4) Let the system evolve and repeat procedures (2) and (3), i.e. each time the error acceptance tolerance is reached specify the initial conditions, $x(\tau_1 + \tau_2),\ x(\tau_1 + \tau_2 + \tau_3), \ldots,$ with precision $\delta_0$. The times $\tau_1, \tau_2, \ldots$ are defined as follows: putting $x'(\tau_1) = x(\tau_1) + \delta_0$, $\tau_2$ is given by the minimum time such that $|x'(\tau_1 + \tau_2) - x(\tau_1 + \tau_2)| \geq \Delta$ and so on.

By following steps (1)–(4) the friend on Mars can reconstruct with a precision $\Delta$ the sequence $\{x(t)\}$ simply iterating on a computer the system (2.50) between 1 and $\tau_1 - 1$, $\tau_1$ and $\tau_1 + \tau_2 - 1$, and so on.

Let us now compute the amount of bits necessary to implement the above transmission (1)–(4). To simplify the notation we introduce the quantities

$$\gamma_i = \frac{1}{\tau_i} \ln \frac{\Delta}{\delta_0} \tag{2.51}$$

which can be considered as sort of *effective* Lyapunov exponents (see the following). The Lyapunov exponent $\lambda$ can be written in terms of $\{\gamma_i\}$ as follows

$$\lambda = \langle \gamma_i \rangle = \frac{\sum_i \tau_i \gamma_i}{\sum_i \tau_i} = \frac{1}{\overline{\tau}} \ln \frac{\Delta}{\delta_0} \tag{2.52}$$

where

$$\overline{\tau} = \frac{1}{N} \sum \tau_i,$$

is the average time after which one has to transmit the new initial condition and $N$ is the number of transmissions (let us observe that to obtain $\lambda$ from the $\gamma_i$ one has to perform the average (2.52) because the transmission time, $\tau_i$, is not constant). If $T$ is large enough $N = T/\overline{\tau} \simeq \lambda T / \ln(\Delta/\delta_0)$. Therefore, noting that in each transmission for the reduction of the error from $\Delta$ to $\delta_0$ one needs to use $\ln_2(\Delta/\delta_0)$ bits, the total amount of bits used in the whole transmission is

$$\frac{T}{\overline{\tau}} \ln_2 \frac{\Delta}{\delta_0} = \frac{\lambda}{\ln 2} T. \tag{2.53}$$

In other words the number of bits per unit time is proportional to $\lambda$.

In more than one dimension, we simply have to replace $\lambda$ with $h_{\mathrm{KS}}$ in (2.53). This point can be understood intuitively by noting that one has to repeat the above transmission procedure in each of the expanding directions and using Pesin's formula. Thus the number of bits to be supplied per unit time typically satisfies relation (2.47).

## 2.3 Beyond the Lyapunov exponents and the Kolmogorov–Sinai entropy

When dealing with the problem of prediction of the behavior of a physical system it is useful to introduce the *predictability time* $T_{\mathrm{p}}$, i.e. the time interval on which one can typically forecast the system. A simple argument previously suggested

$$T_{\mathrm{p}} \sim \frac{1}{\lambda} \ln \left( \frac{\Delta}{\delta_0} \right) . \tag{2.54}$$

However, in any realistic system, relation (2.54) is too naive to be of actual relevance. Indeed, it does not take into account some basic features of dynamical systems.

- The Lyapunov exponent (2.9) is a global quantity: it measures the average rate of divergence of nearby trajectories. In general there exist finite-time fluctuations and the probability distribution function of these fluctuations is important for the characterization of predictability. The *generalized Lyapunov exponents* have been introduced with the aim of taking these fluctuations into account (Fujisaka 1983, Benzi *et al.* 1985).
- The Lyapunov exponent, by definition, involved the linearized dynamics, since it amounts to computing the rate of separation of two infinitesimally close trajectories. On the other hand, with regard to the predictability time (2.54) one is interested in a finite tolerance $\Delta$, because the initial error $\delta_0$ is typically finite. A generalization of the Lyapunov exponent to *finite size* errors extends the study of the perturbation growth to the non-linear regime, i.e. neither $\delta_0$ nor $\Delta$ is infinitesimal (Aurell *et al.* 1996).

### *2.3.1 Characterization of finite-time fluctuations*

Let us consider the linear response, at a delay $t$, to an infinitesimal perturbation $\delta \mathbf{x}(0)$:

$$R(t) = \frac{|\delta \mathbf{x}(t)|}{|\delta \mathbf{x}(0)|} , \tag{2.55}$$

from which the Lyapunov exponent is computed according to (2.9). In order to take into account the finite-time fluctuations, we can compute the different moments $\langle R(t)^q \rangle$ and introduce the so-called generalized Lyapunov exponents (of order $q$)

(Fujisaka 1983, Benzi *et al.* 1985):

$$L(q) = \lim_{t\to\infty} \frac{1}{t} \ln\langle R(t)^q \rangle \tag{2.56}$$

where $\langle \ldots \rangle$ indicates the time average along the trajectory (see Section 2.1.1). It is easy to show that

$$\lambda = \left. \frac{dL(q)}{dq} \right|_{q=0} . \tag{2.57}$$

In the absence of fluctuations (that is, $R(t)$ is deterministic hence $\langle R(t)^q \rangle = \langle R(t) \rangle^q$), $\lambda$ completely characterizes the error growth and we have $L(q) = \lambda q$, while in the general case $L(q)$ is concave in $q$ (i.e. $d^2L/dq^2 > 0$) (Paladin and Vulpiani 1987). Before discussing the properties of the generalized Lyapunov exponents, let us consider a simple example with a non-trivial $L(q)$. The model is the one-dimensional map

$$x(t+1) = \begin{cases} \dfrac{x(t)}{a} & \text{for } 0 \le x \le a \\[2ex] \dfrac{1-x(t)}{1-a} & \text{for } a < x \le 1, \end{cases} \tag{2.58}$$

which for $a = 1/2$ reduces to the tent map. For $a \neq 1/2$ the system is characterized by two different growth rates. The presence of different growth rates makes $L(q)$ non-linear in $q$. Since the map (2.58) is piecewise linear and with a uniform invariant density, by means of ergodicity the explicit computation of $L(q)$ is very easy. The moments of the response after a time $t$ are simply given by

$$\langle R(t)^q \rangle = \left[ a \left(\frac{1}{a}\right)^q + (1-a) \left(\frac{1}{1-a}\right)^q \right]^t . \tag{2.59}$$

From (2.56) and (2.59) we obtain:

$$L(q) = \ln\left[a^{1-q} + (1-a)^{1-q}\right], \tag{2.60}$$

which recovers the non-intermittent limit $L(q) = q \ln 2$ in the symmetric case $a = 1/2$ and, from (2.57), gives for the Lyapunov exponent the known result $\lambda = -a \ln a - (1-a) \ln(1-a)$. In the general case, assuming $0 \le a < 1/2$, we have that for $q \to +\infty$, $L(q)$ is dominated by the less probable, most unstable contributions and $L(q)/q \simeq -\ln(a)$. In the opposite limit, $q \to -\infty$, $L(q)/q \simeq -\ln(1-a)$.

We now show how $L(q)$ is related to the fluctuations of $R(t)$ at finite time $t$. Define an "effective" Lyapunov exponent $\gamma(t)$ at time $t$ by

$$R(t) \sim e^{\gamma(t)t} . \tag{2.61}$$

In the limit $t \to \infty$, the Oseledec (1968) theorem ensures that, for typical trajectories, $\gamma(t) = \lambda = -a \ln a - (1-a)\ln(1-a)$. Therefore, for large $t$, the probability density of $\gamma(t)$ peaks at the most probable value $\lambda$. Let us introduce the probability density $P_t(\gamma)$ of observing a given $\gamma$ on a trajectory of length $t$. Large deviation theory suggests

$$P_t(\gamma) \sim e^{-S(\gamma)t}, \tag{2.62}$$

where $S(\gamma)$ is the Cramer function (Varadhan 1984). The Oseledec theorem implies that $\lim_{t\to\infty} P_t(\gamma) = \delta(\gamma - \lambda)$; this gives constraints on the Cramer function, i.e. $S(\gamma = \lambda) = 0$ and $S(\gamma) > 0$ for $\gamma \neq \lambda$.

The Cramer function $S(\gamma)$ is related to the generalized Lyapunov exponent $L(q)$ through a Legendre transform. Indeed, at large $t$, one has

$$\langle R(t)^q \rangle = \int d\gamma \; P_t(\gamma) e^{q\gamma t} \sim e^{L(q)t}, \tag{2.63}$$

and by a steepest descent estimation one obtains

$$L(q) = \max_{\gamma} \left[ q\gamma - S(\gamma) \right]. \tag{2.64}$$

In other words each value of $q$ selects a particular $\gamma^*(q)$ given by

$$\left. \frac{dS(\gamma)}{d\gamma} \right|_{\gamma^*} = q \tag{2.65}$$

from which $\langle R(t)^q \rangle$ receives the dominant contribution.

We have already discussed how, for negligible fluctuations of the "effective" Lyapunov exponents, the Lyapunov exponent characterizes completely the error growth and $L(q) = \lambda q$. In the presence of fluctuations, the probability distribution for $R(t)$ can be approximated, in a limiting condition, by a log-normal distribution. This can be done assuming weak correlations in the response function so that (2.55) factorizes in several independent contributions and the central limit theorem applies. We can thus write

$$P_t(R) = \frac{1}{R\sqrt{2\pi \mu t}} \exp\left( -\frac{(\ln R - \lambda t)^2}{2\mu t} \right), \tag{2.66}$$

where $\lambda$ and $\mu$ are given by

$$\begin{aligned} \lambda &= \lim_{t\to\infty} \frac{1}{t} \langle \ln R(t) \rangle \\ \mu &= \lim_{t\to\infty} \frac{1}{t} \left( \langle [\ln R(t)]^2 \rangle - \langle \ln R(t) \rangle^2 \right). \end{aligned} \tag{2.67}$$

The log-normal distribution for $R$ corresponds to a Gaussian distribution for $\gamma$, with mean $\lambda$ and variance $\mu/t$, namely

$$S(\gamma) = \frac{(\gamma - \lambda)^2}{2\mu}, \tag{2.68}$$

and a generalized Lyapunov exponent that is quadratic in $q$:

$$L(q) = \lambda q + \frac{1}{2}\mu q^2. \tag{2.69}$$

Let us remark that, in general, the log-normal distribution (2.66) is a good approximation for non-extreme events, i.e. small fluctuations of $\gamma$ around $\lambda$, so that the expression (2.69) is correct only for small $q$ (see Figure 2.1). This is because the moments of the log-normal distribution grow too fast with $q$ (Paladin and Vulpiani 1987). Indeed from (2.65) we have that the selected $\gamma^*(q)$ is given by $\gamma^*(q) = \lambda + \mu q$ and thus $\gamma^*(q)$ is not finite for $q \to \infty$. This is unphysical because $\gamma^*(\infty)$ is the fastest error growth rate in the system, and we may reasonably suppose that it is finite.

Let us consider again the map (2.58). In this case we have $\lambda = L'(0) = -a\ln(a) - (1-a)\ln(1-a)$ and $\mu = L''(0) = a(1-a)(\ln(a) - \ln(1-a))^2$, which are the coefficients of the Taylor expansion of (2.60) around $q = 0$. For large $q$ the log-normal approximation gives $L(q)/q \simeq q\mu/2$ while the correct limit is the constant $L(q)/q \simeq -\ln(a)$.

### 2.3.2 Renyi entropies

Analogous to the generalized Lyapunov exponent, it is possible to introduce a generalization of the Kolmogorov–Sinai entropy in order to take into account the intermittency (in the sense of fluctuation of the "effective" Lyapunov exponent). Let us recall the definition of Kolmogorov–Sinai entropy

$$h_{\mathrm{KS}} = -\lim_{\epsilon\to 0}\lim_{N\to\infty}\frac{1}{N\tau}\sum_{\{W^N(\mathcal{A}_\epsilon)\}} P(W^N(\mathcal{A}_\epsilon))\ln P(W^N(\mathcal{A}_\epsilon)) \tag{2.70}$$

where $\mathcal{A}_\epsilon$ is a partition of the phase space in cells of size $\epsilon$ and $W^N(\mathcal{A}_\epsilon)$ denotes an $N$-step trajectory described, at the level of this partition, by a symbol sequence. The generalized Renyi entropies (Paladin and Vulpiani 1987, Badii and Politi 1997), $K_q$, can be introduced by observing that (2.70) is nothing but the average of $-\ln P(W^N)$ with the probability $P(W^N)$:

$$K_q = -\lim_{\epsilon\to 0}\lim_{N\to\infty}\frac{1}{N\tau(q-1)}\ln\left(\sum_{\{W^N(\mathcal{A}_\epsilon)\}} P(W^N(\mathcal{A}_\epsilon))^q\right). \tag{2.71}$$

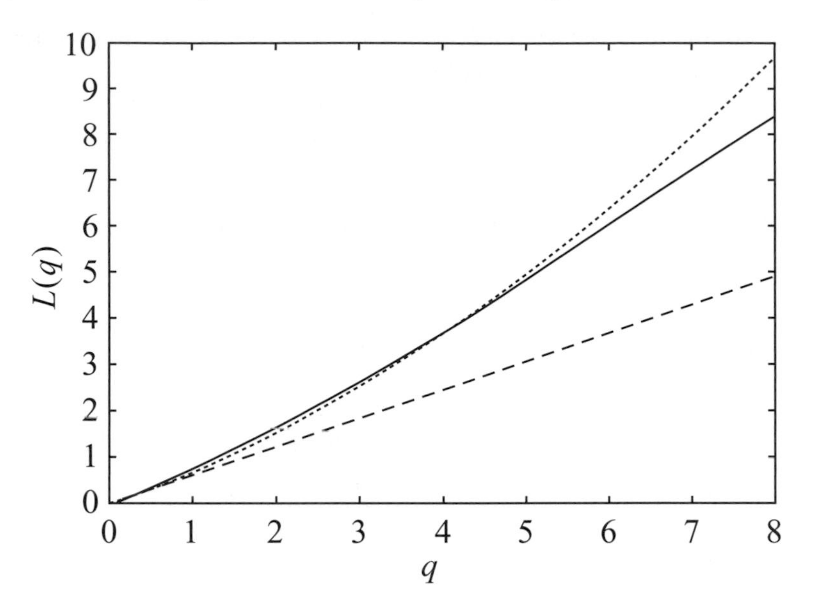


Figure 2.1 Generalized Lyapunov exponent, $L(q)$, for the map (2.58) with $a = 0.3$ (solid line) compared with the linear non-intermittent approximation, $\lambda q$ (dashed line), and with the log-normal one, Eq. (2.69) (dotted line).

As in (2.57) one has $h_{\text{KS}} = \lim_{q\to 1} K_q = K_1$; in addition, from general results of probability theory, one can show that $K_q$ is decreasing monotonically with $q$.

It is not difficult to realize that, by analogy with the $L(q)$, for the Renyi entropies it is also possible to introduce a description in terms of large deviations, connecting the $K_q$ to a suitable Cramer function via a Legendre transform (Paladin and Vulpiani 1987, Badii and Politi 1997).

### 2.3.3 Growth of non-infinitesimal perturbations

In realistic situations, the initial condition of a system is known with limited accuracy. In this case the Lyapunov exponent is of little relevance for the characterization of predictability and new indicators are needed. To clarify the problem, let us consider the following coupled map model:

$$\begin{cases} \mathbf{x}(t+1) = \mathbf{R}\,\mathbf{x}(t) + \varepsilon \mathbf{h}(y(t)) \\ y(t+1) = G(y(t)), \end{cases} \tag{2.72}$$

where $\mathbf{x} \in \mathbb{R}^2$, $y \in \mathbb{R}$, $\mathbf{R}$ is a rotation matrix of an angle $\theta$, $\mathbf{h}$ is a vector function and $G$ is a chaotic map. For simplicity we consider a linear coupling, with both the components of $\mathbf{h}$ equal to $y$: $\mathbf{h}(y) = (y, y)$, and the logistic map: $G(y) = 4y(1-y)$.

For $\varepsilon = 0$ we have two independent systems: a regular and a chaotic one. Thus the Lyapunov exponent of the $\mathbf{x}$ subsystem is $\lambda_x(\varepsilon = 0) = 0$, i.e. it is completely predictable. In contrast, the $y$ subsystem is chaotic with Lyapunov exponent $\lambda_y = \ln 2$.

If we now switch on the (small) coupling ($\varepsilon > 0$) we are confronted with a single three-dimensional chaotic system with a positive global Lyapunov exponent

$$\lambda = \lambda_y + O(\varepsilon).$$

We are interested in making predictions only on the $\mathbf{x}$ subsystem. A direct application of (2.54) would give

$$T_{\mathrm{p}}^{(x)} \sim T_{\mathrm{p}} \sim \frac{1}{\lambda}, \tag{2.73}$$

but this result is clearly unsatisfactory: the predictability time for $\mathbf{x}$ seems to be independent of the value of the coupling $\varepsilon$. Let us underline that this is not due to an artifact of the chosen example. Indeed, one can use the same argument in many physical situations (Boffetta *et al.* 1996). A well-known example is the gravitational three-body problem with one body (asteroid) much smaller than the other two (the Sun and one planet). If one neglects the gravitational feedback of the asteroid on the two massive bodies (restricted problem) one has a chaotic asteroid in the regular field of the two massive bodies. As soon as the feedback is taken into account (i.e. $\varepsilon > 0$ in the example) one has a non-separable three-body system with a positive Lyapunov exponent. Of course, intuition correctly suggests that it should be possible to forecast the motion of the massive bodies for very long times if the asteroid has a very small mass ($\varepsilon \to 0$).

The paradox arises from the use of (2.54), which is valid only for the tangent vectors, also in the non-infinitesimal regime. As soon as the errors become large one has to take into account the full non-linear evolution. The effect is shown for the model (2.3.3) in Figure 2.2. The evolution of $\delta\mathbf{x}$ is given by

$$\delta\mathbf{x}(t+1) = \mathbf{R}\,\delta\mathbf{x}(t) + \varepsilon\,\delta\mathbf{h}(y), \tag{2.74}$$

where, with our choice, $\delta\mathbf{h} = (\delta y, \delta y)$. At the beginning, both $|\delta\mathbf{x}|$ and $\delta y$ grow exponentially. However, the available phase space for $y$ is finite and the uncertainty reaches the saturation value $\delta y \sim O(1)$ in a time $t \sim 1/\lambda$. At larger times the two realizations of the $y$ variable are completely uncorrelated and their difference $\delta y$ in (2.74) acts as a noisy term. As a consequence, the growth of the uncertainty $\delta\mathbf{x}(t)$ on $\mathbf{x}(t)$ becomes diffusive with a diffusion coefficient proportional to $\varepsilon^2$ (Boffetta *et al.* 1996):

$$|\delta\mathbf{x}(t)| \sim \varepsilon t^{1/2} \tag{2.75}$$

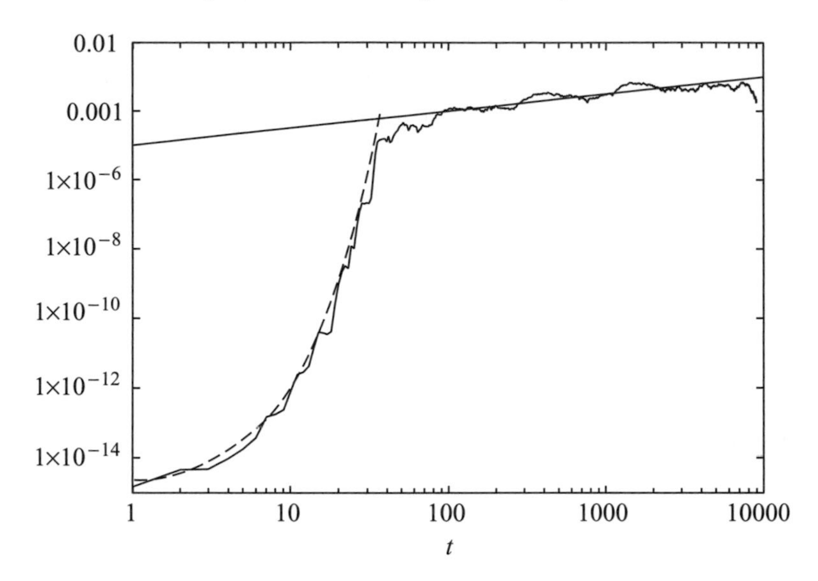


Figure 2.2 Growth of error $|\delta\mathbf{x}(t)|$ for the coupled map (2.72). The rotation angle is $\theta = 0.82099$, the coupling strength $\varepsilon = 10^{-5}$ and the initial error only on the $y$ variable is $\delta y = \delta_0 = 10^{-10}$. The dashed line indicates $|\delta\mathbf{x}(t)| \sim \mathrm{e}^{\lambda t}$ where $\lambda = \ln 2$, the solid line shows $|\delta\mathbf{x}(t)| \sim t^{1/2}$.

so that

$$T_{\mathrm{p}}^{(x)} \sim \varepsilon^{-2}. \tag{2.76}$$

This example shows that, even in simple systems, the Lyapunov exponent can be of little relevance for characterization of the predictability.

In more complex systems, in which different scales are present, one is typically interested in forecasting large-scale motion, while the Lyapunov exponent is related to the small-scale dynamics. A familiar example is weather forecasting: the Lyapunov exponent of the atmosphere is indeed rather large owing to small-scale turbulent motion, but (large-scale) weather prediction is possible for about 10 days (Lorenz 1969). It is thus natural to look for a generalization of the Lyapunov exponent to finite perturbations from which one can obtain a more realistic estimation for the predictability time. It is worth underlining the important fact that finite errors are not confined in the tangent space but are governed by the complete non-linear dynamics. In this sense the extension of the Lyapunov exponent to finite errors will give more information on the system.

With the aim of generalizing the Lyapunov exponent to non-infinitesimal perturbations let us now define the finite size Lyapunov exponent (FSLE) (Aurell *et al.* 1996, Boffetta *et al.* 2002). Consider a reference trajectory, $\mathbf{x}(t)$, and a perturbed one, $\mathbf{x}'(t)$, such that $|\mathbf{x}'(0) - \mathbf{x}(0)| = \delta$ ( $|\ldots|$ being the Euclidean norm, but one can also consider other norms). One integrates the two trajectories and computes the time $\tau_1(\delta, r)$ necessary for the separation $|\mathbf{x}'(t) - \mathbf{x}(t)|$ to increase from $\delta$ to $r\delta$. At time $t = \tau_1(\delta, r)$ the distance between the trajectories is rescaled to $\delta$ and the procedure is repeated in order to compute $\tau_2(\delta, r), \tau_3(\delta, r) \ldots$.

The threshold ratio $r$ must be $r > 1$, but not too large in order to avoid contributions from different scales in $\tau(\delta, r)$. A typical choice is $r = 2$ (for which $\tau(\delta, r)$ is properly a "doubling" time) or $r = \sqrt{2}$. In the same spirit as the discussion leading to Eqs. (2.51) and (2.52), we can introduce an effective finite-size growth rate:

$$\gamma_i(\delta, r) = \frac{1}{\tau_i(\delta, r)} \ln r. \tag{2.77}$$

After having performed $\mathcal{N}$ error-doubling experiments, we can define the FSLE as

$$\lambda(\delta) = \langle \gamma(\delta, r) \rangle_t = \frac{\sum_i \tau_i \gamma_i(\delta, r)}{\sum_i \tau_i} = \frac{1}{\langle \tau(\delta, r) \rangle_e} \ln r, \tag{2.78}$$

where $\langle \tau(\delta, r) \rangle_e$ is

$$\langle \tau(\delta, r) \rangle_e = \frac{1}{\mathcal{N}} \sum_{n=1}^{\mathcal{N}} \tau_n(\delta, r). \tag{2.79}$$

(see Boffetta *et al.* (2002) for details). In the infinitesimal limit, the FSLE reduces to the Lyapunov exponent

$$\lim_{\delta \to 0} \lambda(\delta) = \lambda. \tag{2.80}$$

In practice this limit means that $\lambda(\delta)$ displays a constant plateau at $\lambda$ for sufficiently small $\delta$ (Figure 2.3). For finite value of $\delta$ the behavior of $\lambda(\delta)$ depends on the details of the non-linear dynamics. For example, in the model (2.3.3) the diffusive behavior (2.75), by simple dimensional arguments, corresponds to $\lambda(\delta) \sim \delta^{-2}$.

Since the FSLE measures the rate of divergence of trajectories at finite errors, one might wonder whether it is just another way of looking at the average response $\langle \ln(R(t)) \rangle$ (2.55) as a function of time. A moment of reflection shows that this is not the case. Indeed in the case of $\langle \ln(R(t)) \rangle$ one has to perform an average at fixed time interval, which is not the same as computing the average doubling time at a *fixed scale* $\delta$, as in (2.78).

The FSLE method has been applied successfully to the analysis of geophysical data (Lacorata *et al.* 2001, 2004). The remarkable advantage of the method is the fact that it works at a fixed scale, instead of at a fixed time as in more traditional approaches. This is particularly important in the case of strongly intermittent systems, in which $R(t)$, i.e. the "effective" Lyapunov exponent, can be very different in each realization. In the presence of intermittency, averaging over different realizations

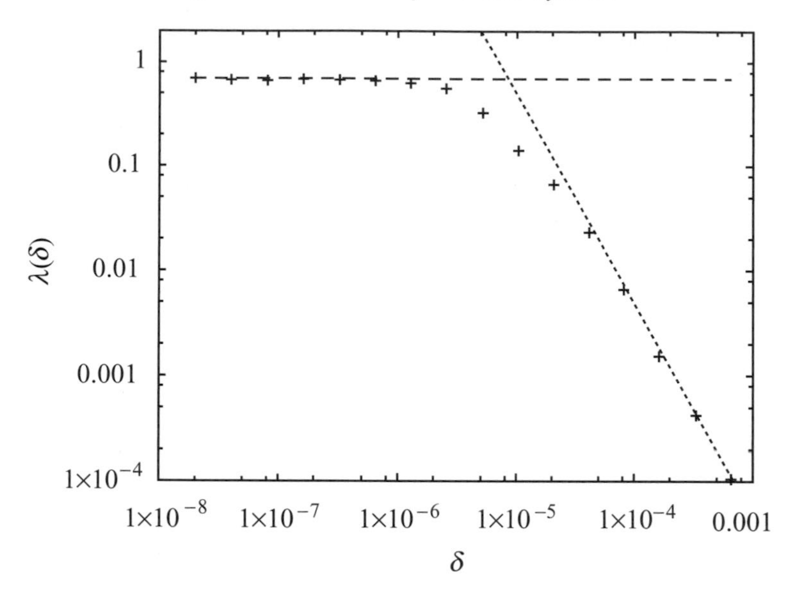


Figure 2.3 $\lambda(\delta)$ as a function of $\delta$ for the coupled map (2.3.3) with $\varepsilon = 10^{-5}$. The perturbation was initialized as in Figure 2.2. For $\delta \to 0$, $\lambda(\delta) \simeq \lambda$ (solid line). The dashed line shows the behavior $\lambda(\delta) \sim \delta^{-2}$.

at fixed times can produce a spurious regime owing to the superposition of exponential and diffusive contributions by different samples at the same time (Artale *et al.* 1997).

### 2.3.4 The $\epsilon$-entropy

The Kolmogorov–Sinai entropy, $h_{KS}$ (2.32), of a system measures the amount of information per unit time that is necessary to record without ambiguity a generic trajectory of a system. Since the computation of $h_{KS}$ involves the limit of arbitrary fine resolution and infinite times, it turns out that, practically, for most systems it is not possible to compute $h_{KS}$. Nevertheless, in the same philosophy as the FSLE, by relaxing the strict requirement of reproducing a trajectory with arbitrary accuracy, one can introduce the $\epsilon$-entropy which measures the amount of information needed to reproduce a trajectory with accuracy $\epsilon$ in phase space. Roughly speaking the $\epsilon$-entropy can be considered the counterpart, in information theory, of the FSLE (as the Kolmogorov–Sinai entropy is for the Lyapunov exponent). This quantity was originally introduced by Shannon (1948), and by Kolmogorov (1956). More recently Gaspard and Wang (1993) and Boffetta *et al.* (2002) made use of this concept (and its space-time extension, the $(\epsilon, \tau)$-entropy) to characterize a large variety of processes.

We start with a continuous (in time) variable $\mathbf{x}(t) \in \mathbb{R}^d$, which represents the state of a $d$-dimensional system, we discretize the time by introducing an interval $\tau$ and we consider the new variable

$$\mathbf{X}^{(m)}(t) = (\mathbf{x}(t), \mathbf{x}(t+\tau), \ldots, \mathbf{x}(t+(m-1)\tau)) . \tag{2.81}$$

Of course $\mathbf{X}^{(m)}(t) \in \mathbb{R}^{md}$ and it corresponds to a piece of trajectory which lasts for a time $T = m\tau$.

In data analysis, the space where the state vectors of the system live is not known. Moreover, usually only a scalar variable $u(t)$ can be measured. In such a situation, as a special case of (2.81), one considers vectors $(u(t), u(t+\tau), \ldots, u(t+m\tau-\tau))$, that live in $\mathbb{R}^m$ and allow a reconstruction of the original phase space (that is, a reconstruction of topologically equivalent geometrical features of the dynamics), known as delay embedding in the literature (Takens 1981, Kantz and Schreiber 1997).

Introduce now a partition of the phase space $\mathbb{R}^d$, using cells of edge $\epsilon$ in each of the $d$ directions. Since the region where a bounded motion evolves contains a finite number of cells, each $\mathbf{X}^{(m)}(t)$ can be coded into a word of length $m$, out of a finite alphabet:

$$\mathbf{X}^{(m)}(t) \longrightarrow W^m(\epsilon, t) = (i(\epsilon, t), i(\epsilon, t+\tau), \ldots, i(\epsilon, t+m\tau-\tau)) , \tag{2.82}$$

where $i(\epsilon, t+j\tau)$ labels the cell in $\mathbb{R}^d$ containing $\mathbf{x}(t+j\tau)$. From the time evolution of $\mathbf{X}^{(m)}(t)$ one obtains, according to the hypothesis of ergodicity, the probabilities $P(W^m(\epsilon))$ of the admissible words $\{W^m(\epsilon)\}$. The $(\epsilon, \tau)$-entropy per unit time, $h(\epsilon, \tau)$, is defined as follows:

$$h_m(\epsilon, \tau) = \frac{1}{\tau}[H_{m+1}(\epsilon, \tau) - H_m(\epsilon, \tau)], \tag{2.83}$$

$$h(\epsilon, \tau) = \lim_{m\to\infty} h_m(\epsilon, \tau) = \frac{1}{\tau} \lim_{m\to\infty} \frac{1}{m} H_m(\epsilon, \tau), \tag{2.84}$$

where $H_m$ is the entropy of blocks of length $m$:

$$H_m(\epsilon, \tau) = - \sum_{\{W^m(\epsilon)\}} P(W^m(\epsilon)) \ln P(W^m(\epsilon)). \tag{2.85}$$

For the sake of simplicity, we ignored the dependence on details of the partition (that is, details other than the cell size $\epsilon$). To make $h(\epsilon, \tau)$ partition independent one has to consider a generic partition of the phase space $\{\mathcal{A}\}$ and to evaluate the Shannon entropy on this partition: $h_{\mathrm{Sh}}(\mathcal{A}, \tau)$. The $\varepsilon$-entropy is thus defined as the infimum over all partitions for which the diameter of each cell is less than $\varepsilon$ (Gaspard and

Wang 1993):

$$h(\varepsilon, \tau) = \inf_{\mathcal{A}:\mathrm{diam}(\mathcal{A})\leq\varepsilon} h_{\mathrm{Sh}}(\mathcal{A}, \tau). \tag{2.86}$$

Note that the time dependence in (2.86) is trivial for deterministic systems, and that in the limit $\epsilon \to 0$ one recovers the Kolmogorov–Sinai entropy

$$h_{\mathrm{KS}} = \lim_{\epsilon\to 0} h(\epsilon, \tau).$$

The $(\epsilon, \tau)$-entropy $h(\epsilon, \tau)$ is well defined also for stochastic processes. Actually the dependence of $h(\epsilon, \tau)$ on $\epsilon$ can give some insight into the underlying stochastic process (Gaspard and Wang 1993, Boffetta *et al.* 2002). For instance, in the case of a stationary Gaussian process with spectrum $S(\omega) \propto \omega^{-(1+2\alpha)}$ with $0 < \alpha < 1$, one has (Kolmogorov 1956):

$$\lim_{\tau\to 0} h(\epsilon, \tau) \sim \epsilon^{-1/\alpha}. \tag{2.87}$$

However, we have to stress that the above behavior may be difficult to observe experimentally, mainly owing to problems related to the choice of $\tau$ (Cencini *et al.* 2000).

In the next chapter we will show how an entropic analysis in terms of $h(\epsilon)$ can be very useful for treating the chaos/noise distinction issue, and more generally for the scale-dependent description of the signal character.

## References

Alekseev, V. M. and Yakobson, M. V. (1981). Symbolic dynamics and hyperbolic dynamic-systems, *Phys. Rep.* **75**, 287.

Artale, V., Boffetta, G., Celani, A., Cencini, M. and Vulpiani, A. (1997). Dispersion of passive tracers in closed basins: beyond the diffusion coefficient, *Phys. Fluids A* **9**, 3162.

Aurell, E., Boffetta, G., Crisanti, A., Paladin, G. and Vulpiani, A. (1996). Growth of non-infinitesimal perturbations in turbulence, *Phys. Rev. Lett.* **77**, 1262.

Badii, R. and Politi, A. (1997). *Complexity. Hierarchical Structures and Scaling in Physics*, Cambridge: Cambridge University Press.

Benettin, G., Galgani, L., Giorgilli, A. and Strelcyn, J. M. (1980). Lyapunov characteristic exponents for smooth dynamical systems and for Hamiltonian systems: a method for computing all of them, *Meccanica* **15**, 9.

Benzi, R., Paladin, G., Parisi, G. and Vulpiani, A. (1985). Characterization of intermittency in chaotic systems, *J. Phys. A* **18**, 2157.

Billingsley, P. (1965). *Ergodic Theory and Information*, New York: Wiley.

Boffetta, G., Paladin, G. and Vulpiani, A. (1996). Strong chaos without butterfly effect in dynamical systems with feedback, *J. Phys. A* **29**, 2291.

Boffetta, G., Cencini, M., Falcioni, M. and Vulpiani, A. (2002). Predictability: a way to characterize complexity, *Phys. Rep.* **356**, 367.

Brudno, A. A. (1983). Entropy and the complexity of the trajectories of a dynamical system, *Trans. Moscow Math. Soc.* **44**, 127.

Bunimovich, L. A. and Sinai, G. (1993). Statistical mechanics of coupled map lattices, in *Theory and Applications of Coupled Map Lattices*, ed. K. Kaneko, p. 169, New York: Wiley.

Cencini, M., Falcioni, M., Kantz, H., Olbrich, E. and Vulpiani, A. (2000). Chaos or noise – sense and nonsense of a distinction, *Phys. Rev. E* **62**, 427.

Chaitin, G. J. (1966). On the length of programs for computing finite binary sequences, *J. Assoc. Comput. Mach.* **13**, 547.

Chaitin, G. J. (1990). *Information, Randomness and Incompleteness*, 2nd edition, Singapore: World Scientific.

Christiansen, F. and Politi, A. (1995). Generating partition for the standard map, *Phys. Rev. E* **51**, R3811.

Christiansen, F. and Politi, A. (1996). Symbolic encoding in symplectic maps, *Nonlinearity* **9**, 1623.

Collet, P. and Eckmann, J.-P. (1980). *Iterated Maps on the Interval as Dynamical Systems*, Boston, MA: Birkhäuser.

Davidchack, R. L., Lai, Y.-C., Bollt, E. M. and Dhamala, M. (2000). Estimating generating partitions of chaotic systems by unstable periodic orbits, *Phys. Rev. E* **61**, 1353.

Dorfman, R. (1999). *Introduction to Chaos in Nonequilibrium Statistical Mechanics*, Cambridge: Cambridge University Press.

Eckmann, J. P. and Ruelle, D. (1985). Ergodic theory of chaos and strange attractors, *Rev. Mod. Phys.* **57**, 617.

Eckmann, J. P. and Ruelle, D. (1992). Fundamental limitations for estimating dimensions and Lyapunov exponents in dynamical systems, *Physica D* **56**, 185.

Ford, J. (1983). How random is a coin tossing?, *Phys. Today* **36**, 40.

Fujisaka, H. (1983). Statistical dynamics generated by fluctuations of local Lyapunov exponents, *Prog. Theor. Phys.* **70**, 1264.

Gaspard, P. and Wang, X. J. (1993). Noise, chaos, and $(\epsilon, \tau)$-entropy per unit time, *Phys. Rep.* **235**, 291.

Giovannini, F. and Politi, A. (1991). Homoclinic tangencies, generating partitions and curvature of invariant manifolds, *J. Phys. A* **24**, 1837.

Goldhirsch, I., Sulem, P.-L. and Orszag, S. A. (1987). Stability and Lyapunov stability of dynamical systems: a differential approach and a numerical method, *Physica D* **27**, 311.

Grassberger, P. (1986). Toward a quantitative theory of self-generated complexity, *Int. J. Theor. Phys.* **25**, 907.

Grassberger, P. and Kantz, H. (1985). Generating partitions for the dissipative Hénon map, *Phys. Lett. A* **113**, 235.

Grassberger, P. and Procaccia, I. (1983). Estimation of the Kolmogorov-entropy from a chaotic signal, *Phys. Rev. A* **28**, 2591.

Grassberger, P., Kantz, H., and Moenig, U. (1989). On the symbolic dynamics of the Hénon map, *J. Phys. A* **22**, 5217.

Hirata, Y., Judd, K. and Kilminster, D. (2004). Estimating a generating partition from observed time series: symbolic shadowing, *Phys. Rev. E* **70**, 016215.

Hohenberg, P. C. and Shraiman, B. I. (1988). Chaotic behavior of an extended system, *Physica D* **37**, 109.

Jaeger, L. and Kantz, H. (1997). Structure of generating partitions for two-dimensional maps, *J. Phys. A* **30**, L567.

Kaneko, K. (1986). Lyapunov analysis and information flow in coupled map lattices, *Physica D* **23**, 436.

Kaneko, K. (1993). The coupled map lattice, in *Theory and Applications of Coupled Map Lattices*, ed. K. Kaneko, p. 1, New York: Wiley.

Kantz, H. and Schreiber, T. (1997). *Nonlinear Time Series Analysis*, Cambridge: Cambridge University Press.

Kaplan, J. L. and Yorke, J. A. (1979). Chaotic behavior of multidimensional difference equations, in *Differential Equations and Approximations of Fixed Points*, ed. H. O. Peitgen and O. Walther, p. 228, New York: Springer.

Kennel, M. B. and Buhl, M. (2003). Estimating good discrete partitions from observed data: symbolic false nearest neighbors, *Phys. Rev. Lett.* **91**, 084102.

Khinchin, A. I. (1957). *Mathematical Foundations of Information Theory*, New York: Dover.

Kolmogorov, A. N. (1956). On the Shannon theory of information transmission in the case of continuous signals, *IRE Trans. Inf. Theory* **1**, 102.

Kolmogorov, A. N. (1958). New metric invariant of transitive dynamical systems and auto-morphism of Lebesgue spaces, *Dokl. Akad. Nauk SSSR* **119**, 861.

Kolmogorov, A. N. (1965). Three approaches to the quantitative definition of information, *Probl. Inf. Transm.* **1**, 1.

Krieger, W. (1970). On entropy and generators of measure preserving transformations, *Trans. Am. Math. Soc.* **149**, 453.

Lacorata, G., Aurell E. and Vulpiani, A. (2001). Drifter dispersion in the Adriatic Sea: Lagrangian data and chaotic model, *Ann. Geophys.* **19**, 1.

Lacorata, G., Aurell, E., Legras, B. and Vulpiani, A. (2004). Evidence for a $k^{-5/3}$ spectrum from the EOLE Lagrangian balloons in the low stratosphere, *J. Atmos. Sci.* **61**, 2936.

Lai, Y.-C., Bollt, E. and Grebogi, C. (1999). Communicating with chaos using two-dimensional symbolic dynamics, *Phys. Lett. A* **255**, 75.

Li, M. and Vitanyi, P. (1997). *An Introduction to Kolmogorov Complexity and its Applications*, Berlin: Springer.

Livi, R., Politi, A. and Ruffo, S. (1986). Distribution of characteristic exponents in the thermodynamic limit, *J. Phys. A* **19**, 2033.

Lorenz, E. N. (1969). The predictability of a flow which possesses many scales of motion, *Tellus*, **21**, 3.

Martin-Löf, P. (1966). The definition of random sequences, *Inf. Contr.* **9**, 602.

Ornstein, D. S. (1974). *Ergodic Theory, Randomness and Dynamical Systems*, New Haven, CT: Yale University Press.

Oseledec, V. I. (1968). A multiplicative ergodic theorem: Lyapunov characteristic numbers for dynamical systems, *Trans. Mosc. Math. Soc.* **19**, 197.

Paladin, G. and Vulpiani, A. (1987). Anomalous scaling laws in multifractal objects, *Phys. Rep.* **156**, 147.

Paladin, G., Serva, M. and Vulpiani, A. (1995). Complexity in dynamical systems with noise, *Phys. Rev. Lett.* **74**, 66.

Pesin, Y. B. (1976). Lyapunov characteristic exponent and ergodic properties of smooth dynamical systems with an invariant measure, *Sov. Math. Dokl.* **17**, 196.

Plumecoq, J. and Lefranc, M. (2000a). From template analysis to generating partitions I: periodic orbits, knots and symbolic encodings, *Physica D* **144**, 231.

Plumecoq, J. and Lefranc, M. (2000b). From template analysis to generating partitions II: characterization of the symbolic encodings, *Physica D* **144**, 259.

Ruelle, D. (1982). Large volume limit of the distribution of characteristic exponents in turbulence, *Commun. Math. Phys.* **87**, 287.

Shannon, C. E. (1948). A mathematical theory of communication, *Bell Syst. Tech. J.* **27**, 379; **27**, 623. This work can be downloaded from: http://cm.bell-labs.com/cm/ms/what/shannonday/paper.html.

Sinai, Y. G. (1959). On the concept of entropy for a dynamic system, *Dokl. Akad. Nauk. SSSR* **124**, 768.

Solomonoff, R. J. (1964). A formal theory of inductive inference, *Inf. Contr.* **7**, 1 (Part I); **7**, 224 (Part II).

Takens, F. (1981). Detecting strange attractors in turbulence, in *Dynamical Systems and Turbulence (Warwick 1980)*, Vol. 898 of *Lecture Notes in Mathematics*, ed. D. A. Rand and L.-S. Young, p. 366, Berlin: Springer.

Varadhan, S. R. S. (1984). *Large Deviations and Applications*, Philadelphia, PA: SIAM.

Welsh, D. (1989). *Codes and Cryptography*, Oxford: Clarendon Press.

White, H. (1993). Algorithmic complexity of points in dynamical systems, *Erg. Theory Dyn. Syst.* **13**, 807.

Young, L.-S. (1982). Dimension, entropy and Lyapunov exponents, *Erg. Theory Dyn. Syst.* **2**, 109.

Zurek, W. H. (ed.) (1990). *Complexity, Entropy and Physics of Information*, Redwood City, CA: Addison-Wesley.

# 3

# Coarse graining, entropies and Lyapunov exponents at work

The meaning of the world is the separation of wish and fact.
*Kurt Gödel*

In the previous chapter we saw that in deterministic dynamical systems there exist well established ways to define and measure the complexity of a temporal evolution, in terms of either the Lyapunov exponents or the Kolmogorov–Sinai entropy. This approach is rather successful in deterministic low-dimensional systems. On the other hand in high-dimensional systems, as well as in low-dimensional cases without a unique characteristic time (as in the example discussed in Section 2.3.3), some interesting features cannot be captured by the Lyapunov exponents or Kolmogorov–Sinai entropy. In this chapter we will see how an analysis in terms of the finite size Lyapunov exponents (FSLE) and $\epsilon$-entropy, defined in Chapter 2, allows the characterization of non-trivial systems in situations far from asymptotic (i.e. finite time and finite observational resolution). In particular, we will discuss the utility of $\epsilon$-entropy and FSLE for a pragmatic classification of signals, and the use of chaotic systems in the generation of sequences of (pseudo) random numbers. In addition we will discuss systems containing some randomness.

## 3.1 Characterization of the complexity and system modeling

Typically in experimental investigations, time records of only few observables are available, and the equations of motion are not known. From a conceptual point of view, this case can be treated in the same framework that is used when the evolution laws are known. Indeed, in principle, with the embedding technique one can reconstruct the topological features of the phase space and dynamics (Takens 1981, Abarbanel *et al.* 1993, Kantz and Schreiber 1997). Nevertheless there are rather severe limitations in high-dimensional systems (Eckmann and Ruelle 1992) and even in low-dimensional systems non-trivial features can appear in the presence of noise (Paladin *et al.* 1995).

### 3.1.1 A simple model

Consider the following map

$$x_{t+1} = [x_t] + F(x_t - [x_t]), \tag{3.1}$$

where $[x_t]$ indicates the integer part of $x_t$ and $F(y)$ is given by:

$$F(y) = \begin{cases} (2+\alpha)y & \text{if } y \in [0, 1/2[ \\ (2+\alpha)y - (1+\alpha) & \text{if } y \in [1/2, 1], \end{cases} \tag{3.2}$$

with $\alpha > 0$. The largest Lyapunov exponent $\lambda$ can be obtained immediately: $\lambda = \ln |F'|$, with $F' = \mathrm{d}F/\mathrm{d}y = 2 + \alpha$. This map generates a diffusive behavior on large scales (Geisel and Nierwetberg 1984) that can be characterized quantitatively by the diffusion coefficient, $D$:

$$\langle (x_t - x_0)^2 \rangle \approx 2Dt \quad \text{for} \quad \text{large} \quad t. \tag{3.3}$$

One expects the following behavior for $h(\epsilon)$:

$$h(\epsilon) \approx h_{\mathrm{KS}} = \lambda \qquad \text{for} \quad \epsilon < 1, \tag{3.4}$$

$$h(\epsilon) \propto \frac{D}{\epsilon^2} \qquad \text{for} \quad \epsilon > 1. \tag{3.5}$$

Relation (3.4) says that, when $\epsilon$ is small enough, the $\epsilon$-entropy reduces to the Kolmogorov–Sinai entropy, given by the Lyapunov exponent in a one-dimensional chaotic system. Relation (3.5) translates the idea that, owing to an efficient time decorrelation, on a large scale the specific dynamical details are not important, and a diffusive behavior dominates giving rise to the $\epsilon$-entropy of Brownian motion (see Section 2.3.4). Consider now as a stochastic system, the following special noisy map

$$x_{t+1} = [x_t] + G(x_t - [x_t]) + \sigma\eta_t, \tag{3.6}$$

where $G(y)$, as shown in Figure 3.1, is a piecewise linear map which approximates the map $F(y)$, and $\eta_t$ is a stochastic process uniformly distributed in the interval $[-1, 1]$, with no correlation in time. When $|\mathrm{d}G/\mathrm{d}y| < 1$, as in the case we consider, the map (3.6), in the absence of noise, gives a non-chaotic time evolution. From a numerical point of view, computation of the FSLE is much less expensive than computation of the $\epsilon$-entropy, moreover one can safely assume that, at least in one dimension, $\lambda(\epsilon)$ and $h(\epsilon)$ are practically the same thing (this has been checked

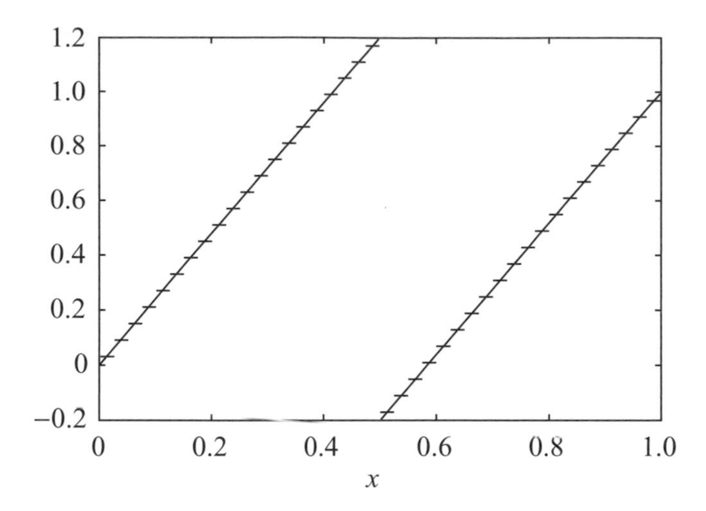


Figure 3.1 The map $F(x)$ (3.2) for $\alpha = 0.4$ is shown together with the superimposed approximating (regular) map $G(x)$ (3.6) obtained using 40 intervals of slope 0.

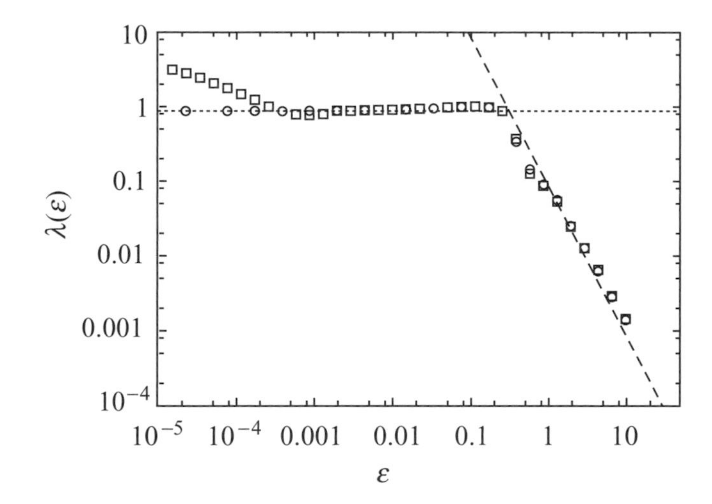


Figure 3.2 $\lambda(\epsilon)$ versus $\epsilon$ obtained with the map $F(y)$ (3.2) with $\alpha = 0.4$ ($\circ$) and with the noisy (regular) map (3.6) ($\Box$) with 10 000 intervals of slope 0.9 and $\sigma = 10^{-4}$. The straight lines indicate the Lyapunov exponent $\lambda = \ln 2.4$ and the diffusive behavior $\lambda(\epsilon) \sim \epsilon^{-2}$.

for $\epsilon$ not too small). Thus we now compare the FSLE for the chaotic map (3.1) and for the noisy map (3.6). In the latter the FSLE has been computed using two different realizations of the noise (see Section 3.3). In Figure 3.2 we show $\lambda(\epsilon)$ versus $\epsilon$ for the two cases. The two curves are practically indistinguishable in the region $\epsilon > \sigma$. The differences appear only at very small scales $\epsilon < \sigma$ where one has a $\lambda(\epsilon)$ which grows with $\epsilon$ for the noisy case, remaining at the same value for the chaotic deterministic case.[1] This analysis shows that we can distinguish three different regimes when observing the dynamics of (3.1) and (3.6) at different length scales. At the large length scales $\epsilon > 1$ we observe diffusive behavior in both models, with FSLE and $\epsilon$-entropy scaling as $\epsilon^{-2}$, as expected for a Wiener process. At length scales $\sigma < \epsilon < 1$ both models show a behavior that is typical of chaotic deterministic systems, because the $\epsilon$-entropy and FSLE are independent of $\epsilon$ and larger than zero. Finally, at the smallest length scales $\epsilon < \sigma$, while the system (3.1) still shows chaotic behavior, we see stochastic behavior for the system (3.6), signaled by the renewal of $h(\epsilon)$ variation. By "stochastic behavior" of a system on a certain interval of $\epsilon$ we mean that its $\epsilon$-entropy or FSLE decreases according to a power law of $\epsilon$ in that interval. The reason is that important stochastic processes display such behavior.

### *3.1.2 On the distinction between chaos and noise*

The above results show that the distinction between chaos and noise can be highly non-trivial, and makes sense only in very peculiar cases, for example, very low-dimensional systems. Nevertheless, even in this case, the entropic analysis can be unable to recognize the "true" character of the system owing to the lack of resolution. Again, the comparison between the diffusive map (3.1) and the noisy map (3.6) makes this point clearer. For $\sigma \leq \epsilon \leq 1$ both the systems (3.1) and (3.6), in spite of their "true" character, will be classified as deterministically chaotic, while for $\epsilon \geq 1$ both can be considered as stochastic as a random walk.

In high-dimensional chaotic systems, with $N$ degrees of freedom, one has typically $h(\epsilon) = h_{\rm KS} \sim O(N)$ for $\epsilon \leq \epsilon_{\rm c}(N)$ (where $\epsilon_{\rm c} \to 0$ as $N \to \infty$) while for $\epsilon \geq \epsilon_{\rm c}$, $h(\epsilon)$ decreases, often with a power law (Gaspard and Wang 1993, Boffetta *et al.* 2002). Since also in some stochastic processes, as noted above, the $\epsilon$-entropy obeys a power law, this can be a source of confusion.

These kinds of problems are not abstract ones, as a recent debate on "microscopic chaos" demonstrates. The detection of microscopic chaos by data analysis has recently been addressed in work by Gaspard *et al.* (1998), see also Grassberger and Schreiber (1999) and Dettmann *et al.* (1999) for technical comments. Gaspard *et al.* (1998), from the entropic analysis of an ingenious experiment on the position of a Brownian particle in a liquid, claim to give empirical evidence for microscopic chaos. In other words, they state that the diffusive behavior observed for a Brownian

[1] Hereafter we use corresponding scale values $\epsilon$ for both the FSLE and the $\epsilon$-entropy in order to make a direct comparison between the two quantities.

particle is the consequence of chaos at a molecular level. Their work can be summarized briefly as follows. From a long ($\approx 1.5 \times 10^5$ data) record of the position of a Brownian particle they compute the $\epsilon$-entropy, from which they obtain:

$$h(\epsilon) \sim \frac{D}{\epsilon^2}, \tag{3.7}$$

where $D$ is the diffusion coefficient. Then, *assuming* that the system is deterministic, and making use of the inequality $h(\epsilon > 0) \leq h_{\rm KS}$, they conclude that the system is chaotic. However, their result does not give direct evidence that the system is deterministic and chaotic. Indeed, the power law (3.7) can be produced with different mechanisms:

(1) a genuine chaotic system with diffusive behavior, such as the map (3.2);
(2) a non-chaotic system with some noise, such as the map (3.6) (Cencini *et al.* 2000);
(3) a deterministic linear non-chaotic system with many degrees of freedom, see for instance Mazur and Montroll (1960);
(4) a "complicated" non-chaotic system such as the Ehrenfest wind-tree model where a particle diffuses in a plane owing to collisions with randomly placed square scatterers, with fixed orientation, as discussed by Dettmann *et al.* (1999) in their comment on Gaspard *et al.* (1998); see also Chapter 6.

It seems to us that the weak points of the analysis in Gaspard *et al.* (1998) are the following:

(a) the explicit assumption that the system is deterministic;
(b) the limited number of data points and therefore limitations in both resolution and block length.

Point (a) is crucial, since without this assumption (even with an enormous data set) it is not possible to distinguish between cases (1) and (2). One has to say that in cases (3) and (4), at least in principle, it is possible to understand that the systems are not chaotic, but for this one has to use a huge number of data. For example Dettmann *et al.* (1999) estimated that in order to distinguish between (1) and (4) using realistic parameters of a typical liquid, the number of data points required is at least $\sim 10^{34}$.

With respect to the discussion about the interpretation of the data of this experiment, in Gaspard *et al.* (1999) and Briggs *et al.* (2001), the authors agree that the data originally presented do not prove that the process generating them must be chaotic. In any case, besides the technical aspects the experiment can be considered an interesting step toward a demonstration that chaos takes place in microscopic systems.

While exposing the difficulties we have set the basis for turning our limitations into an advantage. Certainly, we do not have access to arbitrarily fine resolutions, but the testable ones can be analyzed by powerful tools that allow a useful classification of dynamical behaviors. So we have the apparently paradoxical result that "complexity" helps in the construction of models. Basically, in the case in which one has a variety of behaviors at varying scale of resolution, there is a certain freedom in the choice of the model to be adopted. We have seen that, for some systems, the behavior at large scales can be realized using both chaotic deterministic models and suitable stochastic processes. From a pragmatic point of view, the fact that in certain stochastic processes $h(\epsilon) \sim \epsilon^{-\alpha}$ can indeed be extremely useful for modeling, for example, high-dimensional systems. Perhaps the most relevant case in which one can use this freedom in modeling is fully developed turbulence where the non-infinitesimal (the so-called inertial range) properties can be mimicked successfully in terms of multi-affine stochastic processes (Boffetta *et al.* 2002).

### 3.1.3 Macroscopic chaos in globally coupled maps

The emergence of non-trivial collective behaviors in high-dimensional dynamical systems has attracted much attention (Kaneko 1990, Chaté and Manneville 1992, Pikovsky and Kurths 1994). A limiting case of macroscopic coherence is the global synchronization of all the parts of the system. In addition to synchronization there exist other interesting phenomena (Pikovsky *et al.* 2003), among which we just mention clustering (Konishi and Kaneko 1992), and collective motion in globally coupled maps (Kaneko 1995). The latter behavior can be called *macroscopic chaos* (Shibata and Kaneko 1998, Cencini *et al.* 1999) (see below).

Let us consider a globally coupled map (GCM) defined as follows

$$x_n(t+1) = (1-c)f_a(x_n(t)) + \frac{c}{N}\sum_{i=1}^{N} f_a(x_i(t)), \tag{3.8}$$

where $N$ is the total number of elements and $f_a(x)$ is a non-linear function; in the following we consider the tent map $f_a(x) = a(1/2 - |x - 1/2|)$.

The evolution of a macroscopic variable, for example the center of mass

$$m(t) = \frac{1}{N}\sum_{i=1}^{N} x_i(t), \tag{3.9}$$

upon varying $c$ and $a$ in Eq. (3.8), displays different behaviors (Shibata and Kaneko 1998, Cencini *et al.* 1999):

(a) *standard chaos*: $m(t)$ is almost constant, obeying a Gaussian statistics with a standard deviation $\sigma_N = \sqrt{\langle m(t)^2\rangle - \langle m(t)\rangle^2} \sim N^{-1/2}$ around a fixed value;
(b) *macroscopic periodicity*: $m(t)$ is a superposition of a periodic function and small fluctuations $O(N^{-1/2})$;

(c) *macroscopic chaos*: $m(t)$ displays an irregular but quasi-deterministic motion, as can be seen by looking at the plot of $m(t)$ versus $m(t-1)$ which appears as a structured function (with thickness $\sim N^{-1/2}$), and suggests a chaotic motion for $m(t)$.

In the case of *macroscopic chaos* one expects that the center of mass evolves with typical times longer than the characteristic time of the full dynamics (i.e. the microscopic dynamics); the order of magnitude of the latter time may be estimated as $1/\lambda$, with $\lambda$ the Lyapunov exponent of the microscopic dynamics.

Indeed, conceptually, macroscopic chaos for a GCM can be thought of as analogous to hydrodynamical chaos for molecular motion, where, in spite of a huge microscopic Lyapunov exponent ($\lambda \sim 1/\tau_c \sim 10^{11} s^{-1}$, $\tau_c$ being the collision time), one can have rather different behaviors at a hydrodynamical (coarse-grained) level, i.e. regular motion ($\lambda_{\rm hydro} \leq 0$) or chaotic motion ($0 < \lambda_{\rm hydro} \ll \lambda$). In principle, if one knows the hydrodynamic equations, it is possible to characterize the macroscopic behavior using standard dynamical system techniques. However, in generic coupled map lattices there are no general systematic methods to build up the macroscopic equations, except for in particular cases (Pikovsky and Kurths 1994, Kaneko 1995). Therefore, here we discuss the macroscopic behavior of the system relying upon the full microscopic level of description and its numerical simulation.

Since the microscopic Lyapunov exponents cannot give a characterization of the macroscopic motion, different approaches have been proposed recently based on evaluation of the self-consistent Perron–Frobenius[2] (PF) operator (Perez and Cerdeira 1992, Pikovsky and Kurths 1994, Kaneko 1995) and on the FSLE (Shibata and Kaneko 1998, Cencini *et al.* 1999). Despite the conceptual interest of the former, in some sense the self-consistent PF operator plays a role similar to the Boltzmann equation for gases (Cencini *et al.* 1999); here we shall only discuss the latter which seems to us more appropriate to address the predictability problem.

We recall that for chaotic systems, in the limit of infinitesimal perturbations $\epsilon \to 0$, the FSLE behaves as $\lambda(\epsilon) \to \lambda$, i.e. $\lambda(\epsilon)$ displays a plateau at the value $\lambda$ for sufficiently small $\epsilon$. For non-infinitesimal $\epsilon$, one expects the $\epsilon$-dependence of $\lambda(\epsilon)$ to give information on the existence of other characteristic time scales governing the system, and, hence, that it could be able to characterize the macroscopic motion. In particular, at large scales, i.e. $\epsilon \gg 1/\sqrt{N}$, one expects the (fast) microscopic components to saturate and $\lambda(\epsilon) \approx \lambda_{\rm M}$, where $\lambda_{\rm M}$ can be fairly called the "macroscopic" Lyapunov exponent.

The FSLE of system (3.8) has been determined by looking at the evolution of $|\delta m(t)|$, which has been initialized at the value $\delta m(t) = \epsilon_{\rm min}$ by shifting all the elements of the unperturbed system by the quantity $\epsilon_{\rm min}$ (i.e. $x'_i(0) = x_i(0) + \epsilon_{\rm min}$),

[2] The self-consistent PF operator relates to the evolution of the single variable probability density function.

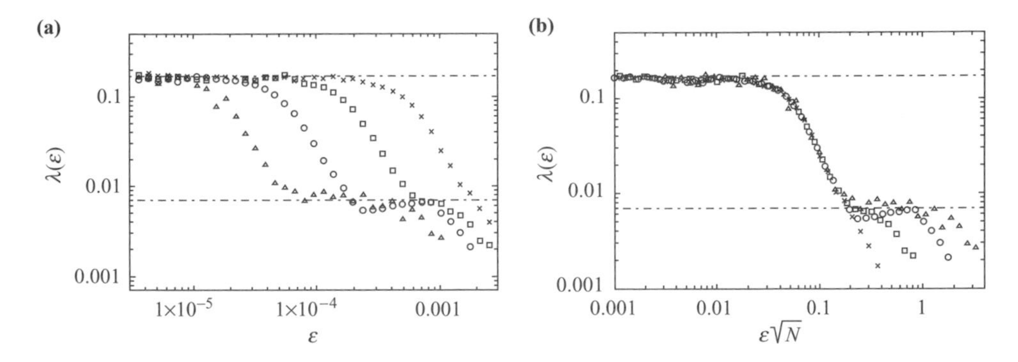


Figure 3.3 (a) $\lambda(\epsilon)$ versus $\epsilon$ for the system (3.8) with $a = 1.7, c = 0.3$ for $N = 10^4$ ($\times$), $N = 10^5$ ($\square$), $N = 10^6$ ($\bigcirc$) and $N = 10^7$ ($\triangle$). The first plateau corresponds to the microscopic Lyapunov exponent $\lambda \approx 0.17$ and the second one corresponds to the macroscopic Lyapunov exponent $\lambda_M \approx 0.007$. The average is over $2 \times 10^3$ realizations for $N = 10^4, 10^5, 10^6$ and 250 realizations for $N = 10^7$. (b) The same as (a) after rescaling the $\epsilon$-axis by $\sqrt{N}$.

for each realization. The computation has been performed by choosing the tent map as local map, but similar results can be obtained for other maps (Shibata and Kaneko 1998, Cencini *et al.* 1999). Figure 3.3(a) shows $\lambda(\epsilon)$ versus $\epsilon$ in the case of macroscopic chaos. One has two plateaus: at small values of $\epsilon (\epsilon \leq \epsilon_1)$, as expected from general considerations, $\lambda(\epsilon) = \lambda$; for $\epsilon \geq \epsilon_2$ one has another plateau, revealing the existence of a "macroscopic" Lyapunov exponent, $\lambda(\epsilon) = \lambda_M$. Moreover, $\epsilon_1$ and $\epsilon_2$ decrease at increasing $N$: indeed, by looking at Figure 3.3(b) one can see that $\epsilon_1, \epsilon_2 \sim 1/\sqrt{N}$. It is important to observe that the macroscopic plateau, which is almost non-existent for $N = 10^4$, becomes more and more resolved and extended at large values of $\epsilon\sqrt{N}$ on increasing $N$ up to $N = 10^7$. Therefore we can argue that the macroscopic motion is well defined in the limit $N \to \infty$ and one can conjecture that in this limit the microscopic signature in the evolution of $\delta m(t)$ disappears completely in favor of the macroscopic behavior. In the case of standard chaos ($\lambda_M < 0$) one has only the microscopic plateau and then a fast decrease in $\lambda(\epsilon)$ (Cencini *et al.* 1999).

We can summarize the main result briefly as follows:

- at small $\epsilon$ ($\ll 1/\sqrt{N}$), where $N$ is the number of elements, one recovers the "microscopic" Lyapunov exponent, i.e. $\lambda(\epsilon) \approx \lambda$;
- at large $\epsilon$ ($\gg 1/\sqrt{N}$) one observes another plateau $\lambda(\epsilon) \approx \lambda_M$ which can be much smaller than the microscopic one.

The emerging scenario is that at a coarse-grained level, i.e. $\epsilon \gg 1/\sqrt{N}$, the system can be described by an "effective" hydrodynamical equation (which in some cases can be low dimensional), while the "true" high-dimensional character appears only at very high resolution, i.e. $\epsilon \leq \epsilon_c = O(1/\sqrt{N})$.

### 3.1.4 Hydrodynamic Lyapunov modes

We saw how in high-dimensional systems one can have coherent behavior of the numerous degrees of freedom. Typically the global quantities which determine the "macroscopic" behavior are rather slow compared with the "microscopic" dynamics, i.e. their typical times are much longer than the inverse of the first Lyapunov exponent. An attempt to connect the macroscopic dynamics with the microscopic dynamics, with a coarse-graining approach (in terms of the FSLE), was discussed in the previous subsection.

Another interesting way is to try to associate the macroscopic behavior of a high-dimensional system with the Lyapunov exponents close to zero. In the last few years, some interest has been devoted to the so-called *hydrodynamic Lyapunov modes* (HLMs), i.e. to the Lyapunov vectors that correspond to the nearly zero Lyapunov exponents (McNamara and Mareschal 2001, Hoover *et al.* 2002, Eckmann *et al.* 2005). Let us review this approach briefly. Write in a formal way the evolution of the tangent vector, induced by the dynamics $\mathbf{x}(0) \to \mathbf{x}(t) = \mathcal{S}^t\mathbf{x}(0)$, as

$$\mathbf{z}(t) = \mathcal{M}(t)\mathbf{z}(0) \tag{3.10}$$

where $\mathcal{M}(t)$ is determined by the evolution of the state between 0 and $t$, see Section 2.1. We know from the Oseledec theorem that, in an ergodic system, the Lyapunov exponents $\lambda_1, \lambda_2, \ldots, \lambda_N$ are determined by the eigenvalues $\alpha_1(t), \alpha_2(t), \ldots, \alpha_N(t)$ of the matrix

$$\mathcal{V}(t) = \left[\mathcal{M}^\dagger(t)\mathcal{M}(t)\right]^{1/2t}$$

since in the limit $t \to \infty$ one has $\alpha_1(t) \to e^{\lambda_1}, \alpha_2(t) \to e^{\lambda_2}, \ldots$ for almost all the initial conditions.

In a similar way one can define the Lyapunov vectors as the eigenvectors (in the limit $t \to \infty$) of the matrix $\mathcal{V}(t)$. As a technical point we note that, at variance with the Lyapunov exponents, the Lyapunov vectors usually depend on $\mathbf{x}(0)$ (Crisanti *et al.* 1993).

Numerical studies of hard-sphere fluid and other systems, for example coupled map lattices and chains of anharmonic oscillators, reveal that Lyapunov exponents close to zero correspond to "hydrodynamic" Lyapunov vectors, i.e. with a structure of weakly perturbed coherent long wavelength waves (McNamara and Mareschal 2001, Hoover *et al.* 2002, Yang and Radons 2006). In order to give an idea of such behavior consider $N_p \gg 1$ disks moving in a two-dimensional box, of sides $L_x$ and $L_y$, with periodic conditions, with elastic scattering (Eckmann and Gat 2000). The

Lyapunov vectors have $4N_p$ components which we indicate as

$$\left(z_j^{(x)}, z_j^{(y)}, z_j^{(p_x)}, z_j^{(p_y)}\right) \qquad j = 1, 2, \ldots, N_p.$$

Let us introduce a vector field

$$\mathbf{v}(\mathbf{x}, t) = \left(v^{(x)}(\mathbf{x}, t), v^{(y)}(\mathbf{x}, t), v^{(p_x)}(\mathbf{x}, t), v^{(p_y)}(\mathbf{x}, t)\right) \in \mathbb{R}^4,$$

defined only by the instantaneous positions $\mathbf{x}_j$ of the particles:

$$v^{(x)}(\mathbf{x}, t) = z_j^{(x)}(t),$$

i.e. one selects the $j$ such that $\mathbf{x} = \mathbf{x}_j(t)$, and similarly for the other components (Eckmann and Gat 2000). Of course in such a treatment the continuous limit is assumed implicitly.

The vector field of the slow Lyapunov vectors, defined by the Lyapunov vectors with small Lyapunov exponents, is well approximated by the long wavelength eigenmodes of a "reverse wave equation" in the considered domain, i.e.

$$\frac{\partial \mathbf{v}(\mathbf{x}, \mathbf{t})}{\partial t} = -\frac{1}{N_p^2} \Delta \mathbf{v}(\mathbf{x}, t);$$

note the unusual sign in front of $\Delta$. One has a long wavelength wave with, say, $n$ nodes in the $x$ direction and $m$ nodes in the $y$ direction and the Lyapunov exponent is proportional to

$$\frac{1}{N_p}\sqrt{\left(\frac{n}{L_x}\right)^2 + \left(\frac{m}{L_y}\right)^2}.$$

In the case of the hard-sphere fluid these collective perturbations are due to the conservation of certain quantities during collisions. These new conservation laws generate new hydrodynamic fields, just as the conservation of mass, momentum, and energy generate the density, velocity, and temperature fields (McNamara and Mareschal 2001).

Such behavior is intriguing because the HLM may be related to hydrodynamic fluctuations. If it were possible to extract the transport coefficients directly from these exponents, it would be rather interesting to find a link between the reversible microscopic dynamics and the irreversible macroscopic behavior.

Unfortunately, progress in such an ambitious project is very slow owing to formidable technical difficulties. On the other hand some interesting results have been obtained. There is numerical and analytical evidence of the existence of HLMs in lattices of coupled Hamiltonian and dissipative maps. The HLMs in Hamiltonian systems are propagating, whereas those of dissipative systems show only diffusive motion. Simulations of various systems confirm that the existence of HLMs is a very general feature of extended dynamical systems with continuous symmetries and that the above-mentioned differences between the two classes of systems are universal to a large extent (Yang and Radons 2006). In a recent work (Ginelli *et al.* 2007) a general method to determine Lyapunov vectors in both discrete- and continuous-time dynamical systems was introduced. For spatially extended systems, it has been shown that the Lyapunov vectors computed using the matrix $\mathcal{V}(t)$ can have properties qualitatively different from those computed from $\mathcal{M}(t)$, composing the orthonormalized basis obtained by the standard procedure used to calculate the Lyapunov exponents.

### 3.1.5 Other high-dimensional systems: convective chaos

A general feature of systems evolving in space and time is that a generic perturbation not only grows in time but also propagates in space. Aiming at a quantitative description of such phenomena, Deissler and Kaneko (1987) introduced a generalization of the Lyapunov exponent to a non-stationary frame of reference: the *comoving* Lyapunov exponent, providing an extension of the notion of convective instability, capturing global and non-linear features. For the sake of simplicity, we consider the case of a one-dimensional coupled map lattice with unidirectional coupling (Rudzik and Pikovsky 1996):

$$x_n(t+1) = (1-c) f_a(x_n(t)) + c f_a(x_{n-1}(t)), \tag{3.11}$$

where $t$ and $n(=1, \ldots, L)$ label the discrete time and space respectively, and $c$ is the coupling strength between near-neighboring lattice sites.

Such a class of models is able to capture the basic phenomenology of many physical systems in which a privileged direction exists, for example boundary layers, thermal convection and wind-induced water waves (Pikovsky 1989, Falcioni *et al.* 1999).

Let us consider an infinitesimally small perturbation initially different from zero only in one site of the lattice. By following the evolution of the perturbation along the sites defined by $j(t) = [vt]$ (where $[\cdots]$ denotes the integer part), one expects:

$$|\delta x_{j(t)}(t)| \approx |\delta x_0(0)| \mathrm{e}^{\lambda(v)t}, \tag{3.12}$$

where $\lambda(v)$ is the largest comoving Lyapunov exponent, defined as

$$\lambda(v) = \lim_{t\to\infty} \lim_{L\to\infty} \lim_{|\delta x_0(0)|\to 0} \frac{1}{t} \ln\left(\frac{|\delta x_{[vt]}(t)|}{|\delta x_0(0)|}\right). \tag{3.13}$$

In this equation the order of the limits is important to obtain a well-defined quantity and avoid finite-size effects. For $v = 0$ one recovers the usual Lyapunov exponent. Moreover, one has that $\lambda(v) = \lambda(-v)$ (and the maximum value is obtained at $v = 0$)

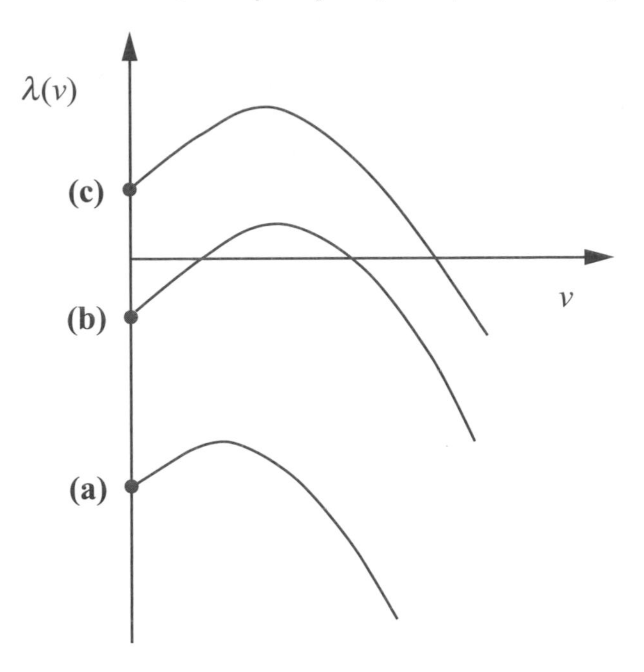


Figure 3.4 Sketch of the behavior of $\lambda(v)$ for (a) absolutely stable and convectively stable flow, (b) absolutely stable but convectively unstable flow, and (c) absolutely unstable flow.

when a privileged direction does not exist, otherwise $\lambda(v)$ can be asymmetric and the maximum can be attained at value $v \neq 0$.

Two other indicators can be defined, that are related to the comoving Lyapunov exponent: the local Lyapunov exponent (Pikovsky 1993) and the specific (or temporal) Lyapunov exponents, see e.g. Cencini and Torcini (2001). A central concept in the study of flow systems is that of *convective* instability, that presents itself when a perturbation grows exponentially along the flow but vanishes locally. We can give a description of the phenomenology of flow systems in terms of the largest Lyapunov exponent and of the comoving Lyapunov exponent (see Figure 3.4). The absolute stability is identified by the condition $\lambda(v) < 0$ for all $v \geq 0$; the convective instability corresponds to $\lambda_1 = \lambda(v = 0) < 0$ and $\lambda(v) > 0$ for some velocities $v > 0$, and finally standard chaos (absolute instability) is present when $\lambda_1 = \lambda(v = 0) > 0$.

The convective instability is conceptually very interesting, because even if the largest Lyapunov exponent is negative the behavior of the system may be very irregular (Falcioni *et al.* 1999).

For the spatial "complexity" associated with convective instability there is no simple and systematic characterization. A first explanation for these features may be found in the sensitivity of convective unstable systems to small perturbations at the beginning of the chain (always present in a physical system), which grow exponentially while they are transmitted along the flow. This simple intuition can be made more quantitative by defining an indicator which measures the degree of sensitivity to the boundary conditions (Falcioni *et al.* 1999). With regard to the state of the initial site, we can have, for instance, $x_0(t) = x^*$ with $x^*$ being an unstable fixed point of the map $f_a(x)$, or more generic time-dependent boundary conditions where $x_0(t)$ is equal to a known function of time $y(t)$, which can be periodic, quasi-periodic or chaotic. Here, following Pikovsky (1989), we consider a quasi-periodic boundary condition $x_0(t) = 0.5 + 0.4\sin(\omega t)$, with $\omega = \pi(\sqrt{5} - 1)$. However, the results we are going to discuss do not depend too much on the details of the boundary conditions, i.e. on $x_0(t)$ being quasi-periodic or chaotic. We wonder how an uncertainty $|\delta x_0(t)|$ of the order of $\delta_0$ in the knowledge of the boundary conditions will affect the system. We consider only the case of infinitesimal perturbations, i.e. $\delta x_n$ evolves according to the tangent space dynamics, and for the moment we do not consider intermittency (i.e. time fluctuations of the comoving Lyapunov exponents).

The uncertainty $\delta x_n(t)$, in the determination of the variable at time $t$ and site $n$, is given by the superposition of the evolved $\delta x_0(t - \tau)$ with $\tau = n/v$:

$$\delta x_n(t) \sim \int \delta x_0(t - \tau) \mathrm{e}^{\lambda(v)\tau} \mathrm{d}v \sim \delta_0 \int \mathrm{e}^{[\lambda(v)/v]n} \mathrm{d}v. \tag{3.14}$$

Since we are interested in the asymptotic spatial behavior, i.e. large $n$, we can write:

$$\delta x_n(t) \sim \delta_0\, \mathrm{e}^{\Gamma n}. \tag{3.15}$$

The quantity $\Gamma$ can be considered as a sort of spatial-complexity-index, an operative definition of which is the following:

$$\Gamma = \lim_{n \to \infty} \frac{1}{n} \left\langle \ln \frac{|\delta x_n|}{\delta_0} \right\rangle, \tag{3.16}$$

where the brackets mean a time average. The previous relation gives a link between the comoving and the "spatial" Lyapunov exponent $\Gamma$, i.e. a relation between the convective instability of a system and its sensitivity to the boundary conditions. In the particular case of a non-intermittent system, a saddle-point estimate of Eq. (3.14) gives

$$\Gamma = \Gamma^* = \max_v \left[ \frac{\lambda(v)}{v} \right]. \tag{3.17}$$

This equation holds exactly only in the absence of intermittency; in general the relation is more complicated. One can introduce the effective comoving Lyapunov exponent, $\tilde{\gamma}_t(v)$, which gives the exponential changing rate of a perturbation, in the frame of reference moving with velocity $v$, on a finite time interval $t$. According to general arguments one has $\langle \tilde{\gamma}_t(v) \rangle = \lambda(v)$. Then, instead of (3.14) one has

$$\delta x_n(t) \sim \delta_0 \int \mathrm{e}^{[\tilde{\gamma}_t(v)/v]n} \mathrm{d}v, \tag{3.18}$$

and therefore:

$$\Gamma = \lim_{n\to\infty} \frac{1}{n} \left\langle \ln \frac{|\delta x_n|}{\delta_0} \right\rangle = \lim_{n\to\infty} \frac{1}{n} \ln \frac{|\delta x_n^{\text{typical}}|}{\delta_0} = \left\langle \max_v \left[ \frac{\tilde{\gamma}_t(v)}{v} \right] \right\rangle. \tag{3.19}$$

Because of the fluctuations, it is not possible to write $\Gamma$ in terms of $\lambda(v)$, although one can obtain a lower bound (Falcioni *et al.* 1999):

$$\Gamma \geq \max_v \left[ \frac{\langle \tilde{\gamma}_t(v) \rangle}{v} \right] = \max_v \left[ \frac{\lambda(v)}{v} \right] \equiv \Gamma^*. \tag{3.20}$$

The difference between $\Gamma$ and $\Gamma^*$ is only due to intermittency, as investigations of a non-intermittent map and computation of the generalized spatial Lyapunov exponents $L_s(q)$ confirm (Falcioni *et al.* 1999).

In conclusion we can say that the spatial complexity displayed by these systems indicates that the unpredictability of a system cannot be completely reduced to the existence of at least one positive Lyapunov exponent.

## 3.2 How random is a random number generator?

In most numerical computations, for example Monte Carlo simulations and molecular dynamics, it is necessary to have a series of independent identically distributed (i.i.d.) continuous random variables $x(1), \ldots, x(n)$ uniformly distributed in the interval [0, 1]. One can produce true random number sequences only by using some non-deterministic physical phenomenon, for example the decay of radioactive nuclei or the arrival on a detector of cosmic rays. A more practical way is to use a computer that produces a "random-looking" sequence of numbers, by means of a recursive rule. Let us comment on this point. It is now well established that deterministic systems may have a time evolution that appears rather "irregular" with the typical features of genuine random processes. Moreover, we know, see Section 3.1.2, that there are rather severe restrictions on the possibility of distinguishing between signals generated by different rules, such as regular (high-dimensional) systems, deterministic chaotic systems, and genuine stochastic processes. However, we know that, although the above result may appear negative, it allows a pragmatic classification of the stochastic or chaotic features of a signal, according to the dependence of the $\epsilon$-entropy on $\epsilon$, and this yields some freedom in modeling systems. As a matter of fact in physical problems one normally uses these similarities to model an "irregular" deterministic behavior by means of a truly stochastic process.

We will see here that one can also use the practical difficulties in the distinction chaos/noise to mimic random processes with deterministic chaotic systems.

We define a pseudo random number generator (PRNG), which is an algorithm, i.e. a deterministic system, designed to mimic a random sequence on a computer. This issue is far from being trivial; in Von Neumann's words: "Anyone who considers arithmetical methods of producing random digits is, of course, in a state of sin" (Von Neumann 1963). Two unavoidable problems are the following.

(a) Since the algorithm is deterministic, the Kolmogorov–Sinai entropy ($h_{\mathrm{KS}}$) is finite. The sequence $\{x(i)\}$ cannot be "really random," i.e. with an infinite Kolmogorov–Sinai entropy, because the deterministic dynamical rule constrains the outputs that are near in time and supplies us with a maximum of $\log_2(e^{h_{\mathrm{KS}}})$ random bits per unit time. This limitation would be present also in a hypothetical computer able to work with real numbers.
(b) Since any deterministic system with a finite number of states is periodic, any sequence produced by an algorithm working with discrete numbers (as happens with a computer) must be periodic, possibly after a transient; therefore, not only $h_{\mathrm{KS}} < \infty$, but also $h_{\mathrm{KS}} = 0$. The computer-implemented system can only be pseudochaotic.

Let us stress the main points to bear in mind when using a deterministic chaotic system as a PRNG.

(1) The outputs $\{x(t)\}$ of a perfect RNG, for small $\epsilon$, have $h(\epsilon) = \ln(1/\epsilon)$. Since in any deterministic $d$-dimensional system $h(\epsilon) \simeq h_{\mathrm{KS}}$ for small $\epsilon$ (i.e. $\epsilon < \epsilon_{\mathrm{c}}$ with $\ln \epsilon_{\mathrm{c}}^d \sim -h_{\mathrm{KS}}$), one should work with a very large $h_{\mathrm{KS}}$. In this way the true (deterministic) nature of the PRNG becomes apparent only below the small scale $\epsilon_{\mathrm{c}}$. Another possibility is to look only at one variable of a high-dimensional system, so that the lower scale is given by $\epsilon_{1\mathrm{c}} \approx \epsilon_{\mathrm{c}}^d \ll \epsilon_{\mathrm{c}}$.
(2) In order to actually observe the behavior $h(\epsilon) \sim \ln(1/\epsilon)$ (i.e. the behavior of independent variables) for $\epsilon \geq \epsilon_{\mathrm{c}}$ in a concrete deterministic algorithm, it is necessary that the time correlation is very weak. We will discuss how this property may be achieved by taking as output a single variable of a high-dimensional chaotic system.

A third point has to be added, dealing with the problem (b). Quantities like $h_{\mathrm{KS}}$ and the $\epsilon$-entropy have an asymptotic nature, i.e. they are related to large time behavior. This allows the existence of situations where the system is, strictly speaking, non-chaotic ($h_{\mathrm{KS}} = 0$) but its features appear irregular to a certain extent. Such a property (denoted by the term pseudochaos (Chirikov and Vivaldi 1999, Mantica 2000, Zaslavsky 2002)) is basically due to the presence of long transient effects. As noted above, the use of a computer discretizes the phase space of a dynamical system, canceling (at least) its asymptotic chaotic properties. However, we may have confidence that, if the period of the realized sequence is long enough, the effects related to points (1) and (2) can survive as a chaotic transient. According to this observation, a third request must be added:

(3) the period of the series generated by the computer (i.e. with a state-discretization of the deterministic system) must be very large.

In the following we discuss some examples of PRNGs.

### 3.2.1 A low-dimensional system with high entropy

A simple and popular PRNG is the multiplicative congruent one (Press *et al.* 1986):

$$\begin{aligned} z_{n+1} &= N_1 z_n \bmod N_2 \\ x_{n+1} &= z_{n+1}/N_2, \end{aligned} \tag{3.21}$$

with an integer multiplier $N_1$ and modulus $N_2$. The $\{z_n\}$ are integer numbers and one hopes to generate a sequence of random variables $\{x_n\}$, which are uncorrelated and uniformly distributed in the unit interval. A first problem one has to face is the periodic nature of (3.21), because of its discrete character. In practice one wants to fix $N_1$ and $N_2$ in such a way as to maximize this period. Note that the rule (3.21) can be interpreted as a deterministic dynamical system, i.e.

$$x_{n+1} = N_1 x_n \bmod 1, \tag{3.22}$$

which has a uniform invariant measure and a Kolmogorov–Sinai entropy $h_{\mathrm{KS}} = \lambda = \ln N_1$. When imposing the integer arithmetics of Eq. (3.21) onto this system, we are, in the language of dynamical systems, considering an unstable periodic orbit of Eq. (3.22), with the particular constraint that, in order to achieve the period $N_2 - 1$ (i.e. all integers $< N_2$ should belong to the orbit of Eq. (3.21)) it has to contain all values $k/N_2$, with $k = 1, 2, \ldots, N_2 - 1$. This results in a condition to sample the uniform distribution, that gives the same weight to each interval $[k/N_2, (k+1)/N_2[$. Since the natural invariant measure of Eq. (3.22) is uniform, such an orbit represents the measure of a chaotic solution in an optimal way. Every sequence of a PRNG is characterized by two quantities: its period $\mathcal{T}$ and its positive Lyapunov exponent $\lambda$, which is identical to the entropy of a chaotic orbit of the equivalent dynamical system.

It is natural to ask how the required (apparent) randomness can be reconciled with the facts that (a) the PRNG is a deterministic dynamical system, and (b) it is a discrete state system (Kantz and Olbrich 2000). If the period is long enough on shorter times one has to face only point (a). Let us discuss this point in terms of the behavior of the $\epsilon$-entropy. As noted above, one should realize the true deterministic chaotic nature of the system when $\ln \epsilon \approx -h_{\mathrm{KS}} = -\ln N_1$. Therefore, when $\epsilon \lesssim 1/N_1, h(\epsilon) \simeq h_{\mathrm{KS}} = \ln N_1$, while for $\epsilon \gtrsim 1/N_1$ one expects to observe the "apparent

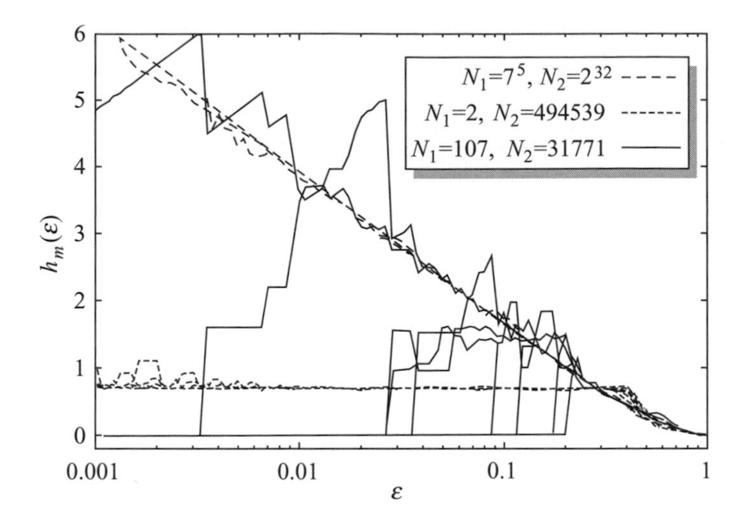


Figure 3.5 The $\epsilon$-entropies, $h_m(\epsilon)$, on varying the embedding dimension $m$ (equivalently the block size) for the multiplicative congruential random number generator Eq. (3.21) for different choices of $N_1$ and $N_2$.

random" behavior of the system, i.e. $h(\epsilon) \sim \ln(1/\epsilon)$. On the other hand, when the spatial resolution is high enough so that every point of this periodic orbit is characterized by its own symbol, then, for arbitrary block length $m$, one has a finite number of $m$-words whose probabilities are different from 0. Therefore, the block entropy $H_m$, for large $m$, is constant and $h_m = 0$.

Figure 3.5 presents the behavior of $h_m(\epsilon)$, computed on sequences of length 60 000 of the PRNG with three different pairs $(N_1, N_2)$ chosen to be $(7^5, 2^{32})$, (2, 494539), and (107, 31771). The first pair is optimal and no deviation from stochastic behavior is visible. The second pair has a small pseudo-entropy, and this is seen in the saturation of all $h_m(\epsilon)$ at $\ln N_1 = \ln 2$, and the last pair has large entropy but a rather short period, so that all $h_m(\epsilon)$ drop to zero for some $\epsilon_m$, where $\epsilon_m$ becomes dramatically larger for increasing $m$ (strong fluctuations arise from the fact that data are confined to a grid of quite large spacing 1/31771 compared to the spacing involved in the other two situations).

### 3.2.2 A system with low entropy and high dimension

As noted in Section 3.1.2 even non-chaotic high-dimensional systems may display a long irregular regime as a transient effect, see also Politi *et al.* (1993). Such a property can be used successfully to generate pseudo-random sequences in systems with a moderate or even vanishing $h_{\mathrm{KS}}$. In these cases, one observes a single variable and its block entropies, $H_m$, which are characterized by a transient with maximal (or almost maximal) value of the slope, i.e. $H_m \simeq m \ln(1/\epsilon)$, and then a crossover when the block size $m$ increases, to a regime with the slope of the true $h_{KS}$ of the system.

The most familiar class of PRNGs using this property is given by the so-called lagged Fibonacci generators (Green *et al.* 1959, Knuth 1981), which correspond to the dynamical system:

$$x(t) = ax(t - \tau_1) + bx(t - \tau_2) \qquad \bmod 1 \tag{3.23}$$

where $a$ and $b$ are $O(1)$ and $\tau_1 < \tau_2$. Equation (3.23) can be written in the form

$$\mathbf{y}(t) = \mathbf{F}\mathbf{y}(t - 1) \tag{3.24}$$

where $\mathbf{F}$ is a $\tau_2 \times \tau_2$ matrix of the form

$$\mathbf{F} = \begin{pmatrix} 0 & \dots & a & \dots & b \\ 1 & 0 & 0 & \dots & 0 \\ 0 & 1 & 0 & \dots & 0 \\ \dots & \dots & \dots & \dots & \dots \\ 0 & \dots & \dots & 1 & 0 \end{pmatrix} \tag{3.25}$$

showing explicitly that the phase space of (3.23) has dimension $\tau_2$. It is easy to prove that this system is chaotic for each value of $a, b \in \mathbb{N}$, with $a, b > 0$. The Kolmogorov–Sinai entropy does not depend on $\tau_1$ or on $\tau_2$, and the only way to obtain high values of $h_{KS}$ is to use large values of $a, b$. Nevertheless, the lagged Fibonacci generators are used with $a = b = 1$: for these values of the parameters $e^{-h_{KS}} \approx 0.618$ and $\epsilon_c$ is not small. This implies that the determinism of the system is detectable also with large graining; however, the generator is a good one. This can be explained because the $m$-words, built up by a single variable (say $y_1$) of the $\tau_2$-dimensional system (3.24), have the maximal allowed block-entropy (Falcioni *et al.* 2005), $H_m(\epsilon) = m \ln(1/\epsilon)$, for $m < \tau_2$, so that:

$$H_m(\epsilon) \simeq \begin{cases} m \ln(1/\epsilon) & \text{for } m < \tau_2 \\ \tau_2 \ln(1/\epsilon) + h_{KS}(m - \tau_2) & \text{for } m \geq \tau_2. \end{cases} \tag{3.26}$$

One may read Eq. (3.26) as follows. Though the "true" $h_{KS}$ is small, it can be revealed only for very large values of $m$. Indeed, by observing the one-variable $m$-words, which corresponds to an embedding procedure, before capturing the dynamical entropy one has to realize that the system has dimension $\tau_2$. This happens only for words longer than $\tau_2$.

The importance of the transient behavior of $H_m$ can be understood in terms of the "effective measure of complexity" $C$, introduced by Grassberger (1986) (see

Section 2.2.1): for large $m$, the block entropies grow as

$$H_m \simeq C + m h_{\mathrm{KS}}. \tag{3.27}$$

In the case of the Fibonacci map, from (3.26) one has that, for small $\epsilon$,

$$C = \tau_2 \left[ \ln\left(\frac{1}{\epsilon}\right) - h_{\mathrm{KS}} \right] \approx \tau_2 \ln\left(\frac{1}{\epsilon}\right). \tag{3.28}$$

For large $\tau_2$ (usually values $O(10^2)$ are used) $C$ is so huge that only an extremely long sequence of the order $\exp(\tau_2)$ (likely outside the capabilities of modern computers) can reveal that the "true" Kolmogorov–Sinai entropy is small. This is so because, recalling that the number of relevant $m$-words is about $\exp(H_m)$, one has that in order to detect a block-entropy $H_m$ one needs a sequence whose length is at least $O(\exp(H_m))$.

### *3.2.3 A system with high entropy and high dimension*

We discuss here a multi-dimensional version of the Arnold map (or cat map, see Section 1.4.2) as a PRNG, showing that this system possesses a high value of $h_{\mathrm{KS}}$ and very good properties from the point of view of correlation functions (Falcioni *et al.* 2005).

The multi-dimensional generalization of the two-dimensional Arnold map can be written in the following way:

$$\begin{pmatrix} \mathbf{x}' \\ \mathbf{y}' \end{pmatrix} = \mathbf{M} \begin{pmatrix} \mathbf{x} \\ \mathbf{y} \end{pmatrix} \mod 1, \tag{3.29}$$

with

$$\mathbf{M} = \begin{pmatrix} \mathbf{I} & \mathbf{A} \\ \mathbf{B} & \mathbf{I} + \mathbf{BA} \end{pmatrix} \tag{3.30}$$

where $\mathbf{M}$ is a $2N \times 2N$ matrix, $\mathbf{x}, \mathbf{y} \in \mathbb{R}^N$, $\mathbf{I}$ is the $N \times N$ identity matrix and $\mathbf{A}, \mathbf{B}$ are symmetric $N \times N$ matrices with integer entries in order to obtain a mapping that is continuous at the boundaries. It easy to see that the evolution law given by the equations above is symplectic, indeed one can write Eq. (3.29) and Eq. (3.30) as a canonical transformation

$$\mathbf{x} = \frac{\partial S(\mathbf{x}', \mathbf{y})}{\partial \mathbf{y}}, \quad \mathbf{y}' = \frac{\partial S(\mathbf{x}', \mathbf{y})}{\partial \mathbf{x}'}, \tag{3.31}$$

where the generating function is

$$S(\mathbf{x}', \mathbf{y}) = \sum_{j=1}^{N} x'_j y_j - \frac{1}{2} \sum_{j,k=1}^{N} (y_j A_{jk} y_k + x'_j B_{jk} x'_k). \tag{3.32}$$

The ordinary Arnold map is obtained with $N = 1$ and $A = B = 1$.

It can be shown that when $\mathrm{Tr}\,\mathbf{M} > 2N$, map (3.29) is a chaotic system with a uniform invariant measure. The Arnold map is an example of what is usually referred to as an Anosov system (see, e.g. Eckmann and Ruelle (1985)). It is not difficult to compute the Kolmogorov–Sinai entropy of the $2N$-dimensional cat map, showing that it grows proportional to the number of dimensions. Detailed numerical computations show that for low enough values of $\epsilon$, i.e. for $\epsilon < \epsilon_c$, $h(\epsilon)$ has a "plateau" around the value of $h_{\mathrm{KS}}$. If one aims to use the map as a generator to produce a random sequence of $\approx 1/\epsilon$ symbols then, when $\epsilon > \epsilon_c$, to a good approximation one is near the value corresponding to a theoretical RNG, i.e. $h(\epsilon) = -\log(\epsilon)$.

If, as the output of the generator one takes a single component of the $2N$-dimensional vector $\mathbf{x}$, one obtains an important advantage compared with other generators. This is the factorization of all $n$-times, with $n < 2N$, correlation functions. Under rather general conditions on the matrices $\mathbf{A}$ and $\mathbf{B}$, it is possible to prove (Falcioni *et al.* 2005) the vanishing of any correlation of up to $n$ ($< 2N$) functions of time-delayed variables:

$$\langle g_1(z(t_1)) \ldots g_n(z(t_n)) \rangle = \langle g_1(z(t_1)) \rangle \ldots \langle g_n(z(t_n)) \rangle \tag{3.33}$$

for every $g_i \in L^2$, where $z$ is one component of the vector $\mathbf{x}$, e.g. $z = x_1$. This is basically due to the high dimensionality of the system and the presence of "hidden variables." This result is rigorously true for the system with continuous states; however, numerical checks show that this property survives in the discrete case.

### 3.2.4 The problem of the period for systems with discrete states

What a computer really calculates is a finite-digit dynamics that can be represented as a dynamics on integers. Therefore one is dealing with a deterministic system with $\mathcal{M}$ discrete states which is periodic. Unfortunately there are no general methods to determine a priori the length of the periodic orbits. A nice result, based on probabilistic considerations, suggests that the period scales as $\mathcal{T} \sim \mathcal{M}^{1/2}$ (Coste and Hénon 1986), although strong fluctuations are present. The use of high-dimensional systems may also be a natural solution for this problem: calling $M$ the number of states along each of the $d$ dimensions, the typical period scales on average as $\mathcal{T} \sim M^{d/2}$ and grows very fast with increasing $M$ and $d$ (note that a lower bound would be sufficient for the present purpose).

In some cases one can estimate the period $\mathcal{T}$. Consider the behavior of the discrete Fibonacci generator:

$$z(t) = az(t - \tau_1) + bz(t - \tau_2) \mod M \tag{3.34}$$

where $z(t) \in [0, M-1]$ is an integer variable, $a, b, \tau_1, \tau_2, M$ are integer parameters and $M \gg \tau_2$. The parameters $\tau_1, \tau_2$ and $M$ are chosen in order to have a period as long as possible. Number-theoretical arguments (Knuth 1981) allow us to choose these parameters such that the period of the orbit is maximum $\mathcal{T} = M^{\tau_2} - 1$.

When the period is maximum, for $\epsilon \geq 1/M$ one has:

$$H_m(\epsilon) \simeq \begin{cases} m \ln(1/\epsilon) & \text{for } m < \tau_2 \\ \tau_2 \ln(1/\epsilon) + h_{\mathrm{KS}}(m - \tau_2) & \text{for } \tau_2 \leq m \leq m^* \\ \tau_2 \ln(M) & \text{for } m > m^* \end{cases} \tag{3.35}$$

where

$$m^* = \frac{\tau_2}{h_{\mathrm{KS}}} \left[ \ln\left(\frac{1}{\epsilon}\right) - \ln M + h_{\mathrm{KS}} \right]. \tag{3.36}$$

When $\epsilon = 1/M$ we have $m^* = \tau_2$, the second regime in Eq. (3.35) disappears and the block-entropy behavior is independent of $h_{\mathrm{KS}}$. Still, as for the continuous case, if $\tau_2$ is large one observes only the pseudo-chaotic transient

$$H_m(\epsilon) \approx m \ln\left(\frac{1}{\epsilon}\right). \tag{3.37}$$

Consider now the discrete multi-dimensional cat map, namely

$$\begin{pmatrix} \mathbf{z}' \\ \mathbf{w}' \end{pmatrix} = \begin{pmatrix} \mathbf{I} & \mathbf{A} \\ \mathbf{B} & \mathbf{I} + \mathbf{BA} \end{pmatrix} \begin{pmatrix} \mathbf{z} \\ \mathbf{w} \end{pmatrix} \quad \mathrm{mod}\ M, \tag{3.38}$$

where, as usual, $\mathbf{A}, \mathbf{B}$ have integer entries, $z_i, w_i \in [0, 1, \ldots, M-1]$.

A peculiar feature of cat maps is that periodic orbits of the continuum system have rational coordinates (Percival and Vivaldi 1987). Consequently, the orbits of the discretized version of the map are completely equivalent to periodic orbits of the continuous system with coordinates $z_i/M, w_i/M$. As a corollary, since the map is invertible, periodic orbits do not have any transient: every state is periodic.

The cat map typically has many orbits, and the great majority of them are of the same length. A theoretical analysis of these orbits has been made in the two-dimensional case (Percival and Vivaldi 1987). The scaling $\mathcal{T} \sim M^N$, suggested by the random map arguments, has been observed numerically for a discrete multi-dimensional cat map (Falcioni *et al.* 2005) and it is rather typical, i.e. it appears for the discrete (state) version of chaotic systems (Grebogi *et al.* 1988).

Unfortunately we have no theoretical control over the period, and wild fluctuations are present when $M$ varies; therefore it is better to choose a value of $M, N, \mathbf{A}, \mathbf{B}$ and check $\mathcal{T}$ or a lower bound directly. With the choice $N = 3$, $M = 1001400791$

and

$$\mathbf{A} = \begin{pmatrix} 1 & 1 & 1 \\ 1 & 3 & 1 \\ 1 & 1 & 5 \end{pmatrix} \qquad \mathbf{B} = \begin{pmatrix} 7 & 1 & 1 \\ 1 & 3 & 1 \\ 1 & 1 & 9 \end{pmatrix}, \tag{3.39}$$

numerically one obtains $\mathcal{T} > 7 \times 10^{12}$, which is a satisfying lower bound for typical simulations.

### *3.2.5 Conclusions on the pseudo random number generators and deterministic chaos*

We have shown how, using some properties of high-dimensional deterministic chaotic systems, it is possible to generate a good approximation of a random sequence. This is in spite of unavoidable constraints of deterministic algorithms running on real computers.

There are two possible mechanisms for obtaining good PRNGs using deterministic systems: very high Kolmogorov–Sinai entropy, and "transient chaos" with a large finite-time $\epsilon$ entropy (due to the high dimensionality of the algorithm). The multi-dimensional cat map introduced above as a PRNG has rather good properties. For example, the generator discussed here has been able to pass all the severe "exams" which have been proposed in order to test "how random" is a given sequence of numbers. These algorithms are available in easy-to-use software packages collecting dozens of different tests, for example, the *DieHard* (Marsaglia 1995), and the NIST batteries.[3]

The main disadvantage of the high-dimensional Arnold map is that one cannot predict analytically the period given the parameters or, equivalently, write a condition on the parameters in order to obtain the maximum period. However, probabilistic arguments (Coste and Hénon 1986), confirmed by numerical checks, show that the period increases exponentially with $N$; therefore with a proper choice of the parameters we have extremely large periods.

## 3.3 Lyapunov exponents and complexity in dynamical systems with noise

We consider here systems containing random perturbations, which are always present in physical systems as a consequence of thermal fluctuations or hidden changes of control parameters, and, in numerical experiments, because of the round-off errors. The combined effect of the noise and of the deterministic part of the

[3] See The National Institute of Standards and Technology, *A Statistical Test Suite for the Validation of Random Number Generators and Pseudo Random Number Generators for Cryptographic Applications*, http://csrc.nist.gov/rng/SP800-22b.pdf.

evolution law can produce highly non-trivial behaviors. Let us mention stochastic resonance, where there is a synchronization with the deterministic periodic forcing of the noise-induced jumps between two stable points (Benzi *et al.* 1982), the phenomena of so-called noise-induced order (Matsumoto and Tsuda 1983) and noise-induced instability (Bulsara *et al.* 1990).

When facing systems with noise, the simplest possibility is to treat the random term as a given time-dependent forcing term, that is to consider the separation of two close trajectories with the same realization of noise. In this way one computes the largest Lyapunov exponent, $\lambda_\sigma$, associated with the separation rate of two nearby trajectories with the same realization of the stochastic term (where $\sigma$ indicates the noise strength). Although $\lambda_\sigma$ is a well-defined quantity, i.e. the Oseledec (1968) theorem holds, sometimes it is not the most useful characterization of complexity. In addition, a moment of reflection shows that it is practically impossible to extract $\lambda_\sigma$ from experimental data.

We will show that, for noisy and random systems, a more natural indicator of complexity can be obtained by computing the separation rate of nearby trajectories evolving with different noise realizations. This measure of complexity, defined in Paladin *et al.* (1995) and Loreto *et al.* (1996a) and inspired by the ideas of information theory, is related to the mean number of bits per unit time that are necessary to specify the sequence generated by a random evolution law.

### *3.3.1 The naive approach: noise treated as a standard function of time*

The approach in which one treats the random term as a usual time-dependent external force can lead to misleading results, as illustrated in the following example. Let us consider a one-dimensional Langevin equation

$$\frac{\mathrm{d}x}{\mathrm{d}t} = -\frac{\partial V(x)}{\partial x} + \sqrt{2\sigma}\,\eta, \tag{3.40}$$

where $\eta(t)$ is a white noise and $V(x)$ diverges for $|\, x \,| \to \infty$, for example the usual double well potential $V = -x^2/2 + x^4/4$.

The Lyapunov exponent $\lambda_\sigma$, associated with the separation rate of two nearby trajectories with the same realization of $\eta(t)$, is defined as

$$\lambda_\sigma = \lim_{t\to\infty} \frac{1}{t} \ln |z(t)| \tag{3.41}$$

where the evolution of the tangent vector is given by:

$$\frac{\mathrm{d}z}{\mathrm{d}t} = -\frac{\partial^2 V(x(t))}{\partial x^2} z(t). \tag{3.42}$$

Since the system is ergodic, with invariant probability distribution $P(x) = Ce^{-V(x)/\sigma}$, in the first place one has that (except for a zero measure set of initial conditions and noise realizations) $\lambda_\sigma$ is independent both of the initial condition and of the particular realization of the noise. Moreover one can write:

$$\lambda_\sigma = \lim_{t\to\infty} \frac{1}{t} \ln |z(t)| = -\lim_{t\to\infty} \frac{1}{t} \int_0^t \partial^2_{xx} V(x(t'))\mathrm{d}t'$$
$$= -C \int \partial^2_{xx} V(x)\mathrm{e}^{-V(x)/\sigma}\,\mathrm{d}x = -\frac{C}{\sigma} \int (\partial_x V(x))^2 \mathrm{e}^{-V(x)/\sigma}\,\mathrm{d}x < 0. \tag{3.43}$$

This result has a rather intuitive meaning: the trajectory $x(t)$ spends most of the time in one of the "valleys" where $-\partial^2_{xx} V(x) < 0$ and only short time intervals on the "hills" where $-\partial^2_{xx} V(x) > 0$, so that the distance between two trajectories evolving with the same noise realization typically decreases. The result obtained for the one-dimensional Langevin equation can easily be generalized to any dimension for gradient systems if the noise is small enough (Loreto *et al.* 1996a).

We have to emphasize that a negative value of $\lambda_\sigma$ implies a fully predictable process only if the realization of the noise is known. In the case of two initially close trajectories evolving under two different noise realizations, after a certain time $T_\sigma$, the two trajectories can be very distant, because they can be in two different valleys. For $\sigma$ small enough, according to the Kramers formula (Chandrasekhar 1943), one has $T_\sigma \sim \exp(\Delta V/\sigma)$, where $\Delta V$ is the difference between the values of $V$ on the top of the hill and at the bottom of the valley.

### 3.3.2 An information theory approach

The main difficulties in defining the notion of "complexity" of an evolution law with a random perturbation already appear in one-dimensional maps. The generalization to $N$-dimensional maps or to ordinary differential equations is straightforward.

Therefore, we consider the model

$$x(t+1) = f[x(t), t] + \sigma w(t), \tag{3.44}$$

where $t$ is an integer and $w(t)$ is an uncorrelated random process, for example the $w$ are independent random variables uniformly distributed in $[-1/2, 1/2]$. To obtain the largest Lyapunov exponent $\lambda_\sigma$, as defined in (3.41), one has to study the equation

$$z(t+1) = \left.\frac{\mathrm{d}f}{\mathrm{d}t}\right|_{x(t)} z(t). \tag{3.45}$$

However, we can follow the approach of Section 2.2.3 to write the Lyapunov exponent of the system. Let $x(t)$ be the trajectory starting at $x(0)$ and $x'(t)$ be the trajectory starting from $x'(0) = x(0) + \delta x(0)$. Let $\delta_0 \equiv |\delta x(0)|$ and indicate by $\tau_1$ the minimum time such that $|x'(\tau_1) - x(\tau_1)| \geq \Delta$. Then, we put $x'(\tau_1) = x(\tau_1) + \delta x(0)$ and define $\tau_2$ as the time such that $|x'(\tau_1 + \tau_2) - x(\tau_1 + \tau_2)| > \Delta$ for the first time, and so on, yielding a sequence $\{\tau_i\}$ of time intervals. In this way we can define the quantity

$$\Lambda_\sigma = \frac{1}{\overline{\tau}} \ln\left(\frac{\Delta}{\delta_0}\right) \tag{3.46}$$

where $\overline{\tau} = \sum \tau_i / \mathcal{N}$, and $\mathcal{N}$ is the number of intervals in the sequence. If the above procedure is applied by considering the same noise realization for both trajectories then, when $\lambda_\sigma > 0$, $\Lambda_\sigma = \lambda_\sigma$. In other words, by considering two different realizations of the noise for the two trajectories, we obtain a new quantity

$$K_\sigma = \frac{1}{\overline{\tau}} \ln\left(\frac{\Delta}{\delta_0}\right), \tag{3.47}$$

which arises naturally in the framework of information theory and algorithmic complexity theory. The times $\tau_1, \tau_2, \ldots$ are the intervals at which it is necessary to repeat the transmission of $x(t)$, with a precision $\delta_0$, and $K_\sigma / \ln 2$ is the number of bits *per* unit time one has to specify in order to transmit the original sequence with a tolerance $\Delta$. If the fluctuations of the effective Lyapunov exponent $\gamma(t)$ (see Section 2.3.1) are very small (i.e. weak intermittency) one has:

$$K_\sigma \simeq \lambda_\sigma. \tag{3.48}$$

The interesting situation occurs for strong intermittency when there are alternations of positive and negative $\gamma$ during long time intervals: this induces a dramatic change in the value of $K_\sigma$. This becomes particularly clear when we consider the limiting case of positive $\gamma^{(1)}$ in an interval $T_1 \gg 1/\gamma^{(1)}$ followed by a negative $\gamma^{(2)}$ in an interval $T_2 \gg 1/|\gamma^{(2)}|$, and again a positive effective Lyapunov exponent and so on. During the intervals with positive effective Lyapunov exponent the transmission has to be repeated rather often with $\simeq T_1/(\gamma^{(1)} \ln 2)$ bits at each time, while during those with negative effective Lyapunov exponent no information has to be sent. Nevertheless, at the end of the contracting intervals one has $|\delta x| = O(\sigma)$, i.e. bounded below (on average) by a finite number, so that, at variance with the noiseless case, it is impossible to use them to compensate for the expanding intervals. This implies that in the limit of very large $T_i$ only the expanding intervals contribute to the evolution of the error $\delta x(t)$ and $K_\sigma$ is given by a time average of the positive effective Lyapunov exponents:

$$K_\sigma \simeq \langle \gamma \theta(\gamma) \rangle. \tag{3.49}$$

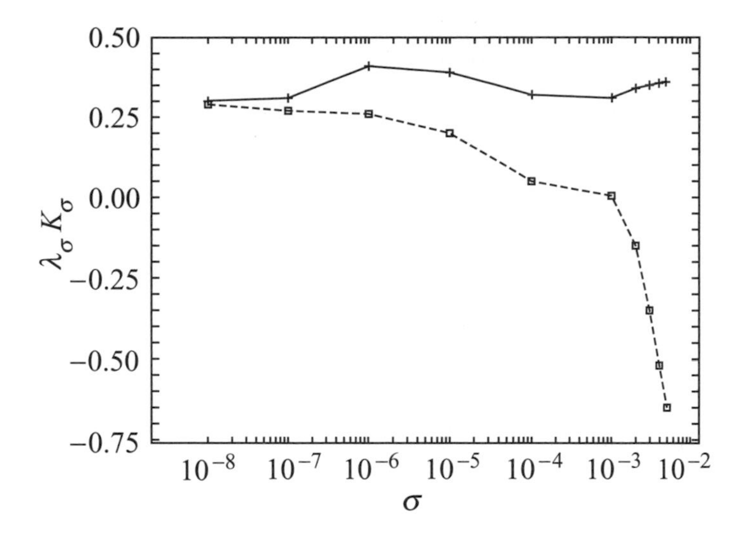


Figure 3.6 $\lambda_\sigma$ (squares) and $K_\sigma$ (crosses) versus $\sigma$ for the map (3.50).

Note that it may happen that $K_\sigma > 0$ with $\lambda_\sigma < 0$. We stress again that (3.49) holds only for strong intermittency, while for uniformly expanding systems or rapid alternations of contracting and expanding behaviors, $K_\sigma \simeq \lambda_\sigma$.

The quantity $K_\sigma$ is a kind of $\epsilon$-entropy, indeed, the complexity we consider is defined for $\delta_0$ not too small ($\delta_0 \gg \sigma$). If $\delta_0$ and $\Delta$ are small enough, but still much larger than $\sigma$, $K_\sigma$ is essentially independent of their values. The relation $K_\sigma \simeq \langle \gamma\, \theta(\gamma) \rangle$ is the time analog of the Pesin relation: $h_{\rm KS} = \sum_i \lambda_i\, \theta(\lambda_i)$. The latter relation expresses the fact that negative Lyapunov exponents do not decrease the value of $h_{\rm KS}$, because the contraction along the corresponding directions cannot be observed for any finite space partition. In the same way the contracting time intervals, if long enough, do not decrease $K_\sigma$. Another important remark is that in the usual treatment of the experimental data, where noise is usually present, one practically computes $K_\sigma$ and the result can be completely different from $\lambda_\sigma$.

Now we discuss briefly some numerical results for the system (3.44) when $f(x)$ gives rise to strongly intermittent behavior. This is the case for the Beluzov–Zhabotinsky map (Matsumoto and Tsuda 1983), introduced to describe the famous chemical reaction:

$$f(x) = \begin{cases} [(1/8 - x)^{1/3} + a]\mathrm{e}^{-x} + b & \text{if } 0 \le x < 1/8 \\ [(x - 1/8)^{1/3} + a]\mathrm{e}^{-x} + b & \text{if } 1/8 \le x < 3/10 \\ c(10\,x\,\mathrm{e}^{-10x/3})^{19} + b & \text{if } 3/10 \le x \end{cases} \tag{3.50}$$

with $a = 0.50607357$, $b = 0.0232885279$, $c = 0.121205692$. The map exhibits a chaotic alternation of expanding and contracting time intervals. In Figure 3.6, one sees that while $\lambda_\sigma$ passes from negative to positive values at decreasing $\sigma$ (Matsumoto and Tsuda 1983), $K_\sigma$ is not sensitive to this transition.

The previous results show that the same system can be regarded either as regular (i.e. $\lambda_\sigma < 0$), when the same noise realization is considered for two nearby trajectories, or as chaotic (i.e. $K_\sigma > 0$), when two different noise realizations are considered. We can say that a negative $\lambda_\sigma$ for relatively large values of $\sigma$ is not an indication that "noise induces order"; a correct conclusion is that noise can induce synchronization (Pikovsky *et al.* 2003).

### 3.3.3 Random dynamical systems

We now discuss dynamical systems where the randomness is not simply given by an additive noise. This kind of system has been of interest in relation to problems involving disorder, such as the characterization of the so-called *on-off intermittency* (Platt *et al.* 1993) and to model transport problems in turbulent flows (Yu *et al.* 1991). In these systems, in general, the random part represents an ensemble of hidden variables believed to be implicated in the dynamics.

For the sake of simplicity we consider only the case of random maps, that exhibit very interesting features ranging from stable or quasi-stable behaviors, to chaotic behaviors and intermittency, see Platt *et al.* (1993). Let us denote by $\mathbf{x}(t)$ the state of a system whose evolution law is given by

$$\mathbf{x}(t+1) = \mathbf{f}(\mathbf{x}(t), J(t)), \tag{3.51}$$

where $J(t)$ is a random variable that, we assume, may take on a discrete set of values. As for the case of additive noise examined previously, the simplest approach is the introduction of the Lyapunov exponent $\lambda_J$ computed considering the separation of two nearby trajectories evolving with the same realization of the random process $J(t) = i_1, i_2, \ldots, i_t$. The Lyapunov exponent $\lambda_J$ generalizes $\lambda_\sigma$ of Section 3.3.1 and can be computed from the tangent vector evolution:

$$\lambda_J = \lim_{t\to\infty} \frac{1}{t} \ln |\mathbf{z}(t)| \tag{3.52}$$

where

$$z_m(t+1) = \sum_n \frac{\partial f_m(\mathbf{x}(t), i_t)}{\partial x_n} z_n(t). \tag{3.53}$$

We note that, assuming ergodicity, $\lambda_J$ does not depend either on the initial condition or on the particular realization of $J(t)$ (apart from zero measure sets of initial conditions and random sequences). On the other hand, for these systems, as in the case of additive noise, it is possible to introduce a measure of complexity, $K_J$, which accounts better for their chaotic properties (Paladin *et al.* 1995, Loreto *et al.* 1996a, 1996b)

$$K_J \simeq h_{\mathrm{Sh}} + \lambda_J \theta(\lambda_J), \tag{3.54}$$

where $h_{\mathrm{Sh}}$ is the Shannon entropy of the random sequence $J(t)$, see Section 2.2.1. The meaning of $K_J$ is rather clear: $K_J / \ln 2$ is the mean number of bits, for each iteration, necessary to specify the sequence $x_1, \ldots, x_t$ with a certain tolerance $\Delta$. Note that there are two different contributions to the complexity: (a) one has to specify the sequence $J(1), J(2), \ldots, J(t)$ which requires $h_{\mathrm{Sh}}/\ln 2$ bits per iteration; (b) if $\lambda_J$ is positive, one has to specify the initial condition $x(0)$ with a precision $\Delta e^{-\lambda_J T}$, where $T$ is the time length of the evolution. This requires $\lambda_J / \ln 2$ bits per iteration; if $\lambda_J$ is negative the initial condition can be specified using a number of bits independent of $T$.

*A toy model: one-dimensional random maps*

The following example may be useful in clarifying the subject since, in spite of its simplicity, it captures some basic features of this kind of system (Platt *et al.* 1993):

$$x(t+1) = a_t x(t)(1 - x(t)), \tag{3.55}$$

where $a_t$ is a random dichotomous variable given by

$$a_t = \begin{cases} 4 & \text{with probability } p \\ 1/2 & \text{with probability } 1-p. \end{cases} \tag{3.56}$$

For $x(t)$ close to zero, one can neglect the non-linear term to obtain

$$x(t) = \prod_{j=0}^{t-1} a_j x(0); \tag{3.57}$$

from the law of large numbers, applied to $\ln x(t)$ and the sequence of i.i.d. variables $\{\ln a_j\}$, one has that the typical behavior is

$$x(t) \sim x(0) e^{\langle \ln a \rangle t}. \tag{3.58}$$

The expression $\langle \ln a \rangle = p \ln 4 + (1-p) \ln 1/2 = (3p-1) \ln 2$ gives a threshold value $p_c = 1/3$. For $p < p_c$, $x(t) \to 0$ for $t \to \infty$ and the linear reasoning is consistent. In contrast, for $p > p_c$, after a certain time $x(t)$ escapes from the fixed point $x = 0$ and the non-linear term becomes relevant. Figure 3.7 shows a typical *on-off intermittency* behavior of the system (3.55)–(3.56) for $p$ slightly larger than $p_c$. Note that, in spite of this irregular behavior, numerical computations show that the Lyapunov exponent $\lambda_J$ is negative for $p < \tilde{p} \simeq 0.5$.

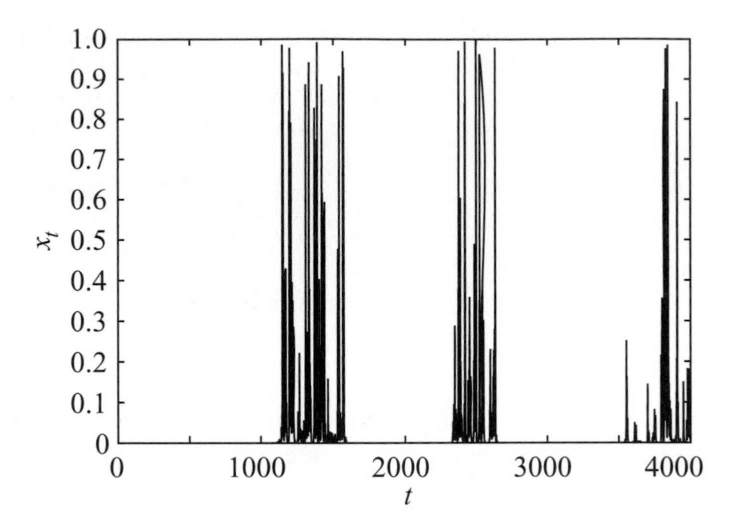


Figure 3.7 $x(t)$ versus $t$ for the random map (3.55)–(3.56), with $p = 0.35$.

By introducing a finite threshold $\epsilon$, in order to discriminate laminar (i.e. $x(t) < \epsilon$) and intermittent phases (i.e. $x(t) \geq \epsilon$), we can define a complexity $K(\epsilon)$. We denote by $l_{\mathrm{L}}$ and $l_{\mathrm{I}}$ the average lifetimes respectively of the laminar and of the intermittent phases. When $p$ is close to $p_{\mathrm{c}}$ we expect $l_{\mathrm{I}} \ll l_{\mathrm{L}}$. The mean number of bits, per iteration, one has to specify in order to transmit the sequence is (Loreto *et al.* 1996a, 1996b)

$$\frac{K(\epsilon)}{\ln 2} \simeq \frac{l_{\mathrm{I}} h_{\mathrm{Sh}}}{(l_{\mathrm{I}} + l_{\mathrm{L}}) \ln 2} \simeq \frac{l_{\mathrm{I}}}{l_{\mathrm{L}}} \frac{h_{\mathrm{Sh}}}{\ln 2}, \tag{3.59}$$

where $h_{\mathrm{Sh}}$ is the Shannon entropy of the sequence $\{a_j\}$. To obtain (3.59) first notice that on an interval $T$ one has approximatively $T/(l_{\mathrm{I}} + l_{\mathrm{L}})$ intermittent bursts and the same number of laminar phases. During a laminar phase $x(t)$ is close to zero, therefore one has to specify only the duration of the phase, for which just a small number of bits ($\mathrm{O}(\ln l_{\mathrm{L}})$) is necessary. However, in the intermittent phase one has to specify the sequence $\{a_j\}$, which requires $h_{\mathrm{Sh}}/\ln 2$ bits per iteration, on average.

### *Sandpile models as random maps*

A class of systems which can be treated in the framework of random maps is represented by the so-called sandpile models that are a paradigmatic example of self-organized criticality (SOC) (Bak *et al.* 1988). This term refers to the tendency of some large dynamical systems to evolve *spontaneously* toward a critical state characterized by spatial and temporal self-similarity. The original sandpile models are probabilistic cellular automata inspired by the dynamics of avalanches in a pile of sand. Dropping sand slowly, grain by grain onto a limited base, one reaches a situation where the pile is critical, i.e. it has a critical slope. This means that a further addition of sand will produce sliding of sand (avalanches) that can be small or cover the entire size of the system. In this case the critical state is characterized by scale-invariant distributions for the size and the lifetime and it is reached without tuning of any critical parameter.

We refer in particular to the model introduced by Zhang (1989), a continuous-state version of the original sandpile model, defined on a $d$-dimensional lattice. The variable on each site $x_i$ (interpretable as energy, sand, heat, mechanical stress etc.) can vary continuously in the range [0, 1] with the threshold fixed at $x_c = 1$. The dynamics is the following:

(a) one chooses a site at random and adds to it an amount $\delta x$;
(b) if at a certain time $t$ the value of the variable in a site, say $i$, exceeds the threshold $x_c$ a relaxation process is triggered defined as

$$\begin{cases} x_{i+nn} \to x_{i+nn} + x_i/2d \\ x_i \to 0, \end{cases} \tag{3.60}$$

where $nn$ indicates the $2d$ nearest neighbors of the site $i$;
(c) one repeats point (b) until all the sites have relaxed;
(d) one goes back to point (a).

Let us discuss the problem of predictability in sandpile models on the basis of some rigorous results (Caglioti and Loreto 1996), which clarify the role of the Lyapunov exponent for this class of systems. It has been proved that the Lyapunov exponent $\lambda_J$ is negative. In fact the dynamics of a small difference between two configurations follows the same rules (a)–(d), i.e. the "error" is redistributed to the nearest-neighbor sites, so that one has

$$\lambda_J \leq -\frac{\text{constant}}{R^2} \tag{3.61}$$

where $R$ is the diameter of the system.

As for the other examples already discussed, the existence of a negative Lyapunov exponent (computed with a given realization of the randomness) does not mean perfect predictability. This can be understood by looking at the growth of the distance, $\delta(t)$, between two initially close trajectories computed with two different realizations of randomness, i.e. by adding sand at different sites. Let us consider the case of the "minimal error": in the reference realization one adds sand on a site $i$ chosen at random. In the perturbed realization, instead, one adds a sand grain at one of the nearest-neighbor sites of $i$. In such a case $\delta(t)$ increases up to a maximal distance in few avalanches (Loreto *et al.* 1996a). Practically, one has the same kind of phenomenon, already discussed, as for the Langevin equation with two noise realizations.

Let us now estimate the complexity $K_J$ of this system. An upper bound can be given by using (3.54), $K_J = h_{\mathrm{Sh}} + \lambda_J \theta(\lambda_J)$, where $h_{\mathrm{Sh}}$ is the entropy of the random sequence of addition of energy. In sandpile models, since each site has the same probability of being selected, one has $h_{\mathrm{Sh}} = \ln V$, where $V$ is the number of sites of the system. Since the Lyapunov exponent is negative, the complexity is just determined by $h_{\mathrm{Sh}}$, i.e. by the nature of the external randomness.

## 3.4 Conclusions

In this chapter we saw that it is possible to analyze different kinds of irregular behaviors of a system by means of different quantities derived from the Lyapunov exponents and the Kolmogorov–Sinai entropy.

At a practical and conceptual level, one has severe difficulty in distinguishing between the deterministic or stochastic nature of systems displaying irregular behavior. Owing to the finiteness of the datasets, it is not possible to perform an entropic analysis with an arbitrary fine resolution, i.e. to compute the $\epsilon$-entropy $h(\epsilon)$ for very small values of $\epsilon$. Although the above result may appear negative, it allows a pragmatic classification of the stochastic or chaotic feature of the signal, according to the dependence of the $\epsilon$-entropy on $\epsilon$, and this yields some freedom in modeling systems. As a relevant example of a representation of a deterministic system in terms of stochastic processes, we mention fully developed turbulence. In addition, one can follow the opposite strategy, i.e. one can mimic noise with deterministic chaotic systems. This point of view is at the basis of the approach that uses a deterministic chaotic system as a PRNG.

Some systems can also manifest irregular spatial properties as happens, for example, in open flows with convective chaos but with negative Lyapunov exponents. In such a case the "complexity" is due to a sort of sensitivity to the boundary conditions. An uncertainty, $\delta x_0$, on the boundary condition is exponentially amplified with the distance, $n$, from the boundary as $\delta x_n \sim \delta x_0 \mathrm{e}^{\Gamma n}$. The "spatial" Lyapunov exponent $\Gamma$ is related to the comoving Lyapunov exponent and gives a characterization of the "spatial complexity."

In the presence of randomness one can introduce two different Lyapunov exponents, $\lambda_\sigma$ (or $\lambda_J$) in the case of trajectories with the same realization of noise (or generic randomness) and $K_\sigma$ (or $K_J$) for different realizations. In general $\lambda_\sigma$ and $K_\sigma$ do not coincide and characterize different aspects of the system. Both quantities have their own relevance, moreover the comparison between $\lambda_\sigma$ and $K_\sigma$ has been found to be useful in the understanding of apparently intricate phenomena, such as noise-induced order and noise-induced instability.

Summarizing, we can say that the study of these different aspects of predictability constitutes a useful method for a quantitative characterization of "complexity," suggesting the following equivalences:

$$\text{COMPLEX} \leftrightarrow \text{INCOMPRESSIBLE} \leftrightarrow \text{UNPREDICTABLE}.$$

## References


Abarbanel, H. D. I., Brown, R., Sidorowich, J. L. and Tsimring, L. Sh. (1993). The analysis of observed chaotic data in physical systems, *Rev. Mod. Phys.* **65**, 1331.

Bak, P., Tang, C. and Wiesenfeld, K. (1988). Self-organized criticality, *Phys. Rev. A* **38**, 364.

Benzi, R., Parisi, G., Sutera, A. and Vulpiani, A. (1982). Stochastic resonances in climatic change, *Tellus* **34**, 10.

Boffetta, G., Cencini, M., Falcioni, M. and Vulpiani, A. (2002). Predictability: a way to characterize complexity, *Phys. Rep.* **356**, 367.

Briggs, M. E., Sengers, J. V., Francis, M. K., Gaspard, P., Gammon, R. W., Dorfman, J. R. and Calabrese, R. V. (2001). Tracking a colloidal particle for the measurement of dynamic entropies, *Physica A* **296**, 42.

Bulsara, A. R., Jacobs, E. W. and Schieve, W. C. (1990). Noise effects in a nonlinear dynamic system – the RF-superconducting quantum interference device, *Phys. Rev. A* **42**, 4614.

Caglioti, E. and Loreto, V. (1996). Dynamical properties and predictability of a class of self-organized critical models, *Phys. Rev. E* **53**, 2953.

Cencini, M. and Torcini, A. (2001). Linear and nonlinear information flow in spatially extended systems, *Phys. Rev. E* **63**, 056201.

Cencini, M., Falcioni, M., Vergni, D. and Vulpiani, A. (1999). Macroscopic chaos in globally coupled maps, *Physica D* **130**, 58.

Cencini, M., Falcioni, M., Kantz, H., Olbrich, E. and Vulpiani, A. (2000). Chaos or noise: difficulties of a distinction, *Phys. Rev. E* **62**, 427.

Chandrasekhar, S. (1943). Stochastic problems in physics and astronomy, *Rev. Mod. Phys.* **15**, 1.

Chaté, H. and Manneville, P. (1992). Collective behaviors in spatially extended systems with local interactions and synchronous updating, *Prog. Theor. Phys.* **87**, 1.

Chirikov, B. V. and Vivaldi, F. (1999). An algorithmic view of pseudochaos, *Physica D* **129**, 223.

Coste, J. and Hénon, M. (1986). Invariant cycles in the random mapping of $N$ integers onto themselves. Comparison with Kauffman binary network, in *Disordered Systems and Biological Organization*, ed. E. Bienenstock, F. Fogelman and G. Weisbuch, p. 360, Berlin: Springer.

Crisanti, A., Paladin, G. and Vulpiani, A. (1993). *Products of Random Matrices in Statistical Physics*, Berlin: Springer.

Deissler, R. J. and Kaneko, K. (1987). Velocity-dependent Lyapunov exponents as a measure of chaos for open-flow systems, *Phys. Lett. A* **119**, 397.

Dettmann, C., Cohen, E. and van Beijeren, H. (1999). Microscopic chaos from Brownian motion?, *Nature* **401**, 875.

Eckmann, J.-P. and Gat, O. (2000). Hydrodynamic Lyapunov modes in translation invariant systems, *J. Stat. Phys.* **98**, 775.

Eckmann, J.-P. and Ruelle, D. (1985). Ergodic theory of chaos and strange attractors, *Rev. Mod. Phys.* **57**, 617.

Eckmann, J.-P. and Ruelle, D. (1992). Fundamental limitations for estimating dimensions and Lyapunov exponents in dynamical systems, *Physica D* **56**, 185.

Eckmann, J.-P., Forster, C., Posch, H. A. and Zabey, E. (2005). Lyapunov modes in hard-disk systems, *J. Stat. Phys.* **118**, 813.

Falcioni, M., Vergni, D. and Vulpiani, A. (1999). Characterization of the spatial complex behavior and transition to chaos in flow systems, *Physica D* **125**, 65.

Falcioni, M., Palatella, L., Pigolotti, S. and Vulpiani, A. (2005). Properties making a chaotic system a good pseudo random number generator, *Phys. Rev. E* **72**, 016220.

Gaspard, P. and Wang, X. J. (1993). Noise, chaos, and $(\epsilon, \tau)$-entropy per unit time, *Phys. Rep.* **235**, 291.

Gaspard, P., Briggs, M. E., Francis, M. K., Sengers, J. V., Gammon, R. W., Dorfman, J. R. and Calabrese, R. V. (1998). Experimental evidence for microscopic chaos, *Nature* **394**, 865.

Gaspard, P., Briggs, M. E., Francis, M. K., Sengers, J. V., Gammon, R. W., Dorfman, J. R. and Calabrese, R. V. (1999). Reply to: Statistical mechanics – Microscopic chaos from Brownian motion?, *Nature* **401**, 876.

Geisel, T. and Nierwetberg, J. (1984). Intermittent diffusion: a chaotic scenario in unbounded systems, *Phys. Rev. A* **29**, 2305.

Ginelli, F., Poggi, P., Turchi, A., Chaté, H., Livi, R. and Politi, A. (2007). Characterizing dynamics with covariant Lyapunov vectors, *Phys. Rev. Lett.* **99**, 130601.

Grassberger, P. (1986). Toward a quantitative theory of self-generated complexity, *Int. J. Theor. Phys.* **25**, 907.

Grassbeger, P. and Schreiber, T. (1999). Statistical mechanics – microscopic chaos from Brownian motion?, *Nature* **401**, 875.

Grebogi, C., Ott, E. and Yorke, J. A. (1988). Roundoff-induced periodicity and the correlation dimension of chaotic attractors, *Phys. Rev. A* **38**, 3688.

Green, B. F., Smith, J. E. K. and Klem, L. (1959). Empirical tests of an additive random number generator, *J. ACM* **6**(4), 527.

Hoover, W. G., Posch, H. A., Forster, C., Dellago, C. and Zhou, M. (2002). Lyapunov modes of two-dimensional many-body systems; soft disks, hard disks, and rotors, *J. Stat. Phys.* **109**, 765.

Kaneko, K. (1990). Globally coupled chaos violates the law of large numbers but not the central-limit theorem, *Phys. Rev. Lett.* **65**, 1391.

Kaneko, K. (1995). Remarks on the mean field dynamics of networks of chaotic elements, *Physica D* **86**, 158.

Kantz, H. and Olbrich, E. (2000). Coarse grained dynamical entropies: investigation of high entropic dynamical systems, *Physica A* **280**, 34.

Kantz, H. and Schreiber, T. (1997). *Nonlinear Time Series Analysis*, Cambridge: Cambridge University Press.

Konishi, T. and Kaneko, K. (1992). Clustered motion in symplectic coupled maps system, *J. Phys. A* **25**, 6283.

Knuth, D. E. (1981). *The Art of Computer Programming*, Vol II: *Seminumerical Algorithms*, 2nd edition, Reading, MA: Addison-Wesley.

Loreto, V., Paladin, G. and Vulpiani, A. (1996a). On the concept of complexity for random dynamical systems, *Phys. Rev. E* **53**, 2087.

Loreto, V., Paladin, G., Pasquini, M. and Vulpiani, A. (1996b). Characterization of Chaos in random maps, *Physica A* **232**, 189.

Mantica, G. (2000). Quantum algorithmic integrability: the metaphor of classical polygonal billiards, *Phys. Rev. E* **61**, 6434.

Marsaglia, G. (1995). *Die Hard: A battery of tests for random number generators*, http://stat.fsu.edu/pub/diehard/.

Matsumoto, K. and Tsuda, I. (1983). Noise-induced order, *J. Stat. Phys.* **31**, 87.

Mazur, P. and Montroll, E. (1960). Poincaré cycles, ergodicity and irreversibility in assemblies of coupled harmonic oscillators, *J. Math. Phys.* **1**, 70.

McNamara, S. and Mareschal, M. (2001). Origin of the hydrodynamic Lyapunov modes, *Phys. Rev. E* **64**, 051103.

Oseledec, V. I. (1968). A multiplicative ergodic theorem: Lyapunov characteristic numbers for dynamical systems, *Trans. Mosc. Math. Soc.* **19**, 197.

Paladin, G., Serva, M. and Vulpiani, A. (1995). Complexity in dynamical systems with noise, *Phys. Rev. Lett.* **74**, 66.

Percival, I. and Vivaldi, F. (1987). A linear code for the sawtooth and cat maps, *Physica D* **27**, 373.

Perez, G. and Cerdeira, H. A. (1992). Instabilities and non statistical behavior in globally coupled systems, *Phys. Rev. A* **46**, 7492.

Pikovsky, A. S. (1989). Spatial development of chaos in non-linear media, *Phys. Lett. A* **137**, 121.

Pikovsky, A. S. (1993). Local Lyapunov exponents for spatiotemporal chaos, *Chaos* **3**, 225.

Pikovsky, A. S. and Kurths, J. (1994). Collective behavior in ensembles of globally coupled maps, *Physica D* **76**, 411.

Pikovsky, A., Rosenblum, M. and Kurths, J. (2003). *Synchronization: A Universal Concept in Nonlinear Sciences*, Cambridge: Cambridge University Press.

Platt, N., Spiegel, E. A. and Tresser, C. (1993). On-off intermittency – a mechanism for bursting, *Phys. Rev. Lett.* **70**, 279.

Politi, A., Livi, R., Oppo, G. L. and Kapral, R. (1993). Unpredictable behavior in stable systems, *Europhys. Lett.* **22**, 571.

Press, W. H., Teukolsky, S. A., Vetterling, W. T. and Flannery, B. P. (1986). *Numerical Recipes*, Cambridge: Cambridge University Press.

Rudzick, O. and Pikovsky, A. (1996). Unidirectionally coupled map lattices as a model for open flow systems, *Phys. Rev. E* **54**, 5107.

Shibata, T. and Kaneko, K. (1998). Collective chaos, *Phys. Rev. Lett.* **81**, 4116.

Takens, F. (1981). Detecting strange attractors in turbulence, in *Dynamical Systems and Turbulence (Warwick 1980)*, Vol. 898 of *Lecture Notes in Mathematics*, ed. D. A. Rand and L.-S. Young, p. 366, Berlin: Springer.

Von Neumann, J. (1963). Various techniques used in connection with random digits, in *John Von Neumann Collected Works*, Vol. V, ed. A. H. Taub, New York: MacMillan.

Yang, H. L. and Radons, G. (2006a). Hydrodynamic Lyapunov modes in coupled map lattices, *Phys. Rev. E* **73**, 016202.

Yang, H. L. and Radons, G. (2006b). Hydrodynamic Lyapunov modes and strong stochasticity threshold in Fermi–Pasta–Ulam models, *Phys. Rev. E* **73**, 066201.

Yu, L., Ott, E. and Chen, Q. (1991). Fractal distribution of floaters on a fluid surface and the transition to chaos for random maps, *Physica D* **53**, 102.

Zaslavsky, G. M. (2002). Chaos, fractional kinetics, and anomalous transport, *Phys. Rep.* **371**, 461.

Zhang, Y.-C. (1989). Scaling theory of self-organized criticality, *Phys. Rev. Lett.* **63**, 470.

# 4

# Foundation of statistical mechanics and dynamical systems

> To know that you know when you do know, and know that you do not know when you do not know: that is knowledge.
>
> *Confucius*

Statistical mechanics was founded by Maxwell, Boltzmann and Gibbs to account for the properties of macroscopic bodies, systems with a very large number of particles, without very precise requirements on the dynamics (except for the assumption of ergodicity).

Since the discovery of deterministic chaos it is now well established that statistical approaches may also be unavoidable and useful, as discussed in Chapter 1, in systems with few degrees of freedom. However, even after many years there is no general agreement among the experts about the fundamental ingredients for the validity of statistical mechanics.

It is quite impossible in a few pages to describe the wide spectrum of positions ranging from the belief of Landau and Khinchin in the main role of the many degrees of freedom and the (almost) complete irrelevance of dynamical properties, in particular ergodicity, to the opinion of those, for example Prigogine and his school, who consider chaos as the basic ingredient.

For almost all practical purposes one can say that the whole subject of statistical mechanics consists in the evaluation of a few suitable quantities (for example, the partition function, free energy, correlation functions). The ergodic problem is often forgotten and the (so-called) Gibbs approach is accepted because "it works." Such a point of view cannot be satisfactory, at least if one believes that it is not less important to understand the foundation of such a complex issue than to calculate useful quantities.

The present chapter is meant to serve as a global introduction to the problem of the connection between dynamical behavior (mainly ergodicity and chaos) and statistical laws.

## 4.1 The ergodic problem: a brief random walk among an intricate history

Sometimes Boltzmann and Gibbs are viewed as the champions of different points of view about statistical mechanics. According to this vulgata Gibbs was the founder of the ensemble approach while Boltzmann was the supporter of a dynamical theory based on ergodicity. In modern textbooks we use Gibbs's terminology for the ensembles (i.e. microcanonical, canonical and grand canonical). Actually, both ergodicity and ensembles are inventions of Boltzmann, though he is not always credited with them. As a matter of fact a statistical ensemble was called a *monode* by Boltzmann who was able to prove the equivalence, in the thermodynamic limit, of the *ergode* and the *holode* (respectively the microcanonical and canonical ensemble in Gibbs's terminology). Indeed this explains the existence of the *ergodic hypothesis* and of the consequent *ergodic theory*.

For a discussion about Boltzmann, Gibbs and the origin of statistical mechanics we strongly recommend the wonderful book by Cercignani (1998) and the papers by Gallavotti (1995) and Klein (1973).

### *4.1.1 Statistical mechanics as a form of statistical inference?*

Let us open a short parenthesis. Beyond the historical events and the opinions of the founding fathers, there exists an "extremistic anti-dynamical" point of view which considers statistical mechanics as a form of statistical inference rather than a description of objective physical reality. In this approach the probabilities are interpreted as measures of the degree of "trueness" of a logical proposition, rather than as quantities which can be measured physically.

Jaynes (1967, 1989) proposed the maximum entropy principle (MEP) as a rule to find the probability of a given event, in circumstances where only partial information is available. If the mean values of $m$ independent functions $f_i(\mathbf{X})$ are given,

$$c_i = \langle f_i \rangle = \int f_i(\mathbf{X})\rho(\mathbf{X})\,\mathrm{d}\mathbf{X} \quad i = 1, \ldots, m, \tag{4.1}$$

the prescribed rule of the MEP to determine the probability density function $\rho(\mathbf{X})$ is to maximize the entropy $-\int \rho(\mathbf{X}) \ln \rho(\mathbf{X})\,\mathrm{d}\mathbf{X}$ under the constraints $c_i = \langle f_i \rangle$. Using the Lagrangian multipliers one easily obtains

$$\rho(\mathbf{X}) = \frac{1}{Z} \exp \sum_{i=1}^{m} \lambda_i f_i(\mathbf{X}) \tag{4.2}$$

where $\lambda_1, \lambda_2, \ldots, \lambda_m$ depend on $c_1, c_2, \ldots, c_m$.

This rule, when applied to statistical mechanics, with a fixed number of particles and the unique constraint on the mean value of the energy, leads to the usual canonical distribution in a very simple way. In an analogous way, also imposing the constraint of the mean value of the particles, one obtains the grand canonical distribution. The most frequent objection to this point of view can be summarized with the motto *Ex nihilo nihil*, i.e. it is not possible that because we are ignorant of the matter we can know something about it.

The reader interested in this issue can consult the extended literature, see for example Jaynes (1989), Buck and Macaulay (1991), and Uffink (1995).

### 4.1.2 From Boltzmann to Birkhoff

Let us come back to our main aim. Since there is a certain confusion, mainly because of the terminology introduced in the celebrated monograph of Paul and Tatiana Ehrenfest (1956) on Boltzmann's work, it seems to us that a short discussion is appropriate.

The ergodic theory begins with Boltzmann's attempt at justifying the determination of average values in kinetic theory. His original "ergodic hypothesis" was the following: the energy surface consists of finitely many cells, that can be labeled and counted, and during the time evolution a trajectory will pass through all the cells. This raises the possibility of replacing a time average by a far simpler phase average (see below).

An interpretation, due to the Ehrenfests, in the context of classical mechanics of point particles evolving in a continuum, reformulated the original "ergodic hypothesis" into the hypothesis that a trajectory moving on the energy surface visits all its points. It is well documented that Boltzmann had a resolutely finitist point of view: for him, as in calculus, the concepts without any discrete representation are purely metaphysical. Therefore it is certain that the ergodic hypothesis as formulated by the Ehrenfests cannot be attributed to Boltzmann.

The (rather obvious) impossibility of a single phase trajectory visiting every point of the energy surface led the Ehrenfests to formulate the so-called "quasi-ergodic hypothesis" which basically proposed that each evolution on the energy surface (but a zero measure set of initial points) covers densely the surface itself.

Modern ergodic theory can be viewed as a branch of the abstract theory of measure and integration. The aim of this field goes far beyond its original problem as formulated by Boltzmann in the statistical mechanics context. We can now formulate the ergodic problem in the following terms. Consider a dynamical system, i.e. a deterministic evolution law in the phase space $\Omega$

$$\mathbf{X}(0) \to \mathbf{X}(t) = U^t \mathbf{X}(0) \tag{4.3}$$

and a measure $\mathrm{d}\mu(\mathbf{X})$ invariant under the evolution given by $U^t$, i.e. $\mathrm{d}\mu(\mathbf{X}) = \mathrm{d}\mu(U^{-t}\mathbf{X})$. The dynamical system $(\Omega, U^t, \mathrm{d}\mu(\mathbf{X}))$ is called ergodic, with respect to the measure $d\mu(\mathbf{X})$, if, for every integrable function $A(\mathbf{X})$ and for almost all initial conditions $\mathbf{X}(t_0)$, with respect to $\mu$, one has:

$$\overline{A} \equiv \lim_{\mathcal{T}\to\infty} \frac{1}{\mathcal{T}} \int_{t_0}^{t_0+\mathcal{T}} A(\mathbf{X}(t))\mathrm{d}t = \int A(\mathbf{X})\mathrm{d}\mu(\mathbf{X}) \equiv \langle A \rangle, \tag{4.4}$$

where $\mathbf{X}(t) = U^{t-t_0}\mathbf{X}(t_0)$.

Let us now discuss the relation between this issue and the foundation of equilibrium statistical mechanics.

Macroscopic systems contain a very large number (of the order of Avogadro's number) of particles; this implies the practical necessity of a statistical description. Denoting by $\mathbf{q}_i$ and $\mathbf{p}_i$ the position and momentum vectors of the $i$th particle, the state of an $N$-particle system at time $t$ is described by the vector $\mathbf{X}(t) \equiv (\mathbf{q}_1(t), \ldots, \mathbf{q}_N(t), \mathbf{p}_1(t), \ldots, \mathbf{p}_N(t))$ in a $6N$-dimensional phase space. The evolution law is given by Hamilton's equations. If $V(\{\mathbf{q}_j\})$ is the interaction potential the Hamiltonian is

$$H = \sum_{i=1}^{N} \frac{\mathbf{p}_i^2}{2m} + V(\{\mathbf{q}_j\}), \tag{4.5}$$

and the evolution equations are

$$\begin{aligned} \frac{\mathrm{d}\mathbf{q}_i}{\mathrm{d}t} &= \frac{\partial H}{\partial \mathbf{p}_i} = \frac{\mathbf{p}_i}{m}, \\ \frac{\mathrm{d}\mathbf{p}_i}{\mathrm{d}t} &= -\frac{\partial H}{\partial \mathbf{q}_i} = -\frac{\partial V}{\partial \mathbf{q}_i}, \end{aligned} \tag{4.6}$$

with $i = 1, \ldots, N$.

Here it is important to note that the macroscopic time scale (the time scale at which we observe the system) is much larger than the microscopic dynamics time scale over which the molecular changes take place. This means that an experimental measurement is actually the result of a single observation during which the system goes through a very large number of microscopic states. If the measurement refers to an observable $A(\mathbf{x})$, the result can be considered as an average taken over a very long time (from the microscopic point of view):

$$\overline{A}^{\mathcal{T}} = \frac{1}{\mathcal{T}} \int_{t_0}^{t_0+\mathcal{T}} A(\mathbf{X}(t))\mathrm{d}t\,. \tag{4.7}$$

The calculation of the time average $\overline{A}^{\mathcal{T}}$, in principle, requires both knowledge of the complete microscopic state of the system at a given time and the determination of its trajectory. These are evidently impossible requirements so that, beyond the difficulty of integrating the system (4.6), if $\overline{A}^{\mathcal{T}}$, for the system in equilibrium, depends too strongly on the initial conditions, not even statistical predictions can be made. The ergodic hypothesis allows us to overcome this obstacle. A rather natural candidate for the invariant measure $d\mu(\mathbf{X})$ is the microcanonical measure on the constant energy surface $H = E$:

$$d\mu_{\mathrm{mc}}(\mathbf{X}) = \frac{d\sigma(\mathbf{X})}{|\nabla H|} \tag{4.8}$$

where $d\sigma$ is the surface element and $\nabla H = (\partial_{\mathbf{q}_1} H, \ldots, \partial_{\mathbf{q}_N} H, \partial_{\mathbf{p}_1} H, \ldots, \partial_{\mathbf{p}_N} H)$.

The ergodic hypothesis is satisfied if for sufficiently large $\mathcal{T}$ the average $\overline{A}^{\mathcal{T}}$ depends only on the energy and hence it has the same value for (almost) all the trajectories on the same constant energy surface and therefore:

$$\overline{A} \equiv \lim_{\mathcal{T}\to\infty} \frac{1}{\mathcal{T}} \int_{t_0}^{t_0+\mathcal{T}} A(\mathbf{X}(t))dt = \int A(\mathbf{X})d\mu_{\mathrm{mc}}(\mathbf{X}) \equiv \langle A \rangle. \tag{4.9}$$

The validity of such an equality eliminates the necessity both of determining a detailed initial state of the system and of solving Hamilton's equations. Whether (4.9) is valid or not, i.e. whether it is possible to substitute the temporal average by an average in phase space, constitutes the main question of the ergodic problem. The crucial role of this issue for statistical mechanics rests also in the following fact. If the statistical properties of a large isolated system in equilibrium can be properly described in terms of the microcanonical ensemble, then it is not difficult to show that the equilibrium properties of a small subsystem (but still large at microscopic level) is properly described by the canonical ensemble. Therefore a proof of the validity of (4.9) provides the dynamical justification of the statistical ensembles. On the other hand, one has to remember that, in this framework, the ensemble is just a useful mathematical tool, but in reality one is considering only a single physical system.

It is rather natural, both from a mathematical and from a physical point of view, to wonder under which conditions a dynamical system is ergodic. The problem, at an abstract level, i.e. for a dynamical system given by $(\Omega, U^t, d\mu(\mathbf{X}))$, was tackled by Birkhoff (1931). He proved the following two theorems.

**Theorem I** *For almost every initial condition $\mathbf{X}_0$ the infinite time average*

$$\overline{A}(\mathbf{X}_0) \equiv \lim_{\mathcal{T}\to\infty} \frac{1}{\mathcal{T}} \int_0^{\mathcal{T}} A(U^t\mathbf{X}_0)dt \tag{4.10}$$

*exists.*

**Theorem II** *A necessary and sufficient condition for the system to be ergodic, i.e. the time average $\overline{A}(\mathbf{X}_0)$ does not depend on the initial condition (for almost all $\mathbf{X}_0$), is that the phase space $\Omega$ be metrically indecomposable. This property means that $\Omega$ cannot be subdivided into two invariant (under the dynamics $U^t$) parts each of positive measure.*

(Sometimes instead of metrically indecomposable the equivalent term metrically transitive is used.) Theorem I is rather general and not very stringent, in fact the time average $\overline{A}(\mathbf{X}_0)$ can depend on the initial condition. The result of Theorem II is more interesting although practically inconclusive as regards statistical mechanics, since, in general, it is not possible to decide whether a given system satisfies the condition of metrical indecomposability. So, at a practical level, Theorem II is only a shift of the problem.

## 4.2 Beyond abstract ergodic theory

When one speaks of equilibrium of a physical system one must take into consideration the problem of the observation time, with respect to the considered system. This time cannot be too large, otherwise the equilibrium can become meaningless (Ma 1985). As an example consider a cold cup filled with boiling water. After a few minutes the water and the cup will reach the same temperature and within five or ten minutes we can consider the cup and water systems to be in equilibrium. After a few hours the temperature of the water (and cup) will be equal to room temperature: another equilibrium state. However, water molecules will evaporate, so if the observation time is over four or five days we will have another equilibrium state. On the other hand, strictly speaking, this is not an absolute unchanging state; in fact the molecules of the cup will also evaporate over a very long time.

For a physicist the ergodic problem is surely interesting, but for the foundation of statistical mechanics one has to consider more specific issues. First we note that the mathematical infinite time limit $\overline{A}$ needs a physical interpretation. A more relevant question is how long do we have to take $\mathcal{T}$ to be in such a way that $\overline{A}^{\mathcal{T}}$ is close to $\langle A \rangle$. It is easy to realize that the answer to this question must depend both on the observable $A$ and on the number of particles $N$. As a simple example let us consider a region $G$ of the phase space $\Omega$, and the observable

$$A(\mathbf{X}) = \begin{cases} 1 & \text{if} \quad \mathbf{X} \in G \\ 0 & \text{otherwise.} \end{cases} \tag{4.11}$$

To be more specific we assume that the region $G$ is a $6N$-dimensional hypercube whose edge is $\epsilon$. Of course $\langle A \rangle \propto \epsilon^{6N}$ is just the probability of remaining in $G$ and $\overline{A}^{\mathcal{T}}$ is the fraction of time the trajectory $\mathbf{X}(t)$ spent in $G$ during the interval $[0, \mathcal{T}]$. A crude estimation of the equilibration time $\mathcal{T}_{\text{eq}}$, i.e. the time necessary to give a fair agreement between $\overline{A}^{\mathcal{T}}$ and $\langle A \rangle$, is the following. Consider the projection $\mu_n$ of $G$ on the 6-dimensional space spanned by the variables $(\mathbf{q}_n, \mathbf{p}_n)$ describing the $n$th particle. If this particle starts from $\mu_n$, it needs a certain time $\tau_n(\epsilon)$ to come back to $\mu_n$. A rough estimate gives $\tau_n(\epsilon) \sim \epsilon^{-\gamma}$, the precise value of $-\gamma$ is not particularly relevant. If one assumes for the $n$th particle a random walk dynamics among all the $\mathcal{N} \sim \epsilon^{-6}$ cells of volume $\epsilon^6$, the number of different cells visited grows as $t^{1/2}$, therefore the typical time required to pass again to the original cell is $\tau \sim \epsilon^{-12}$. As a first approximation, which is reasonable in diluted gases, we can assume that for different particles the return mechanisms are independent of each other. Therefore in order to have $\overline{A}^{\mathcal{T}} \simeq \langle A \rangle$ one has to wait until $\mathcal{T}_{\mathrm{eq}} \sim \epsilon^{-\gamma N}$, i.e. a time exponentially large in $N$.

Note that the variable defined in (4.11) can also be written:

$$A(\mathbf{X}) = \prod_n \chi_{\mu_n}(\mathbf{q}_n, \mathbf{p}_n),$$

where $\chi_{\mu_n}(\mathbf{q}_n, \mathbf{p}_n)$ is the characteristic function of the subset $\mu_n$, i.e. $\chi_{\mu_n}(\mathbf{q}_n, \mathbf{p}_n) = 1$ if $(\mathbf{q}_n, \mathbf{p}_n) \in \mu_n$ and $\chi_{\mu_n}(\mathbf{q}_n, \mathbf{p}_n) = 0$ otherwise. This makes it clear that such a function is not very interesting from a macroscopic point of view, since the exiting of a single particle from the tagged region $\mu_n$ causes a variation of $A(\mathbf{X})$ of the same order as its value. This is why we call $A(\mathbf{X})$ a *microscopic* function. The above discussion shows that the equilibration time for a microscopic function typically is closely related to the Poincaré recurrence time. On the other hand the previous argument is nothing but a reformulation of the original answer of Boltzmann to criticisms of the validity of his $\mathcal{H}$ theorem (Cercignani 1998). Because of the enormous number of particles in a macroscopic system, the time of recurrence of a non-equilibrium state is astonishingly large. For instance, Boltzmann estimated that, in a sphere of air of radius 1 cm at temperature 300 K and standard pressure, for a state of fluctuation in which the concentration of molecules will differ from the average value by 1% one has to wait $10^{10^{14}}$ seconds!!

The exponentially large (in $N$) times, appearing in these two examples, have a common origin in the exponential smallness of the phase space regions involved: the region where $A(\mathbf{X})$ is different from zero and the region where the state of the system is slightly out of equilibrium. However, they convey very different information. In the first case such a large time is not helpful, since it means that Eq. (4.9) is practically useless. In the second case it is welcome, since it allows the notion of equilibrium of a macroscopic system to be introduced in statistical mechanics. In this respect, it is necessary to stress that the relevant observables for thermodynamics, those by which equilibrium states are characterized, are not generic functions. They are few and mainly of a special kind, so that the physically interesting question is whether the equilibration time can be short enough for these special phase-functions.

### 4.2.1 Possibility of ergodicity without metrical indecomposability

Khinchin (1949) in his celebrated book *Mathematical Foundations of Statistical Mechanics* presents some important results on the ergodic problem which do not need the metrical transitivity of the Birkhoff theorem. The general idea of this approach is based on the following facts:

(a) in the systems which are of interest to statistical mechanics the number of degrees of freedom is very large;
(b) in statistical mechanics the important observables are not generic (in the mathematical sense) functions, so it is enough to show the validity of (4.9) for the relevant observables;
(c) one can allow that equation (4.9) does not hold for the initial conditions $\mathbf{X}_0$ in a region of small measure (which goes to zero as $N \to \infty$).

Therefore there is hope of obtaining some interesting results beyond the Birkhoff theorems which hold for generic dynamical systems (i.e. also for the low-dimensional case), for non-specific observables, and for almost all initial conditions.

Kinchin considers a separable Hamiltonian system, i.e.

$$H = \sum_{n=1}^{N} H_n(\mathbf{q}_n, \mathbf{p}_n), \tag{4.12}$$

and a special class of observables (called *sum functions*) of the form

$$f(\mathbf{X}) = \sum_{n=1}^{N} f_n(\mathbf{q}_n, \mathbf{p}_n) \tag{4.13}$$

where $f_n = O(1)$. Interesting examples of sum functions are given by the pressure, the kinetic energy, the total energy and the single-particle distribution function. Notice that, at variance with the function $A(\mathbf{X})$ in (4.11), a change $O(1)$ in a single $f_n$ results in a relative variation $O(1/N)$ in $f(\mathbf{X})$: the sum functions are "good" macroscopic functions, since they are not so sensitive to microscopic details.

Using the fact that the Hamiltonian is separable one has:

$$\langle f \rangle = O(N) \quad \text{and} \quad \sigma^2 = \langle (f - \langle f \rangle)^2 \rangle = O(N). \tag{4.14}$$

We recall that $\langle\ \rangle$ indicates the microcanonical ensemble average. Consider the time average $\overline{f}(\mathbf{X})$ of the observables $f$ along a trajectory starting from $\mathbf{X}$. Under quite general hypothesis (without invoking metrical transitivity) one has:

$$\langle \overline{f}(\mathbf{X}) \rangle = \langle f \rangle \quad \text{and} \quad \langle [\overline{f}(\mathbf{X})]^2 \rangle \leq \langle f^2 \rangle \tag{4.15}$$

so that

$$\langle(\overline{f} - \langle f\rangle)^2\rangle \leq \langle(f - \langle f\rangle)^2\rangle = O(N). \tag{4.16}$$

Now we can use the Markov inequality

$$\text{Prob}\left(\frac{|\overline{f} - \langle f\rangle|}{|\langle f\rangle|} \geq a\right) \leq \frac{\sigma}{a|\langle f\rangle|} = \frac{1}{a}O(N^{-1/2}) \tag{4.17}$$

to obtain, taking $a = O(N^{-1/4})$,

$$\text{Prob}\left(\frac{|\overline{f} - \langle f\rangle|}{|\langle f\rangle|} \geq K_1 N^{-1/4}\right) \leq K_2 N^{-1/4} \tag{4.18}$$

where $K_1$ and $K_2$ are $O(1)$.

Therefore we have that for the class of sum functions the set of points for which time and phase averages differ more than a given amount, which goes to zero as $N \to \infty$, has a measure which goes to zero as $N \to \infty$.

Note that, as already stressed, this result also holds for systems which are not metrically transitive, so it is rather different from the Birkhoff theorem. The price one has to pay in order to avoid metrical transitivity is

- the system must have a separable Hamiltonian;
- only special observables are concerned, i.e. sum functions;
- the number of degrees of freedom must be very large, $N \gg 1$;
- a region of small (but finite, for finite $N$) measure exists where time and phase averages do not coincide.

However, now we see that the physical interest is shifted from the ergodicity of the system to that of some functions, those for which the time and phase averages coincide for almost all trajectories. But this is not yet enough, because all these results still concern infinite time averages.

Finally, since the dynamics in Khinchin's approach plays a rather marginal role, a much more interesting result, not concerning time averages, can be obtained. From the assumptions (4.12)–(4.13) and the result (4.14) one can show that:

$$\text{Prob}\left(\frac{|f - \langle f\rangle|}{|\langle f\rangle|} \geq K_1 N^{-1/4}\right) \leq K_2 N^{-1/4}, \tag{4.19}$$

in other words, the physically relevant observables are self-averaging, i.e. they are practically constant (except in a region of small measure) on a constant-energy surface. This implies that the time average operation is not so important for assigning a physical meaning to the quantity $\langle f\rangle$, which turns out to be the near constant value of the observable. It also means, and this can be viewed as the essence of Khinchin's results, that the ensemble based statistical mechanics can work independently of the validity of ergodicity (in the mathematical sense). Actually this was also (and already) the point of view of Boltzmann himself (Gallavotti 1999).

A weak aspect, from the physical point of view, of Khinchin's approach concerns the no-interaction assumption (4.12). In contrast, an essential requisite for thermodynamic behavior is the possibility of an exchange of energy among the particles. Of course Khinchin noted the problem and argued that the actual Hamiltonian is indeed only approximated by the separable Hamiltonian. The feeling of Khinchin was that the interaction among the particles contributes very little to evaluating the averages and for the majority of computations in statistical mechanics one can neglect these terms.

The undesirable restriction to the separable structure of the Hamiltonian, i.e. (4.12), was removed by Mazur and van der Linden (1963). They extended the result to systems of particles interacting through a short-range potential, showing that the intuition of Khinchin, that the interaction among the particles is of little relevance, was basically correct. The physical interpretation of the result is that, owing to the short range of the interactions, a many-particle system behaves as if it consists of a large number of non-interacting components. As Mazur and van der Linden write: "One might think of subsystems consisting of large numbers of particles; the interaction between these subsystems is then a surface effect and very small compared to the energy content of the subsystems themselves." Their calculations then imply that "the energies of these subsystems behave as almost independent random variables, so that a central limit theorem still applies." It is interesting to note that the result obtained is valid for all but a finite number of values of the temperature, where the system may undergo a phase transition, at variance with the non-interacting case considered by Khinchin.

Let us stress again that in the Khinchin result, as well as in the generalization of Mazur and van der Linden, basically the dynamics plays no role and the existence of good statistical properties is due to the fact that $N \gg 1$.

### 4.2.2 A note on the ergodic hypothesis

We introduced the ergodic problem by arguing that a macroscopic observation on a thermodynamic system necessarily involves a huge number of microscopic states of the system, so that, in order to relate theory with experiment, the computation of a time average is mandatory. At this point of the reasoning the ergodic hypothesis is introduced, freeing us from that impossible task, but replacing it with the need to explain how thermodynamic finite time properties can be linked to infinite time averages. The discussion above, about the sum functions and the relevant concept of ergodic functions, gives a possible explanation.

Since we put forward the ergodic hypothesis giving no detailed justification, it may be interesting to arrive at it by following a route that is likely to be close to the path followed by Boltzmann himself (Gallavotti 1999) and that, moreover, may help in understanding why the hypothesis was formulated.

The starting point is a theorem initially studied by Boltzmann, and further also elaborated by Helmholtz. It was called the *heat theorem*, not to be confused with the heat theorem of Nernst. It can be stated as follows. Consider a mechanical system belonging to the special class of *monocyclic* systems that, by definition, possess motions that are periodic and non-degenerate, i.e. such that only one trajectory corresponds to a given energy value. Assume that the potential energy, $\Phi$, of the system depends on a parameter, $V$, and put

- $T$ the time average of the kinetic energy, $K$,
- $U = K + \Phi$ the total energy of the system,
- $P$ the time average of $-\partial_V \Phi$.

Identify a state with a complete motion of given energy $U$ and given $V$. If the parameters defining a state vary by the infinitesimal quantities $\mathrm{d}U$ and $\mathrm{d}V$, then one has

$$\text{Helmholtz's theorem:} \qquad \textit{the differential} \quad \frac{\mathrm{d}U + P\,\mathrm{d}V}{T} \quad \textit{is exact.}$$

Helmholtz's theorem is intriguing since it states that, at least in these systems, purely mechanical quantities can be found that satisfy a relation *formally* identical to the one that allows the existence of the entropy function in thermodynamics.

Essentially, monocyclic Hamiltonian systems are one-dimensional. In order to use these ideas in the case of high-dimensional systems, we can think that the lesson of thermodynamics is just that macroscopic systems enjoy a kind of monocyclicity property because, for example for a gas in a container, a unique behavior corresponds to a given energy and a given volume. We may assume, as Boltzmann did within a discrete view of nature, that an energy surface consists of finitely many cells, that can be labeled and counted, and during the time evolution a motion will visit all the cells. This is the "ergodic hypothesis," as formulated by Boltzmann, and it corresponds to the assumption that a generic Hamiltonian system can be considered as monocyclic, i.e. all motions with a given energy are periodic and differ, at most, by a time shift. Thus the ergodic hypothesis allows us to apply the *heat theorem* to generic Hamiltonian systems. Moreover, at this point, it is clear that the calculation of time averages over a complete period can be replaced by the calculation of phase averages over the whole energy surface, with respect to the uniform distribution, since the cells are all equal.

Now, let us consider a gas in a container, whose volume, $V$, is the parameter on which the potential energy of the system depends, since a piston allows it to be varied. If we calculate, by substituting time averages with microcanonical averages, the mean value of $-\partial_V \Phi$ (see, for instance, Gallavotti (1999)) the quantity we obtain, that we named $P$, is just the average momentum communicated by the gas molecules to the walls, per unit time and unit surface, i.e. the physical pressure; this also satisfies the perfect gas law, once the average kinetic energy is taken as proportional to the absolute temperature.

Summing up, we see that the ergodic hypothesis translates the requirement that the system be monocyclic. The latter property suggests that mechanical analogs of thermodynamical quantities may exist, it allows the microcanonical distribution to be introduced, and it guarantees that the mechanically defined quantities having the physical meaning of pressure and temperature satisfy, via Helmholtz's theorem, the correct properties that thermodynamics imposes. Thus we have a "mechanical model" of thermodynamics.

However, the physical content of the model does not yet have a direct thermodynamic interpretation. We are confronted again with the problem of infinite times. Indeed the quantities appearing in Helmholtz's theorem are obtained by performing averages over one period that, in macroscopic systems, is practically infinite. In contrast, the physical quantities involved in the (true) thermodynamic relations are obtained from measures localized in time. The remaining task is to find the relation between infinite time properties and finite time laws. This is a problem whose solution can be found, as already suggested by Boltzmann, in the peculiarities of the physical observables, that are studied in a systematic way in Khinchin's book, as discussed above. The result is that the huge number of particles provides a solution, since it is the reason why the physical observables involved are almost constant, so that a short time average is the same as a long time average.

## 4.3 The connection between analytical mechanics and the ergodic problem

The issue of ergodicity is entangled with the problem of the existence of non-trivial integrals (i.e. conserved quantities) in Hamiltonian systems. Given a Hamiltonian $H(\mathbf{q}, \mathbf{p})$, with $\mathbf{q}, \mathbf{p} \in \mathbb{R}^N$, if there exists a canonical transformation (i.e. a change of variables which does not change the Hamiltonian structure of the system) from the variables $(\mathbf{q}, \mathbf{p})$ to the action-angle variables $(\mathbf{I}, \phi)$, such that the new Hamiltonian depends only on the action $\mathbf{I}$,

$$H = H_0(\mathbf{I}), \tag{4.20}$$

then the system is called integrable. In this case the time evolution of the system is

$$\begin{cases} I_i(t) = I_i(0) \\ \phi_i(t) = \phi_i(0) + \omega_i(\mathbf{I}(0))\, t, \end{cases} \tag{4.21}$$

where $\omega_i = \partial H_0 / \partial I_i$ and $i = 1, \ldots, N$.

Note that in integrable systems there are $N$ independent first integrals, since all the actions $I_i$ are conserved and the motion evolves on $N$-dimensional tori. The Solar System provides an important example: if the planetary interactions are neglected one has the two-body problem (Sun–planet), for which the integrability can be easily proved.

It is fairly natural to wonder about the effect of perturbations on (4.20), i.e. to study the Hamiltonian

$$H(\mathbf{I}, \phi) = H_0(\mathbf{I}) + \epsilon H_1(\mathbf{I}, \phi), \tag{4.22}$$

also called a near-integrable Hamiltonian.

For the Solar System, this would imply accounting for the interactions between planets, leading to $\epsilon \sim 10^{-3}$, which is the ratio between the masses of Jupiter (the largest planet) and the Sun. Do the perturbed system (4.22) trajectories end up "close" to those of the integrable system (4.20)? Does the introduction of the perturbation term $\epsilon H_1(\mathbf{I}, \phi)$ still allow for the existence of integrals of the motion besides the energy?

These questions are of obvious interest to celestial mechanics, and also relevant to the ergodic problem. Of course, if there are first integrals, beyond the energy, the system cannot be ergodic: choosing one of the first integrals as the observable $A$, one has $\overline{A} = A(\mathbf{X}(0))$, which depends on the initial condition $\mathbf{X}(0)$ and therefore, in general, it cannot coincide with the phase average $\langle A \rangle$.

Curiously, in statistical mechanics and in celestial mechanics there are contrasting expectations or, better, wishes: in statistical mechanics one wishes for "irregular" dynamical behavior, in order to justify the ergodic hypothesis; conversely, in celestial mechanics "regular" behavior is desired so that accurate predictions can be made.

### *4.3.1 The great Poincaré result*

In a very important work Poincaré showed that generally a system like (4.22), with $\epsilon \neq 0$, does not allow analytic first integrals, except for energy (Poincaré 1892). The existence of first integrals is equivalent to the possibility of finding a change of variables $(\mathbf{I}, \phi) \to (\mathbf{I}', \phi')$ preserving the Hamiltonian nature of the system (canonical transformation) such that the Hamiltonian is a function only of the (new) actions $\mathbf{I}'$. Practically, one has to find a generating function $S(\mathbf{I}', \phi)$, which links $(\mathbf{I}, \phi)$ to $(\mathbf{I}', \phi')$:

$$I_n = \frac{\partial S(\mathbf{I}', \phi)}{\partial \phi_n}, \quad \phi'_n = \frac{\partial S(\mathbf{I}', \phi)}{\partial I'_n}, \tag{4.23}$$

in such a way that the perturbed Hamiltonian

$$H(\mathbf{I}, \phi) = H_0\left(\frac{\partial S}{\partial \phi}\right) + \epsilon H_1\left(\frac{\partial S}{\partial \phi}, \phi\right) \tag{4.24}$$

is a function only of $\mathbf{I}'$. One approach is to look for a solution in the form of a power series in $\epsilon$:

$$S = S_0 + \epsilon S_1 + \epsilon^2 S_2 + \cdots. \tag{4.25}$$

For $S_0$ one has $S_0 = \mathbf{I}' \cdot \phi$ (this corresponds to the identity transformation for $\epsilon = 0$). Substituting the series (4.25) for $S$ in (4.24) one obtains an equation for $S_1$:

$$\frac{\partial H_0(\mathbf{I}')}{\partial \mathbf{I}'} \cdot \frac{\partial S_1(\mathbf{I}', \phi)}{\partial \phi} = -H_1(\mathbf{I}', \phi). \tag{4.26}$$

If one expresses $H_1$ and $S_1$ as Fourier series in the angle vector $\phi$:

$$H_1 = \sum_{\mathbf{m}} h_{\mathbf{m}}^{(1)}(\mathbf{I}')\mathrm{e}^{\mathrm{i}\mathbf{m}\cdot\phi}, \quad S_1 = \sum_{\mathbf{m}} s_{\mathbf{m}}^{(1)}(\mathbf{I}')\mathrm{e}^{\mathrm{i}\mathbf{m}\cdot\phi}, \tag{4.27}$$

where $\mathbf{m}$ is an $N$-component vector of integers, one obtains

$$S_1 = i \sum_{\mathbf{m}} \frac{h_{\mathbf{m}}^{(1)}(\mathbf{I}')}{\mathbf{m} \cdot \omega_0(\mathbf{I}')} \mathrm{e}^{\mathrm{i}\mathbf{m}\cdot\phi}, \tag{4.28}$$

where $\omega_0(\mathbf{I}) = \partial H_0(\mathbf{I})/\partial \mathbf{I}$ is the unperturbed $N$-dimensional frequency vector for the torus corresponding to action $\mathbf{I}$.

One understands immediately the origin of the non-existence of first integrals: this is the celebrated *problem of small denominators*. Clearly (4.28) does not work for the values of $\mathbf{I}$ for which $\mathbf{m} \cdot \omega_0(\mathbf{I}') = 0$ for some value of $\mathbf{m}$. Also, in the case where the unperturbed frequencies $\omega_0(\mathbf{I})$ are rationally independent, the denominator $\mathbf{m} \cdot \omega_0(\mathbf{I}')$ can be arbitrarily small, therefore one has to conclude that first integrals (beyond energy) cannot exist.

### 4.3.2 Does non-integrability imply ergodicity?

Poincaré' s result sounds rather positive for statistical mechanics. In 1923 the young Fermi first generalized Poincaré's result, showing that in a Hamiltonian system which is a perturbation of an integrable system, if $N > 2$, for a generic perturbation $H_1$, there cannot exist even a single surface of dimension $2N - 2$ embedded in the $2N - 1$ dimensional constant energy surface, and analytical in the variable $(\mathbf{I}, \phi)$ and $\epsilon$, that contains all the trajectories starting on it (Fermi 1923). This (correct) result induced Fermi to argue that Hamiltonian systems (apart from the integrable ones, which must be considered atypical) in general, are ergodic, as soon as $\epsilon \neq 0$. This conclusion has been generally accepted by the physics community.

Following Fermi's 1923 work, even in the absence of a rigorous demonstration, the ergodicity problem seemed, at least to physicists, essentially solved. There was a general consensus that the non-existence theorems for regular first integrals implied ergodicity. It seems that Fermi was not very worried at the lack of rigor of his "proof," likely the main reason was his (and more generally the large part of the physics community) interest in the development of quantum physics. In the 1930s the ergodic problem thus became a subject studied mainly by mathematicians, who tackled it in a rather general and abstract way, without particular interest in its connections with statistical mechanics.

## 4.4 An unexpected result revitalizes interest in the ergodic problem

After the Second World War Fermi continued periodically to visit the Los Alamos Laboratories. As Ulam wrote in the introduction to the celebrated paper *Studies of non-linear problems* (Fermi *et al.* 1955) for *Note e Memorie* (Collected Papers) (Fermi 1965), he quickly became interested in the development of computers and in their use for scientific research. That paper, which is often referred to by the acronym FPU (from the names of the authors: Fermi, Pasta and Ulam), was completed in May 1955, but appeared for the first time in 1965 as a contribution to *Note e Memorie*, an anthology of Fermi's papers. This work had a unique role in the development of different fields of research such as dynamical chaos and numerical simulations. In FPU Fermi and collaborators studied the time evolution of $N+2$ particles of mass $m$, interacting with non-linear springs (i.e. Hooke's law is not exactly valid). The Hamiltonian of the system is

$$H = \sum_{i=0}^{N} \left[ \frac{p_i^2}{2m} + \frac{K}{2}(q_{i+1} - q_i)^2 + \frac{\epsilon}{r}(q_{i+1} - q_i)^r \right] \tag{4.29}$$

where $q_0 = q_{N+1} = 0 = p_0 = p_{N+1}$ and $r = 3$ or $r = 4$.

If $\epsilon = 0$ the system is integrable, since it is equivalent to $N$ independent harmonic oscillators. In such a case using the normal modes,

$$a_k = \sqrt{\frac{2}{N+1}} \sum_i q_i \sin \frac{i k \pi}{N+1} \qquad (k = 1, \ldots, N), \tag{4.30}$$

the system reduces to $N$ non-interacting harmonic oscillators whose angular frequencies are

$$\omega_k = 2\sqrt{\frac{K}{m}} \sin \frac{k\pi}{2(N+1)}$$

and whose energies are

$$E_k = \frac{1}{2}\left(\dot{a}_k^2 + \omega_k^2 a_k^2\right) .$$

The $E_k$, in this case, are clearly constant during the time evolution and they are proportional to the action variables $E_k = \omega_k I_k$. The Hamiltonian (4.29) is, therefore, a typical example of a perturbed integrable system. For small values of $\epsilon$ it is not difficult to compute all the thermodynamically relevant quantities in the framework of equilibrium statistical mechanics, i.e. in terms of averages over a statistical ensemble (e.g. the canonical or microcanonical ones). In particular, it can be shown that

$$\langle E_k \rangle \simeq \frac{E_{\text{tot}}}{N}. \tag{4.31}$$

Equation (4.31) is just one way of writing the equipartition law. It must be noted that equipartition can be valid for $\epsilon = 0$, or small; however $\overline{E_k}$, the time average computed along a trajectory, can coincide with $\langle E_k \rangle$, only if $\epsilon$ is different from zero, so that the normal modes interact, loosing memory of their initial conditions.

What happens if an initial condition is chosen in such a way that all the energy is concentrated in a few normal modes, for instance $E_1(0) \neq 0$ and $E_k(0) = 0$ for $k = 2, \ldots, N$?

Before FPU, the general expectation would have been (based on the discussion of the previous section) that the first normal mode would have progressively transferred energy to the others and that, after some relaxation time, every $E_k(t)$ would fluctuate around the common value given by (4.31). Even though there is no specific evidence (Ulam does not mention this explicitly), it is reasonable to think that Fermi shared this expectation. Likely Fermi was interested in a numerical simulation, not so much to verify his "demonstration" of the ergodic hypothesis, as to investigate the thermalization times, i.e. the times necessary for the system to go from a non-equilibrium state (all energy concentrated in only one mode) to the equipartition expected by statistical mechanics. In FPU a numerical simulation was performed with $N = 16, 32, 64$, $\epsilon \neq 0$ and all the energy concentrated initially in the first normal mode. Unexpectedly, no tendency toward equipartition was observed, even for long times. In other words, a violation of ergodicity and mixing was found. In Figure 4.1 we show the time behavior of the quantities $E_k/E_{\text{tot}}$, for several values of $k$, for $N = 32$ and $r = 3$. Instead of a loss of memory of the initial condition, we see an almost periodic mode: after a long time, $E_1$ reverts almost back to its initial value. The non-equipartition of energy can be clearly observed in Figure 4.2, which shows the time-average energy of mode $k$, as a function of the

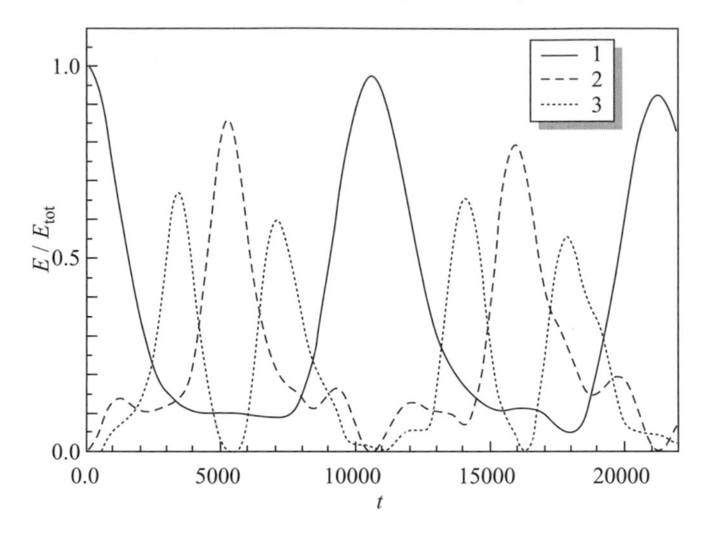


Figure 4.1 $E_1(t)/E_{tot}$, $E_2(t)/E_{tot}$, $E_3(t)/E_{tot}$ for the FPU system, with $N = 32$, $r = 3$, $\epsilon = 0.1$ and energy density $\mathcal{E} = E_{tot}/N = 0.07$. Courtesy of G. Benettin.

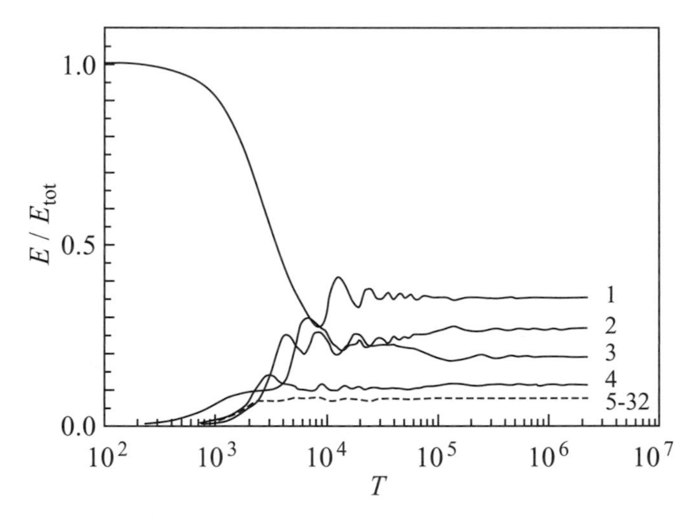


Figure 4.2 Time-averaged fraction of energy, in modes $k = 1, 2, 3, 4$ (bold lines, from top to bottom) and $\sum_{k=5}^{32} E_{(av)k}(\mathcal{T})/E_{tot}$ (dashed line). The parameters of the system are the same as in Figure 4.1. Courtesy of G. Benettin.

observation time $\mathcal{T}$:

$$\overline{E_k}(\mathcal{T}) = \frac{1}{\mathcal{T}} \int_0^{\mathcal{T}} E_k(t)\mathrm{d}t, \qquad \text{with} \qquad k = 1, \ldots, N. \tag{4.32}$$

The FPU results contrasted strongly with expectations, and Fermi himself, according to Ulam, said that he was rather surprised and that they were dealing with an important discovery which showed unambiguously how the prevalent opinion (at that time) on the generality of mixing and thermalization properties of non-linear systems might not always be justified.

## 4.5 Some modern developments

The existence of non-ergodic behavior in non-integrable Hamiltonian systems had (paradoxically) already been found by the soviet mathematician Kolmogorov (1954), one year before the FPU paper. This fact was surely unknown to the authors of FPU.

### 4.5.1 The KAM theorem

Kolmogorov proposed (without a detailed proof, but clearly expressing the basic idea) an important theorem, which was subsequently completed by Arnold (1963) and Moser (1962). The theorem, now known as KAM, reads as follows.

**Theorem (KAM)** *Given a Hamiltonian* $H(\mathbf{I}, \phi) = H_0(\mathbf{I}) + \epsilon H_1(\mathbf{I}, \phi)$, *with* $H_0(\mathbf{I})$ *sufficiently regular and* $\det |\partial^2 H_0(\mathbf{I})/\partial I_i \partial I_j| \neq 0$, *if* $\epsilon$ *is small enough, then on the constant-energy surface, invariant tori survive in a region whose measure tends to* 1 *as* $\epsilon \to 0$. *These tori, called KAM tori, result from a small deformation of those present in the integrable system* ($\epsilon = 0$).

At first glance, if the theorems on the non-existence of non-trivial first integrals were not known, the KAM theorem might seem obvious. Actually instead, as a result of the small denominators, the existence of the KAM tori is a rather subtle and strongly counterintuitive fact. In fact for every value (even very small) of $\epsilon$, some tori of the perturbed system, the so-called resonant tori, are destroyed, and this forbids analytic first integrals. In spite of that, for small $\epsilon$ most tori survive, even if slightly deformed; thus the perturbed system (at least for "non-pathological" initial conditions) behaves similarly to the integrable system.

In a nutshell, the idea of the KAM theorem is the following. Poincaré's result shows that, because of the small denominators, it is not possible to find a canonical transformation such that in the new variables the system is integrable. On the other hand one can try to obtain a weaker result: i.e. not for the whole phase space but for a region of non-zero measure. This is possible if the Fourier coefficients of $S_1$ in (4.28) are small. Assuming that $H_1$ is an analytic function, the $h^{(1)}_{\mathbf{m}}$ decrease exponentially with $m = |m_1| + |m_2| + \cdots + |m_N|$. On the other hand, there exist tori, with frequencies $\omega_0(\mathbf{I})$, such that the denominator is *not too small*, i.e.

$$|\mathbf{m} \cdot \omega_0(\mathbf{I}')| > K(\omega_0) m^{-(N+1)}, \tag{4.33}$$

for all integer vectors $\mathbf{m}$ (except the zero vector). The set of $\omega_0$ for which Eq. (4.33) holds has a non-zero measure in the $\omega_0$-space, and thus one can build the $S_1$ in a suitable non-zero measure region (around the non-resonant tori). Then one has to repeat the procedure for $S_2, S_3, \ldots$ and to control the convergence.

The FPU results can be seen (a posteriori) as a "verification" of the KAM theorem and above all, of its physical relevance, i.e. of the fact that the tori survive for physically significant values of the non-linear parameter $\epsilon$.

In the case of Hamiltonian systems with two degrees of freedom, the KAM tori have dimension 2 and separate regions of the three-dimensional surface of constant energy. Therefore chaotic disjoint regions separated by invariant surfaces (KAM tori) can coexist. Let us note that in general the KAM tori have dimension $N$ while the available phase space has dimension $2N - 1$. Thus in the case $N = 2$, the tori are able to separate regions that, in principle, can exhibit different behaviors. In contrast, for $N \geq 3$ the complement of the set of invariant tori is connected. This fact allows for so-called Arnold diffusion: the system can move on the whole surface of constant energy, by diffusing among the unperturbed tori (Arnold 1964). The presence of Arnold diffusion has been proved for particular systems, but it is believed to hold in a generic system. Unfortunately it is not easy to give theoretical estimates of the time scale of Arnold diffusion in the general case.

Let us note that the KAM theorem gives a result (i.e. the existence of the invariant tori) which is valid at any time, but only for a part of the phase space. If one is simply interested in times smaller than a given (large) $T_{\max}$ and for generic initial conditions, the results of the KAM theorem are somehow too restrictive (with regard to the times) and not completely satisfactory (with regard to the initial conditions). From an important theorem due to Nekhoroshev (1977) it follows that for a near-integrable Hamiltonian system any of the actions remains close to its initial value.

**Theorem (Nekhoroshev)** *Given a Hamiltonian* $H(\mathbf{I}, \phi) = H_0(\mathbf{I}) + \epsilon H_1(\mathbf{I}, \phi)$, *with* $H_0(\mathbf{I})$ *satisfying the same assumptions as for the KAM theorem, positive constants* $A, B, C, \alpha, \beta$, *exist such that any motion* $(\mathbf{I}(t), \phi(t))$ *satisfies the inequality*

$$|I_n(t) - I_n(0)| \leq A\epsilon^{\alpha} \quad n = 1, \ldots, N \tag{4.34}$$

*for*

$$t \leq B \exp[C\epsilon^{-\beta}]. \tag{4.35}$$

Both the KAM and Nekhoroshev theorems show clearly that both ergodicity and integrability are not generic properties for Hamiltonian systems which are perturbations of integrable systems. On the other hand it is not at all easy to master a priori, even at a qualitative level, important physical aspects such as the dependence on $N$ of the constants $A, B, C, \alpha, \beta$ in the Nekhoroshev theorem or the behavior of the measure of the KAM tori as a function of $N$ and $\epsilon$.

### 4.5.2 A parenthesis on solitons, FPU and KAM

In the 1960s, Zabusky and Kruskal (1965) developed the idea that the regular behavior of the Hamiltonian system (4.29) could be attributed to some solutions, called solitons, of a partial differential equation for which the Hamiltonian (4.29), is a discrete approximation.

Since this (often accepted) interpretation can generate some confusion we want to discuss this point briefly.

The equations which govern the evolution of the FPU system are

$$m\,\frac{\mathrm{d}^2 q_n}{\mathrm{d}t^2} = f(q_{n+1} - q_n) - f(q_n - q_{n-1}), \tag{4.36}$$

where, for $r = 3$, $f(y) = Ky + \epsilon y^2$. Assuming that $q_n(t)$ is the value of a spatially continuous variable, the field $\psi(x, t)$ at $n\,\Delta x$, where $\Delta x$ is the spacing of the lattice with which one approximates a continuous interval, it is easy to write a partial differential equation for $\psi(x, t)$:

$$\frac{1}{c^2}\frac{\partial^2\psi}{\partial t^2} = a\frac{\partial^2\psi}{\partial x^2} + 2\,g\,\frac{\partial\psi}{\partial x}\frac{\partial^2\psi}{\partial x^2}, \tag{4.37}$$

where, in the limit $\Delta x \to 0$, a proper rescaling of $K$, $m$ and $\epsilon$ has been performed.

It can be shown that the solutions of (4.37) develop spatial discontinuities after a finite time $t_c \sim 1/|\psi_0|$, where $\psi_0$ is the maximum field amplitude at $t = 0$ (Cercignani 1977).

However, we can look for a solution of (4.36) which, in the continuous limit, will be slowly varying with $t$, if $x - ct$ is fixed. In such a limit, for the variable $v = \partial\psi/\partial\xi$ we obtain the equation

$$\frac{\partial v}{\partial \tau} + \epsilon v\frac{\partial v}{\partial \xi} + \frac{1}{24}\frac{\partial^3 v}{\partial \xi^3} = 0, \tag{4.38}$$

where the variables $\xi$ and $\tau$ are proportional to $x - ct$ and to $t$ respectively.

Equation (4.38) is another way of writing the Korteweg–de Vries (KdV) equation, which was introduced in 1895 to describe the propagation of surface waves in shallow water. Equation (4.38) admits a "solitonic wave" solution of the type $v = F(\xi - V\tau)$, where $V$ is a constant and $F(z)$ is a localized function that decays to zero at large values of $|\xi - V\tau|$. Solitary waves have been considered for a long time as a mere mathematical curiosity. Since the work of Zabusky and Kruskal on vibrations in anharmonic crystals and plasma waves, solitonic properties have turned out to be fundamental to many physical phenomena (Cercignani 1977).

The original Zabusky and Kruskal explanation of the regularity of the FPU system in terms of solitary waves originating from the KdV equation, however, is not totally convincing. In fact the passage from Eq. (4.36), with variables $q_n(t)$, i.e. an ordinary differential equation, to a partial differential equation for the field $\psi(x, t)$ is very delicate, since similar assumptions can lead to very different systems. For instance, one can obtain an equation such as (4.37) which develops singularities in a finite time, or an equation such as (4.38) which has very regular behavior. On the other hand, the feature observed in FPU is not pathological, mainly because of its relation with the KAM theorem. Therefore the Zabusky and Kruskal interpretation of the regular behavior of FPU does not seem completely appropriate.

### *4.5.3 Numerical results and physical questions*

Since it is difficult to control the dependence on $N$ of the constants involved in the KAM theorem and the Nekhoroshev theorem, computer investigations are practically unavoidable. The great merit of numerical computations is that they allow experimental tests, and thus serve as a guide toward future theories. Let us discuss some simulations on the FPU. Izrailev and Chirikov (1966) first noted that for high values of $\epsilon$, where the effects of the KAM theorem are switched off, there is good statistical behavior, as can be seen in Figure 4.3. The energy, initially concentrated in the lowest frequency normal modes, can be seen to spread equally on all normal modes, therefore the time averages are in agreement with those of equilibrium statistical mechanics. For a fixed number of particles $N$, at least for large but finite

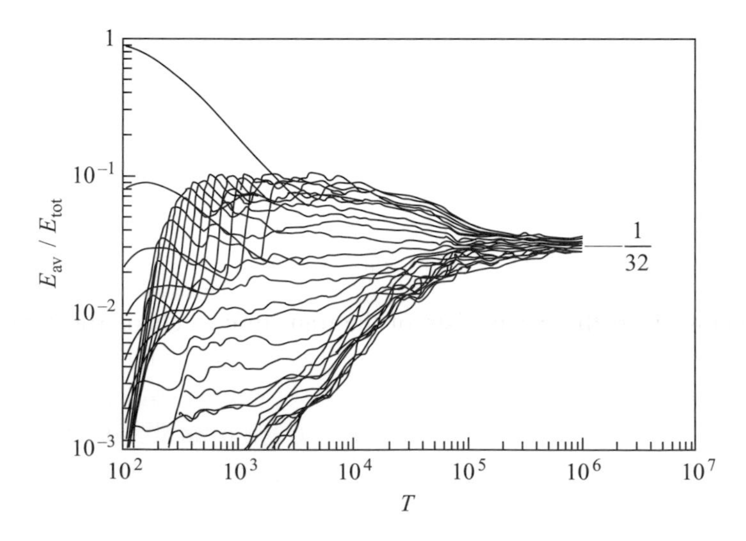


Figure 4.3 Time-averaged fraction of energy, in all the modes $k = 1, \ldots, 32$. The parameters of the system are $N = 32$, $r = 3$, $\epsilon = 0.1$ and energy density $\mathcal{E} = E_{tot}/N = 1.2$. Courtesy of G. Benettin.

times, the following scenario essentially holds (Livi *et al.* 1985, Ruffo 2001, Carati *et al.* 2005).

For a given energy density $\mathcal{E} = E/N$ there is a threshold $\epsilon_c$ for the strength of the perturbation such that:

(a) if $\epsilon < \epsilon_c$ the KAM tori play a major role and the system does not follow equipartition, even after a very long time;
(b) if $\epsilon > \epsilon_c$ the KAM tori have a minor effect, the system follows equipartition and there is agreement with standard statistical mechanics.

It is easy to realize that if the value of the perturbation $\epsilon$ is given, as happens in actual physical situations, the energy density could play the role of a control parameter and a threshold $\mathcal{E}_c$ would exist which separates regular from irregular behavior.

At variance with dissipative systems, in the Hamiltonian cases there is not a sharp transition from regular to chaotic behavior, and regular and chaotic motions can coexist.

Since the habitual aim of statistical mechanics is the study of systems with a large number of degrees of freedom, it is of genuine physical interest to understand whether the results observed in low-dimensional systems can be extended to cases with $N \gg 1$. Several physical questions arise:

(1) whether the regular behavior for small non-linearities, and irregular behavior for large non-linearities, is peculiar to the FPU Hamiltonian;
(2) what is the dependence of $\epsilon_c$ on $N$ (at fixed $\mathcal{E}$) or, equivalently, what is the dependence of $\mathcal{E}_c$ on $N$ (at fixed $\epsilon$);
(3) what are the characteristic times of the equipartition process.

More generally (i.e. for generic Hamiltonian systems) open questions are:

(4) how small is the part of the phase space with regular behavior;
(5) how does it depend on the number of degrees of freedom;
(6) what is the behavior of the relaxation times as functions of $N$ and $\epsilon$.

Point (1) is clear: the mechanism of the transition to chaos for increasing $\epsilon$ is standard for all systems which (like FPU) are obtained by perturbing harmonic systems. Furthermore, this behavior is present not only in one-dimensional lattices, but also in multi-dimensional lattices, for instance in Lennard-Jones two-dimensional systems at low energy, where the Hamiltonian can be written in form (4.22), i.e. a harmonic part plus an anharmonic perturbation (Benettin and Tenenbaum 1983). For points (2) and (3), the answers are less clear. The dependence of $\epsilon_c$ (or, equivalently, of $\mathcal{E}_c$ at fixed $\epsilon$) on $N$ is obviously very important: if $\epsilon_c \to 0$ when $N \to \infty$, the traditional point of view, i.e. that preceding FPU, is re-established. On the contrary, if $\epsilon_c$ were not dependent on $N$, there would be a serious discrepancy with the results expected from equilibrium statistical mechanics. Detailed numerical simulations and analytic computations have been performed to answer points (2) and (3). In spite of great effort, owing to technical and numerical difficulties, there is still no general agreement. In the following we discuss briefly some recent results.

Casetti *et al.* (1997) show clearly that in FPU with cubic non-linearity ($r = 3$), for an energy density smaller than $\mathcal{E}_c = E_c/N \sim 1/N^2$ the motion is very regular, even in the limit $N \to \infty$, with solitonic behavior, in agreement with the Zabusky and Kruskal interpretation. Above this threshold the system has good statistical behavior. However, the time $\tau_R(\mathcal{E})$ necessary to reach equipartition, starting from a far-from-equilibrium initial condition (for instance all the energy is concentrated in a few normal modes) may be very long: $\tau_R \sim \mathcal{E}^{-3}$. Similar results for quartic ($r = 4$) non-linearities have been obtained by De Luca *et al.* (1999). The relaxation time $\tau_R$ might also depend on the number of degrees of freedom $N$; for instance, if the initially excited normal modes are always between $k_1$ and $k_2$ (with fixed $k_1$ and $k_2$), on increasing $N$ we have $\tau_R \sim N^{1/2}\mathcal{E}^{-1}$ (Ruffo 2001). Since Hamiltonian systems do not have an attractor, the choice of initial conditions (particularly for $N \gg 1$) is an important technical aspect and it may have a non-trivial influence, even at a qualitative level, on the relaxation to statistical equilibrium. Without going into detail we note that, even starting from initial conditions which are typical of statistical equilibrium, partially regular behavior is also observed above the stochasticity threshold ($\mathcal{E} > \mathcal{E}_c$) (Livi *et al.* 1987). For a recent review on the FPU see Berman and Izrailev (2005) and Carati *et al.* (2005).

## 4.6 On the role of chaos in statistical mechanics

From the results discussed above one could be tempted to say that chaos (in the sense of positive Lyapunov exponent) is a necessary ingredient for the validity of statistical mechanics. Unfortunately this scenario seems to be much more complicated than originally expected. Even if the system turns out to be chaotic and most KAM tori are destroyed, the automatic validity of ordinary statistical mechanics is, in fact, not obtained, at least over long but finite times (Livi *et al.* 1987). This behavior is not restricted to systems similar to FPU, i.e. anharmonic perturbation of harmonic chains. Let us discuss briefly the results of numerical studies of high-dimensional symplectic coupled maps of the form

$$\begin{cases} \phi_n(t+1) = \phi_n(t) + I_n(t) & \mod 2\pi \\ I_n(t+1) = I_n(t) + \epsilon \nabla F(\phi(t+1)) & \mod 2\pi \end{cases} \tag{4.39}$$

where $n = 1, \ldots, N$ and $\nabla = (\partial/\partial\phi_1, \ldots, \partial/\partial\phi_N)$. It is easy to see that the above symplectic map is just a canonical transformation from the "old" variables $(\mathbf{I}, \phi)$, i.e. those at time $t$, to the "new" variables $(\mathbf{I}', \phi')$, at time $t+1$, via the generating function:

$$S(\mathbf{I}, \phi') = \sum_n \phi'_n I_n - \frac{1}{2} \sum_n I_n^2 + \epsilon F(\phi'). \tag{4.40}$$

The map (4.39) can be seen as the Poincaré section of a Hamiltonian system with $N+1$ degrees of freedom. When the coupling constant $\epsilon$ vanishes the system is integrable, and the term $\epsilon F(\phi)$ plays the role of the non-integrable perturbation of the Hamiltonian. In Falcioni *et al.* (1991) and Hurd *et al.* (1994) one can find clear evidence that the irregular behavior becomes dominant as $N$ becomes large. Specifically, one observes that the volume of the phase space occupied by the KAM tori decreases exponentially as $N$ increases. Although this sounds good for the foundation of statistical mechanics, one finds that very long time scales are involved: individual trajectories forget their initial conditions and invade a non-negligible part of the phase space only after an extremely long time. Also, in very large systems Arnold diffusion is very weak and, even with a high value of the Lyapunov exponent, different trajectories maintain some of their own features for a very long time.

Livi *et al.* (1987) studied the relevance of chaos in non-linear Hamiltonian systems with respect to the predictions of statistical mechanics. The canonical ensemble is the most suitable for computing averages analytically. From a conceptual point of view one can consider the canonical ensemble as describing the fluctuations of a small part of a large conservative system. This naturally suggests how to simulate a canonical ensemble, avoiding any noise source modeling the heat bath. Indeed the dynamics of a canonical ensemble of the FPU system can be simulated by subdividing a chain of $N$ particles into $N_1$ subsystems of $N_2 = N/N_1$ particles each, with $N_1 \gg 1$ and $N_2 \gg 1$.

In such a way one can compute the time average of observables defined in the subsystems and compare them with the values computed according to the canonical ensemble. For example one can define the internal energy $U$ as the mean value (over time) of the energy $E_j$ in the $j$th subsystem: $U = \overline{E_j}/N_2$. In an analogous way for the specific heat $C_V$ one has:

$$C_V = \frac{\overline{E_j^2} - \overline{E_j}^2}{N_2 T^2} \tag{4.41}$$

where the temperature is defined as $T = \overline{p^2}$.

Detailed numerical computations show that both the internal energy $U$ and the specific heat $C_V$, as functions of the temperature, are close to the predictions of the canonical ensemble. It is non-trivial that this agreement also holds in the region at small energy (i.e. small temperature) where the system behaves regularly (the KAM tori are dominant).

The above result seems to confirm Khinchin's approach (although the observables are not in the class of sum functions) with the small role played by dynamics.

However this conclusion does not seem to be general, in fact in other non-linear systems one can find rather different behavior (Livi *et al.* 1987). Consider a system of coupled rotators

$$H = \sum_{i=0}^{N} \left[ \frac{p_i^2}{2m} + \gamma(1 - cos(q_{i+1} - q_i)) \right] \tag{4.42}$$

where the variables $\{q_i\}$ are defined in $(0, 2\pi)$. For fixed $\gamma$ this system has two different integrable limits:

(a) for very small energy one has perturbation of a harmonic chain of oscillators;
(b) for very large energy, because the potential is bounded, one has perturbation of independent rotators.

For the system (4.42), though $U$ is very close to the canonical prediction, at high temperatures $C_V$ disagrees strongly with the results of the canonical ensemble.

The different behavior of $C_V$ in the two near-integrable regimes of low and high temperature can be understood as follows. For the FPU system and for the low-temperature rotators the "natural" variables are the normal modes, which, even in a statistical analysis, are able to show regular behavior and where the energy of the system is resident. However, even if the normal modes are almost decoupled, when observing the energy of a subsystem, identified by some set of "local" variables $\{p_j, q_j\}$, non-negligible fluctuations of the "local" energy can be seen. On the other hand, for the chain of rotators at large energy the normal modes, i.e. the carriers of the energy, are the "local" variables $\{p_j, q_j\}$ themselves, and therefore the fluctuations of the local energy are strongly depressed, as is the exchange of energy among the subsystems.

## 4.7 Some general remarks

Let us conclude this chapter with some remarks on ergodicity and chaos with respect to the foundation of statistical mechanics.

First we note that the ergodic approach can be seen as a natural way to introduce probabilistic concepts in a deterministic context. It seems to us that the ergodic theory provides support for the frequentistic interpretation of probability in the foundation of statistical mechanics. The other way (which is not in disagreement with the point of view of Boltzmann) to introduce probability is to assume an amount of uncertainty in the initial conditions. This approach is due to Maxwell who considers that there are *a great many systems the properties of which are the same, and that each of these is set in motion with a different set of values for the coordinates and momenta* (Maxwell 1879).

Since one is forced to deal with a unique system (although with many degrees of freedom) it seems natural to assume that the purpose of statistical mechanics, for equilibrium phenomena, is to calculate time averages according to the temporal evolution of the system. Therefore the ensemble theory should be seen only as a practical mathematical tool and the ergodic theory (or a "weak" version, such as that of Khinchin and Mazur and van der Linden) is an unavoidable step. Of course there is no complete consensus on this; for example Jaynes's opposite opinion is that ergodicity is simply not relevant for the Gibbs method (Jaynes 1967).

The ergodicity is, at the same time, an extremely demanding property (i.e. the time and phase averages must be equal for almost all the initial conditions), and not very conclusive at a physical level (because of the average over an infinite time). On the other hand, in the quasi-integrable limit the analytical results (KAM and Nekhoroshev) give only qualitative indications and do not allow for quantitative aspects. Therefore it is not possible to avoid detailed numerical investigations.

There are also opposing answers to the question of whether the systems which are described by statistical mechanics must have a large number of degrees of freedom, and it is possible to find eminent scientists with opposite opinions. For instance Grad (1967) writes explicitly that "the single feature which distinguishes statistical mechanics is the large number of degrees of freedom." One can read rather similar sentences in the well known textbook of Landau and Lifshitz. In contrast Gibbs (1902) believed that "the laws of statistical mechanics apply to conservative systems of any number of degrees of freedom, and are exact."

Extended simulations on high-dimensional Hamiltonian systems show in a clear way that chaos is not necessarily a fundamental ingredient for the validity of equilibrium statistical mechanics: the naive idea that chaos implies good statistical properties is inconsistent. Indeed sometimes, even in the absence of chaos (in agreement with Khinchin's ideas), one can have good agreement between the time averages and their values predicted by equilibrium statistical mechanics.

## References

Arnold, V. I. (1963). Proof of a theorem of A. N. Kolmogorov on the preservation of conditionally-periodic motions under small perturbations of the Hamiltonian, *Russ. Math. Surv.* **18**, 9.

Arnold, V. I. (1964). Instability of dynamical systems with many degrees of freedom, *Dokl. Akad. Nauk SSSR* **18**, 85.

Benettin, G. and Tenenbaum, A. (1983). Ordered and stochastic behavior in a two-dimensional Lennard-Jones system, *Phys. Rev. A* **28**, 3020.

Berman, G. P. and Izrailev, F. M. (2005). The Fermi–Pasta–Ulam problem: fifty years of progress, *Chaos* **15**, 015104.

Birkhoff, G. D. (1931). Proof of the ergodic theorem, *Proc. Natl. Acad. Sci.* **18**, 279.

Buck, B. and Macaulay, V. A. (eds.) (1991). *Maximum Entropy in Action*, Oxford: Clarendon Press.

Carati, A., Galgani, L. and Giorgilli, A. (2005). The Fermi–Pasta–Ulam problem as a challenge for the foundations of physics, *Chaos* **15**, 015105.

Casetti, L., Cerruti-Sola, M., Pettini, M. and Cohen, E. G. D. (1997). The Fermi–Pasta–Ulam problem revisited: stochasticity thresholds in nonlinear Hamiltonian systems, *Phys. Rev. E* **55**, 6566.

Cercignani, C. (1977). Solitons: theory and application, *Riv. Nuovo Cimento* **7**, 429.

Cercignani, C. (1998). *Ludwig Boltzmann: the Man who Trusted Atoms*, Oxford: Oxford University Press.

De Luca, J., Lichtenberg, A. J. and Ruffo, S. (1999). Finite times to equipartition in the thermodynamic limit, *Phys. Rev. E* **60**, 3781.

Ehrenfest, P. and Ehrenfest, T. (1956). *The Conceptual Foundation of the Statistical Approach in Mechanics*, New York: Cornell University Press, original edition in German 1912.

Falcioni, M., Marconi, U. M. B. and Vulpiani, A. (1991). Ergodic properties of high-dimensional symplectic maps, *Phys. Rev. A* **44**, 2263.

Fermi, E. (1923). Dimostrazione che in generale un sistema meccanico normale è quasi ergodico, *Nuovo Cimento* **25**, 267.

Fermi, E. (1965). *Note e Memorie (Collected Papers)*, Rome: Accademia Nazionale dei Lincei and The University of Chicago Press.

Fermi, E., Pasta, J. and Ulam, S. (1955). Studies of non linear problems, *Los Alamos Sci. Lab. Rep.* LA-1940.

Gallavotti, G. (1995). Ergodicity, ensemble, irreversibility in Boltzmann and beyond, *J. Stat. Phys.* **78**, 1571.

Gallavotti, G. (1999). *Statistical Mechanics. A Short Treatise*, Berlin: Springer.

Gibbs, J. W. (1902). *Elementary Principles in Statistical Mechanics*, New York: Scribner's.

Grad, H. (1967). Levels of description in statistical mechanics and thermodynamics, in *Delaware Seminar in the Foundations of Physics*, ed. M. Bunge, p. 49, Berlin: Springer.

Hurd, L., Grebogy, C. and Ott, E. (1994). On the tendency toward ergodicity with increasing number of degrees of freedom in Hamiltonian systems, in *Hamiltonian Mechanics*, ed. J. Siemenis, p. 123, New York: Plenum.

Izrailev, F. M. and Chirikov, B. V. (1966). Statistical properties of a nonlinear string, *Sov. Phys. Dokl.* **166**, 57.

Jaynes, E. T. (1967). Foundations of probability theory and statistical mechanics, in *Delaware Seminar in the Foundations of Physics*, ed. M. Bunge, p. 77, Berlin: Springer.

Jaynes, E. T. (1989). *Papers on Probability, Statistics and Statistical Physics*, Dordrecht: Kluwer.

Khinchin, A. I. (1949). *Mathematical Foundations of Statistical Mechanics*, New York: Dover.

Klein, M. J. (1973). The development of Boltzmann's statistical ideas, in *The Boltzmann Equation: Theory and Application*, ed. E. G. D. Cohen and W. Thirring, p. 53, Berlin: Springer.

Kolmogorov, A. N. (1954). On conservation of conditionally-periodic motions for a small change in Hamilton's function, *Dokl. Akad. Nauk SSSR* **98**, 525.

Livi, R., Pettini, M., Ruffo, S., Sparpaglione, M. and Vulpiani, A. (1985). Equipartition threshold in nonlinear large Hamiltonian systems: the Fermi–Pasta–Ulam model, *Phys. Rev. A* **31**, 1039.

Livi, R., Pettini, M., Ruffo, S. and Vulpiani, A. (1987). Chaotic behavior in nonlinear Hamiltonian systems and equilibrium statistical mechanics, *J. Stat. Phys.* **48**, 539.

Ma, S. K. (1985). *Statistical Mechanics*, Singapore: World Scientific.

Maxwell, J. C. (1879). On the Boltzmann's theorem on the average distribution of energy in a system of material points, *Proc. Cambridge Philos. Soc.* **12**, 547.

Mazur, P. and van der Linden, J. (1963). Asymptotic form of the structure function for real systems, *J. Math. Phys.* **4**, 271.

Moser, J. K. (1962). On invariant curves of area-preserving mapping of an annulus, *Nachr. Akad. Wiss. Göttingen Math. Phys. Kl.* **2**, 1.

Nekhoroshev, N. N. (1977). An exponential estimate of the time of stability of nearly integrable Hamiltonian systems, *Usp. Mat. Nauk* **32**, 6.

Poincaré, H. (1892). *Les Méthodes Nouvelles de la Mécanique Céleste*, Paris: Gauthier-Villars.

Ruffo, S. (2001) Time-scales for the approach to thermal equilibrium, in *Chance in Physics*, ed. J. Bricmont *et al.*, p. 243, Berlin: Springer.

Uffink, J. (1995). Can the maximum entropy principle be explained as a consistency requirement?, *Stud. Hist. Philos. Mod. Phys.* **26**, 223.

Zabusky, N. J. and Kruskal, M. D. (1965). Interaction of solitons in a collisionless plasma and the recurrence of initial states, *Phys. Rev. Lett.* **15**, 240.

# 5
# On the origin of irreversibility

> Since in the differential equations of mechanics themselves there is absolutely nothing analogous to the Second Law of thermodynamics the latter can be mechanically represented only by means of assumptions regarding initial conditions.
>
> *Ludwig Boltzmann*

By irreversibility here we mean a well evident fact which is part of the everyday experience of everybody: there are a lot of phenomena of which we do not see the reverse order evolution, and we do not expect to see it. For instance, if we put a hot coffee on the table then, after a while it gets colder, transferring some heat to the environment. However, if the coffee, at room temperature, has become too cold, and we want to warm it up, then we put it in the microwave oven. This is because we are pretty sure that the required heat will not come into the coffee from the surroundings, even if the reverse process has just taken place, as always. This certainty allows us to judge time ordering. Given the two states of the coffee on the table, hot and cold, we know that, without external intervention, the hot state cannot come after the cold state: it always comes before. As a consequence, once the cooling has occurred, we say that something (spontaneously) irreversible has happened. So, irreversibility is the asymmetric time evolution of certain macroscopic systems. The theoretical frame where this kind of irreversibility is accommodated is thermodynamics, where the Second Principle dictates the prohibitions.

## 5.1 The problem

As a fact within the coherent theoretical description of thermodynamics, irreversibility is not a problem. It becomes a problem when we adopt the atomistic point of view, and we pretend to explain the behavior of the macroscopic world starting from the laws of motion of its microscopic constituents. The origin of the problem lies in the fact that the laws of mechanics, that are assumed to govern the evolution of the individual particles, are time-reversal invariant.

We are inclined to believe that the time-symmetry breaking process, leading to decay of neutral kaons (Cronin 1981), cannot have any physically meaningful effect in determining the behavior of the usual macroscopic systems, that exhibit irreversibility under standard conditions of temperature and pressure. Moreover, we can safely assume that the relevant features of an irreversible behavior do not depend on the internal structure of the molecules forming the gas; this allows us to restrict our attention to the translational degrees of freedom, for which a classical mechanics description is fully adequate. So we can consider only classical mechanics, leaving aside a possible discussion based on the (time-reversal invariant) Schrödinger equation.

Assuming that the interparticle force, $\mathbf{F}$, depends only on particle positions, Newton's equation of motion for $N$ particles

$$m_i \frac{\mathrm{d}^2 \mathbf{x}_i}{\mathrm{d}t^2} = \mathbf{F}_i(\mathbf{x}_1, \ldots, \mathbf{x}_N) \qquad (i = 1, 2, \ldots, N) \tag{5.1}$$

is such that, if $\{\mathbf{x}_i(t)\}$ is an allowed solution, describing a possible evolution of $N$ particles, then $\{\mathbf{x}_i(-t)\}$, describing the reverse motion, with all velocities inverted, is also a solution. Essentially, this is because if one lets $t \to -t$ and $\mathbf{v}_i \to -\mathbf{v}_i$ while the $\mathbf{x}_i$ do not change, Eq. (5.1), which depends only on second derivatives with respect to time, stays unaltered. More precisely, one considers an initial condition, with particles at positions $(\mathbf{x}_1(0), \ldots, \mathbf{x}_N(0))$ possessing velocities $(\mathbf{v}_1(0), \ldots, \mathbf{v}_N(0))$, and lets it evolve, according to Eq. (5.1), until a time $t$. At this time one inverts the velocities, and one takes as a new initial condition that with particles at $(\mathbf{x}_1(t), \ldots, \mathbf{x}_N(t))$ and velocities $(-\mathbf{v}_1(t), \ldots, -\mathbf{v}_N(t))$. The evolution of the system from the new condition, during another time interval $t$, will bring it back to the points $(\mathbf{x}_1(0), \ldots, \mathbf{x}_N(0))$ with inverted velocities $(-\mathbf{v}_1(0), \ldots, -\mathbf{v}_N(0))$.

It is nearly unavoidable to mention here the so-called *echo phenomena*, where the time reversal of microscopic degrees of freedom (e.g. nuclear spins and positions of impurities) is realized; see, for instance, Hahn (1950) and Rhim *et al.* (1971). An interesting discussion of the subject can be found in Chapter 24 of Ma (1985).

The time invariance of classical dynamics implies the existence of the backward running of every process. Thus, referring to the above example, together with spontaneous cooling of the coffee, spontaneous warming up must also exist. That is to say, if a molecular evolution takes energy away from the coffee (making it colder, until its temperature is that of the room), then the inverted molecular motion must exist which, starting from the cold coffee, realizes a transfer of energy from the environment to the coffee, making its temperature increase. No one ever saw this: where is it?

Let us introduce another example that is more easily treated. Consider a gas in a box, that is initially compressed (say, by a piston) in one half of the box. When the piston is released the gas expands and fills the whole box uniformly. Together with the motion that starts soon after the removal of the piston and takes the molecules everywhere in the container, an inverted evolution must exist that brings all the molecules back into the smaller region. Once again we wonder where all the backward motions have gone.

## 5.2 Toward the solution

A possible answer to the question is to assume the drastic attitude of assigning to irreversibility a "fundamental status." As a consequence, the elementary laws of motion must be reformulated in a larger theoretical structure, that includes, as a genuine new kind of solution, the time-symmetry breaking solutions. We do not go further along this path. For the state of the art of this program see Prigogine (1999) and references therein; for a detailed critical analysis of this point of view see Bricmont (1995).

Here we adhere to the more conservative point of view that does not consider irreversibility as a fundamental behavior, that must play a part in the construction of dynamical theory. On the contrary, we assume that the laws of thermodynamics, that rule irreversibility, are not exactly valid, i.e. for all systems and on all physical scales. We argue that thermodynamics is a very accurate "effective theory" to describe the behavior of the macroscopic observables of large systems, whose microscopically detailed evolution is governed by the known (quantum or classical) dynamics. This thinking dates back to the founding fathers of statistical mechanics and, for instance, motivated the work of Einstein on Brownian movement, also viewed as a means to test the strict validity of the Second Law of thermodynamics. As Einstein writes in the introduction of his first article (May, 1905): "If the movement discussed here can actually be observed (together with the laws relating to it that one would expect to find), then classical thermodynamics can no longer be looked upon as applicable with precision to bodies even of dimension distinguishable in a microscope. . . " The experimental confirmation, provided by Perrin, of the Brownian motion fluctuation theory of Einstein and Smoluchowski, seems to us good support of the position we take (Mehra 2001).

In our treatment of the irreversibility problem we will not discuss the chaotic properties, if any, of the underlying dynamics; this because, in agreement with Bricmont (1995), we believe that the main ingredients for irreversibility are the large number of degrees of freedom and the initial conditions. At the end of the chapter we will discuss why, in spite of some claims, chaos plays a rather marginal role in irreversibility.

### *5.2.1 Some preliminary considerations*

We will try to understand whether the reversible Newton equations of motion can generate, in suitable conditions, practically irreversible behavior. We take upon ourselves the task of explaining how certain kinds of motions can "disappear." So we have to characterize the favorable settings that could allow time inversion to become a hidden symmetry.

*About the size of the system*

A first remark is that if Newton's equations are applied to describe the evolution of a small number of degrees of freedom, then direct and reversed motions are perceived as equivalent, and one (usually) does not distinguish significantly between them. For example, consider a billiard table with no pockets and two balls: neglecting friction, whatever the motion of these "big molecules" can be, one realizes easily its reverse motion; and the labeling of the motions, as direct or reversed, in these cases, has no particular meaning. However, in this simple setting one can also get in trouble with reversibility. Suppose that there are 16 balls on the table: 15 are numbered and clustered at rest in a (triangular) region in the middle of one of the two halves of the table; the 16th (and white) one is far apart in the middle of the other half of the table. One then gives a high enough speed to the latter, that it hits the group of balls, spreading them all around the table. We can call this motion, beginning with the white ball approaching the cluster, the direct motion. We are accustomed to this kind of behavior, which is a typical beginning of a game. But now, the reversed motion, where 15 balls cluster together and stop, transferring all their energies to only one ball, is so complicated to realize, and thus so unusual, that we feel a clear substantial difference between the two processes. If we had a movie of the direct event, that revealed to us the correct positions and velocities the balls must possess, nonetheless the implementation of a starting configuration for the reversed process would be a formidable task, involving the synchronized and calibrated action of 16 persons, or the rapid execution of 16 "right" shots by one person. This is to be contrasted with the ease of setting up the initial condition of what we called the direct motion. We are faced with a sort of "practical irreversibility," given by the extreme difficulty of preparing the initial state for the inverse motion, a motion that, on the other hand, we are able to imagine clearly, and that dynamical laws do not forbid. We can try to give a quantitative measure of the "strangeness" of the inverse motion, imagining realizing it by chance. A very rough estimation could be the following. In order to give an idea of the order of magnitude of the typical numbers involved in the problem, we disregard the rotational degrees of freedom and suppose that each ball can be located in $\mathcal{R} = 1000$ different positions on the table. If we admit that the speed of a ball ranges from 0 to 10 m/s with a step of

0.1 m/s, and that the orientation of a vector in the plane can be selected between 50 different possibilities, then the velocity vector can be chosen in, about, $\mathcal{V} = 5000$ ways. The number $\mathcal{N}$ of different states of $\mathcal{B} = 16$ distinguishable billiard balls on the table is

$$\mathcal{N} = \frac{\mathcal{R}!}{(\mathcal{R} - \mathcal{B})!} \mathcal{V}^{\mathcal{B}}.$$

Among all the $\mathcal{N}$ states we must select those, $\mathcal{N}_0$, leading to the (reversed) initial configuration, described above, with the white ball going away with maximum speed. Since we are not so fussy, we do not pretend to see just the 15 numbered balls stopping in exactly the initial positions, so that $\mathcal{N}_0 = \mathcal{B}!$ (not 1). At this point we can try to "guess the state." The fraction of good shots is

$$\frac{\mathcal{N}_0}{\mathcal{N}} = \left( \binom{\mathcal{R}}{\mathcal{B}} \mathcal{V}^{\mathcal{B}} \right)^{-1} = \frac{\mathcal{B}!(\mathcal{R} - \mathcal{B})!}{\mathcal{R}!} \mathcal{V}^{-\mathcal{B}},$$

a quantity that we can also interpret as the probability of success. Using the values in the text, the estimate for the fraction of propitious events is something like $10^{-94}$, which is comparable with the probability of guessing the *exact* sequence of heads and tails in 310 throws of a coin.

A process more within reach, conceptually analogous to the one described above, is a drop of water falling from a faucet, hitting the sink and splashing around. The time-reversed event would require an extremely fine tuning of the scattered tiny masses, to be gathered together so as to rebuild the drop, with the correct vertical velocity to jump back into the faucet. As in the previous example, the incomparable degree of difficulty in the preparation of the two initial states has to be noted: everyone is able to let a drop fall, but (we think) nobody has ever been able to realize the "antisplashing."

It appears that a great number of degrees of freedom in a system has the effect of making it hard both to realize the very particular "good" initial states and to determine them, among too many possibilities. The backward running states are there, but are almost inaccessible. The complication originates in the enlargement of the number of available states of the system, as a consequence of the relaxation of some constraint, used to prepare the "direct" initial state. After the enlargement, it becomes difficult to arrange all the degrees of freedom in order to produce, by evolution, some very particular state belonging to the small class as the constrained states.

At this point, we may notice that, from a theoretical point of view, the preparation of states evolving toward a (very) small region of phase space results in a hard task not only for high-dimensional systems but also for low-dimensional chaotic systems. In the former (if non-chaotic) one has to control few digits for each one of a great number of degrees of freedom, in the latter one has to control very many digits for each one of a small number of degrees of freedom. So, this is not the whole story.

*About the monitoring of the system*

To proceed further, a second point to be stressed, as a relevant fact, is that our bare perception is a direct and immediate witness of these irreversible behaviors.

In a system with many degrees of freedom one can decide to keep under observation only a few of them. In this case, as a consequence of their interaction with the rest of the system, one would see a very irregular, chaotic looking, behavior. The paradigmatic example is, of course, a Brownian particle. In the case of $N$ identical particles, we can refer to the work of Mazur and Montroll (1960) where the one-particle impulse in a system of harmonically coupled particles, at equilibrium, was studied. It is shown that, in spite of the quasi-periodicity of the dynamics, in the high $N$ limit, almost all of the time the impulse undergoes fluctuations, well described by an equilibrium Maxwellian distribution. On the other hand, it is known that in an ideal gas, one obtains an approximately Maxwellian distribution of the single-particle velocity, not only from the microcanonical uniform density but also for almost any distribution on the energy sphere.

However, this single-particle stochasticity is not a characterizing property of the irreversibility we are interested in. It is present in the states of macroscopic systems, either in equilibrium or in non-equilibrium, and indeed, as remarked above, it is an indication that thermodynamics, somehow, is not universally applicable. As a fact, the irreversibility we are discussing shows itself in the evolution of collective variables of macroscopic systems, that is, the variables underlying the image of the system that is manifest to us and that we are able to observe in usual situations. It is not concerned with properties of single molecules or of microscopic regions of space.

Indeed, suppose that our sensory faculties were able to capture the microscopic details of the external world. In the case of the coffee, we may suppose that one is able to follow the motion of one particular molecule. If one is given the values of the kinetic energy of the molecule at two different times, corresponding to the two states of the coffee (hot and cold), one could hardly perceive a substantial difference between the two cases. Because of the large variability of the possible energy values, one does not have a clear characterization of the one-molecule events, allowing for their time ordering. In the case of the gas in a box, we may think that we are able to take under control microscopic regions of space (on the scale, say, of the mean molecular separation). Then one is given two snapshots of a fixed microscopic region, contained in the initially filled half-box, taken before and after the piston is released. Also in this case, because of the large fluctuations in the number of molecules in the tagged region, one is not able to distinguish unambiguously two different situations, so as to give an ordering. What we mean is the following. One can decide that a higher kinetic energy and a higher number of particles "come before." However, upon repeating the tests, this will not always be the correct answer. If one is driven by a microscopic view one does not recognize a definite asymmetric time evolution.

It is only when one gathers information on a human scale and with human ability (i.e. involving a macroscopic number of particles, macroscopic regions, and with a coarse resolution) that one can notice a clear-cut distinction between two situations, and one learns that one situation always comes after the other. In fact, our lips come into contact with a macroscopic number of molecules of the coffee and they experience the effect of a macroscopic kinetic energy, a quantity which is very robust with respect to perceptible fluctuations, and is definitively different in the two states of the system (as the temperature is). Upon repeating the experiment we will always give the right answer. Also the density of the gas in the initially filled half-box is a quantity that (in non-critical settings) is not plagued with macroscopic fluctuations and is different before and after the free expansion of the gas.

The consideration that, in a dynamical system, the details of the states may be not all equally relevant, or that the states are observed with bounded accuracy, in information theory led to the notion, due to Shannon (1948) of "transmission to within a given tolerance" and to the birth of rate distortion theory (Berger 1971). The great value of Shannon's original idea was immediately recognized by Kolmogorov, who developed the concept of $\epsilon$-entropy for general dynamical systems and functional analysis (Kolmogorov 1956). See Chapters 2 and 3 for the use of the $\epsilon$-entropy to treat some aspects of predictability and complexity in dynamical systems.

### 5.2.2 *Along the Boltzmann–Khinchin way*

*The role of the law of large numbers*

To see a well-defined quantity, to within a certain accuracy, we must observe, as we really do in everyday life, a global variable, one that refers additively to a macroscopic number of particles. The contributions of all the molecules add up to produce a very stable result, with respect to our capability of resolution. This may be seen as an effect of central-limit kind. In fact, the observables one typically encounters in statistical mechanics (e.g. temperature, energy, pressure, number of particles in a given region) are of sum-function type (see Chapter 4). In the case of the perfect gas, Khinchin (1949) showed that, over a surface of given energy, the mean square deviations of these kinds of functions are so small, with respect to their mean values (macroscopic, of order $N$), that the functions can be regarded as practically (i.e. macroscopically) constant. One may say that the sum functions are subject to the law of large numbers. This fact, as Khinchin writes, *establishes the "representability" of the mean values of the sum functions, and permits us to identify them with the time averages which represent the direct results of any physical measurement.* Mazur and van der Linden (1963) explored the case of molecules with a binary interaction given by the Herzfeld potential. This is a piecewise constant potential possessing the characteristic properties of a realistic short-range potential function with a hard core. Their results, on the structure function of the system (that is, a quantity proportional to the measure of the hypersurface with given energy $E$: $\int \mathrm{d}^N\mathbf{x}\, \mathrm{d}^N\mathbf{p}\, \delta[H(\mathbf{x}^N, \mathbf{p}^N) - E]$), show that in this case the law of large numbers is also at work. One finds that, owing to the short range of interactions, the system behaves as if it were split into a large number of (almost) non-interacting components. So, we can safely retain that nice feature of the sum functions as valid also in the case of interacting molecules.

### *The macrostates*

The above remarkable property may be rephrased as follows. Given a system at a fixed energy, one classifies the microscopic states according to a macroscopic variable, $F$ (of sum type), i.e. one puts in the same class all the states giving rise to the same value of $F$. Then one measures the size of each class (in a continuous phase space this means, for instance, computing the Liouville volume occupied by the states in the class). It appears that there is one class containing the overwhelming majority of the states. The $F$ value of this class can be seen as the (practically constant) value of $F$ in the system. The grouping by classification of microscopic states realizes partitions of the phase space, and that is useful when dealing with macrostates of a macroscopic system. A macrostate specifies the values of a suitable number of global variables (mostly additive), resulting in a very crude description of the system, with respect to a detailed microscopic view, but which is considered exhaustive from a macroscopic point of view. For instance, the macrostates of a gas in a volume $V$ can be defined by the number of molecules, their total energy, and by the fraction of particles contained in each half of the available volume. Every macrostate identifies a class of microscopic states: those on which the relevant macroscopic variables assume, within a given tolerance, the correct values. The above kind of reasoning leads us to believe that a class of microstates exists that almost exhausts the available phase space, so defining an equilibrium macrostate. It is worth stressing that equilibrium is a property of a macrostate, i.e. of the variables pertaining to the way the system appears to us: the microstate of the system is changing continuously, while the relevant macroscopic observables maintain practically constant values.

A very simple example where a similar state of affairs is at work is the following. Consider the set of all $N$-long symbol sequences $S^N = (s_1, s_2, \ldots, s_N)$ where the variable $s$ takes values from a finite alphabet $\mathcal{A} = a_1, a_2, \ldots, a_L$. The *type* of a sequence, $P(S^N)$, is defined as its empirical distribution of the symbols. That is, $P(S^N)$ is given by the ensemble of frequencies, $(\nu_1, \nu_2, \ldots, \nu_L)$, with which each symbol occurs in $S^N$: we indicate a particular type by $P_\alpha^{(N)}$. In general several sequences (related by a mere permutation of symbols) possess the same ensemble of frequencies, so that a given $P_\alpha^{(N)}$ characterizes a set of sequences, $C(P_\alpha^{(N)})$, called the *type class* of $P_\alpha^{(N)}$.

Let us make explicit the simple case $N = 3$ with $L = 2$, say $a_1 = a$, $a_2 = b$. We must consider the following three-symbol sequences:

$$\begin{array}{llll} \omega_1 = a\,a\,a & \omega_2 = a\,a\,\theta, & \omega_3 = a\,b\,a & \omega_4 = b\,a\,a \\ \omega_5 = b\,b\,a & \omega_6 = b\,a\,b & \omega_7 = a\,b\,b & \omega_8 = b\,b\,b. \end{array} \tag{5.2}$$

The types of the sequences are:

$$\begin{aligned} &P(\omega_1) = (\nu_a = 1; \nu_b = 0) \quad P(\omega_2) = \left(\frac{2}{3}; \frac{1}{3}\right) = P(\omega_3) = P(\omega_4) \\ &P(\omega_5) = \left(\frac{1}{3}; \frac{2}{3}\right) = P(\omega_6) = P(\omega_7) \quad P(\omega_8) = (0; 1). \end{aligned} \tag{5.3}$$

So we have four types: $P_1 = P(\omega_1)$ (we omit the exponent 3), $P_2 = P(\omega_2)$, $P_3 = P(\omega_5)$, $P_4 = P(\omega_8)$ and the corresponding type classes contain $1, 3, 3, 1$ elements, respectively.

One may imagine $S^N$ as the analog of the microstate of a system with $N$ particles, each one having access to $L$ states, and $P_\alpha^{(N)}$ as the analog of a macrostate (a one-particle distribution), represented by the microstates in $C(P_\alpha^{(N)})$. One can show, (see Cover and Thomas 1991), that the number of different types is only polynomial in the length $N$ (it is at most $(N+1)^L$), while the number of sequences is exponential in $N$ (it is $L^N$), so there must be at least one type with exponentially many sequences in its type class. Indeed one has that, when $N$ is not small, to first order in the exponent, the number of sequences in $C(P_\alpha^{(N)})$ is given by

$$\|C(P_\alpha^{(N)})\| \approx \mathrm{e}^{NH(P_\alpha^{(N)})}, \tag{5.4}$$

where

$$H(P_\alpha^{(N)}) = -\sum_{i=1}^{L} \nu_i \ln \nu_i. \tag{5.5}$$

This implies that, when $N$ is very large, the largest type class essentially contains the entire set of sequences: it is the analog of the equilibrium macrostate. It is important to note that these results follow from simple combinatorial considerations in the counting of symbol arrangements, plus the Stirling approximation for factorial coefficients.

To summarize, when the number of particles of a system grows larger and larger, the full microscopic state of the system goes very rapidly out of control, while the variables pertaining to a small number of particles show irregular behavior, with large fluctuations relative to their mean values. But, at the same time, global observables, whose average values are of order $N$, become more and more well defined, because their relative fluctuations become smaller and smaller. We may suspect that the irreversible behavior is an emerging property of the evolution of global variables in systems composed of a huge number of molecules.

### 5.2.3 A proposal for the solution

We set out on a path leading us to explore the following scenario for the appearance of an asymmetric time behavior.

(a) The microscopic state of a system rapidly enters (if it is not already there) into the class of states representing the largest class with common values of the relevant macroscopic variables (we are, maybe, observing). These values can be identified as the equilibrium values of the variables.
(b) Once the equilibrium values have been attained, a change is almost impossible, since the microstates of the other classes represent an irrelevant fraction of the total: the system, practically, has "no chance" to enter into one of them, on its own.

Since it is not difficult to prepare a system in a non-equilibrium macrostate (it is enough to remove a constraint), point (a) says that it is easy to observe the spontaneous evolution of a macrostate toward equilibrium. This would account for the existence of what we called the direct motions. At the same time point (b) says that it is not at all easy to observe spontaneously the inverse motions.

We have stated qualitative expectancies, based on general considerations about very-high-dimensional phase spaces. The hard task is to substantiate this description by means of Newtonian dynamics, which drives each and every real macroscopic system. In particular, we have to show that (a) and (b) are true for essentially all the microstates belonging to a set of points describing a (initial) non-equilibrium situation. That is, we have to show that irreversibility is true for all macroscopic systems, that are in the same macroscopic initial conditions. One can say that this must be so, because thermodynamics has been developed independently from an atomistic view of matter, and therefore the details of the microscopic description of the system cannot have a part in thermodynamic irreversibility. We are sure to observe a certain irreversible behavior in a system, when some $O(10)$ (macro)variables are properly initialized, while the remaining $O(10^{23})$ are completely neglected.

## 5.3 Some results

### *5.3.1 The freely expanding Knudsen gas*

*Quantitative aspects*

So, let us analyze from a quantitative point of view the case of an isolated gas performing a free expansion. We start with the simplest possible approximation of non-interacting molecules (ideal gas) with no internal structure. This is also known as the Knudsen gas, i.e. a gas in which the mean free path of the molecules is much longer than the linear dimension of the container, so that the collisions between particles may be neglected. When the system is at equilibrium (before and after the expansion process) its state can be characterized macroscopically as having $N$ (of order $10^{23}$) identical molecules, with energy $U$, confined in a volume $V$. We begin by posing the question: how many microscopic states belong to this macrostate. A microscopic state is specified by the values of position ($\mathbf{x}$) and impulse ($\mathbf{p}$) of all the $N$ particles.

It is necessary here to make explicit an assumption already present in the reasoning above. In the scenario we are following, we consider, a priori, on an equal footing all the microstates of a system that satisfy the constraints defining a macrostate (e.g. fixed $N$, $U$, $V$ in an isolated gas). That is, we assume that in a single realization of the experiment the thermodynamic system can be found in any one of the allowed microstates, with the same probability. Then, when a particular phenomenon has to be explained, we feel satisfied if we are able to show that it takes place for all the microstates but a set of very small phase space volume. Thus we are choosing the Liouville measure in phase space as the "natural" tool to estimate the physical relevance of sets of points. This means that, in this approach to the problem, a particular behavior will seldom appear if it is pertinent to microstates represented by a set of phase space points with a small Liouville measure. This was the important part of the remark. Now, if we prefer dimensionless quantities, so as to avoid arbitrariness in the values of volume (dependent on the units chosen for mass, length and time) we have to introduce a reference phase space volume. A "natural" choice is $h^3$ for the elementary volume of the one-particle phase space, where $h$ is Planck's constant. Of course, since we are dealing in a classical context, the actual value of $h$ is not relevant at all. At this point, if one is willing to think of $h^3$ as the minimum phase space extension of a one-particle state (because of the Uncertainty Principle) then one can also think about the number of states instead of phase space volume. In the present context this is a matter of taste.

According to these considerations, in order to answer the posed question about the macrostate size, we have to compute the integral:

$$\Gamma(U, V, N) = \int_{U-\delta U/2 \leq H \leq U+\delta U/2} \frac{\mathrm{d}^N \mathbf{x}\, \mathrm{d}^N \mathbf{p}}{N!\, h^{3N}}. \tag{5.6}$$

In Eq. (5.6) $H$ is the Hamiltonian function (the energy) of the system, $U$ is its actual value, known to within an uncertainty $\delta U$, and $N!$ takes into account the physical indistinguishability of the particles. In the perfect gas approximation, and considering only the translational degrees of freedom ($H = \sum_{i=1}^{N} \mathbf{p}_i^2/2m$), to the leading order in $N$ we have

$$\Gamma(U, V, N) \approx \left[ \left(\frac{V}{N}\right) \left(\frac{U}{N}\right)^{3/2} \left(\frac{4\pi m}{3h^2}\right)^{3/2} \mathrm{e}^{5/2} \right]^N, \tag{5.7}$$

where, owing to the very high dimension of the phase space, we do not consider the small role of the uncertainty $\delta U$ (the volume is concentrated in a very thin layer, smaller than any macroscopic uncertainty). One could check here that macrostates of homogeneous particle density, with respect to every splitting up of $V$ into macroscopic regions, always contain the great majority of states of $\Gamma(U, V, N)$: at equilibrium the gas is found in microstates of uniform density.

If now, by a free expansion, the volume accessible to the gas is doubled (with no change in energy), then we see that the number of states after ($\Gamma_\mathrm{a}$) and before ($\Gamma_\mathrm{b}$) the expansion are in the ratio

$$\frac{\Gamma_\mathrm{a}}{\Gamma_\mathrm{b}} \approx 2^N. \tag{5.8}$$

The relaxing of the initial constraint gives each molecule twice the initial spatial freedom, but multiplies by an astronomically huge factor the number of available states of the macroscopic system. Let us assume, for a while, that all the microscopic states of the gas in the volume $V$ ("initial states") lead to the final macroscopic state of the gas filling the volume $2V$. It is evidently very difficult (practically impossible) to invert the velocities of all the molecules coming from any one of the initial states. So, if we want to see the antidiffusion of the gas we can only hope for the chance that, repeating the preparation of the state in the volume $2V$, we hit a "good" one: relation (5.8) says that this happens once every $2^{10^{23}}$ times, i.e. practically never. If we identify the initial volume as $V_\mathrm{L}$ and we put $2V - V_\mathrm{L} \equiv V_\mathrm{R}$, then during the expansion a suitable macrostate description of the gas is obtained giving $N_\mathrm{R}$, the number of molecules in $V_\mathrm{R}$ (or $N_\mathrm{L}$, the number of molecules in $V_\mathrm{L}$). Before lifting the constraint, $N_\mathrm{R} = 0$ (or $N_\mathrm{L} = N$).

We come back now, when considering the initial states, to stress that not all of them can evolve into the space-filling states: some must exist that concentrate the gas in a smaller volume, say $V/2$. The reason for this is the same reason that ensures that, among all the states of the gas in the volume $2V$, those from the above "initial states" are also included, together with their time-reversed states, bringing the gas back to $V$. The aim of this observation is not to underline that the chance of antidiffusion is smaller than we assumed (it is already small enough), but to introduce another conceptual point, that will allow us to distinguish clearly between the microscopic time direction and the macroscopic time-arrow. So far, by counting the states, we have gained some confidence about the following. After we prepared the gas in the volume $V$, because of relation (5.8), it will not become concentrated into a smaller volume on its own (with a probability about $1 - 2^{-N}$). Moreover, if we relax the constraint keeping the gas in $V$, then its microscopic state will wander in the new accessible states, whose macroscopic characteristic, concerning the distribution of molecules, is a uniform density in the volume $2V$. In the macrostate picture, the final configuration is expected to be $N_L = N_R = N/2$, within macroscopic accuracy. The given values of the macrovariables $N_L$ and $N_R$ are those shared by the very great majority of microscopic states. In contrast, the microstates with the initial values ($N_L = N$, $N_R = 0$) are only an infinitesimal fraction in the final phase space, whereas they practically exhausted the initially accessible phase space.

So we are brought to believe that practically all the microscopic states of the system in $V$ will evolve toward the uniform density in $2V$, while only an infinitesimal fraction will lead the gas to occupy a smaller volume. This may appear to generate a problem. We could think that the application of the time-reversal operation to all these states would produce an opposite situation: a great majority of concentrating evolutions and an infinitesimal fraction of expanding evolutions. In the end we stumbled over the paradox connected with the time-invariance of Newton's equations. However, this would be so only if the time reversal of every expanding state were a contracting one, and vice versa. But it is not so.

To proceed further we cannot avoid introducing some dynamical considerations. Following Hurley (1980, 1981), one can show that, in the simple system we are considering, practically all the microscopic states leading to a correct increase in $N_R$ possess a time-reversed state again with increasing $N_R$.

One would like to characterize the time behavior, with respect to $N_R$, of microstates belonging to the class selected by a macrostate with $N_R < N/2$. To this end one computes $\Gamma_{(+,+)}(N_R)$, the number of states which at the present time, say $t = 0$, have $N_R(0) = N_R$ particles in the right half of the box, and for which, a time $\Delta t$ before, $N_R$ was smaller ($N_R(-\Delta t) < N_R$) and a time $\Delta t$ after, $N_R$ will be greater ($N_R(\Delta t) > N_R$). Thus, $\Gamma_{(+,+)}(N_R)$ is the number of states for which the number of particles in the right side of the box is an increasing quantity both before and after now. Here $\Delta t$ is a short time, much less than the mean transit time of a particle in the box. Then one computes $\Gamma_{(-,-)}(N_R)$, the number of states with negative derivatives of $N_R$; these are the states with $N_R(-\Delta t) > N_R$ and $N_R(\Delta t) < N_R$. In the case at hand ($N_R < N/2$) the microstates contained in $\Gamma_{(+,+)}(N_R)$ have expanding histories, going toward equilibrium, while those in $\Gamma_{(-,-)}(N_R)$ are their time-reversed states with contracting histories, moving away from equilibrium. These are the kinds of states causing our troubles with time invariance. At this point, however, one observes that different kinds of time behavior are possible. Let $\Gamma_{(-,+)}(N_R)$ count the states with $N_R(-\Delta t) > N_R$ and $N_R(\Delta t) > N_R$, i.e. states that are nearer to the equilibrium value for $N_R$ in both time directions. Finally, let us denote by $\Gamma_{(+,-)}(N_R)$ the number of states with $N_R(-\Delta t) < N_R$ and $N_R(\Delta t) < N_R$, i.e. states that are farther from equilibrium in both time directions. Note that, while the set $\Gamma_{(+,+)}$ is transformed into $\Gamma_{(-,-)}$ by time reversal, $\Gamma_{(-,+)}$ and $\Gamma_{(+,-)}$ are left invariant by the transformation: they contain together with a state its time-reversed state too. From the above quantities one computes $G_{(+,+)}(N_R)$, $G_{(-,-)}(N_R)$, $G_{(-,+)}(N_R)$ and $G_{(+,-)}(N_R)$, the fractions of microstates, with given $N_R$, each subclass contains. These quantities can be evaluated explicitly (Hurley 1980, 1981) in the thermodynamic limit and one can write:

$$\begin{aligned} G_{(+,+)} &= g(1-g) \\ G_{(-,-)} &= (1-g)g \\ G_{(-,+)} &= (1-g)^2 \\ G_{(+,-)} &= g^2 \end{aligned} \tag{5.9}$$

where $g(N_R)$ is a function that in this limit and for this condition ($N_R < N/2$) goes rapidly to zero, at increasing $N$. From this result one obtains the following.

The diffusive states with antidiffusive time-reversed partners, $\Gamma_{(+,+)}$, are indeed a near zero fraction of the total. This result is in agreement with the above observation on the extreme smallness in $V$ of the fraction of states coming from $V/2$, leading us into trouble; however, here we can see the solution of the problem. Virtually all the microscopic evolutions, belonging to a macrostate with $N_R < N/2$, drive the gas toward a more uniform density *in both microscopic time directions*, because the fraction of states with this property, $G_{(-,+)} \simeq 1 - 2g$, is practically 1. So, the latter microstates are the carriers of the typical behavior of the macrostate, while the former are very untypical. A macroscopic system in a macrostate $N_R < N/2$ is almost certainly found in a microstate of the class $\Gamma_{(-,+)}$ and its evolution will be diffusive, independently of possible time reversal (in the example we are considering, practically all the states of the gas in $V$ are ready to diffuse toward $2V$ in both time directions). A system can be found in a $\Gamma_{(+,+)}$ microstate, that under time reversal would generate an antidiffusive evolution, only if it comes from a lesser uniform macrostate. Moreover, a system coming from a lesser uniform macrostate can also be found in a $\Gamma_{(+,-)}$ microstate, with incorrect antidiffusive future behavior, but Eq. (5.9) tells us that this is an almost impossible event. A macroscopic system that is expanding will continue the expansion, with a probability near 1. Indeed we can identify

$$\frac{G_{(+,+)}}{G_{(+,+)} + G_{(+,-)}} = 1 - g$$

with the relative probability that a state with the right diffusion property in the past will continue to expand in the future, and we can identify

$$\frac{G_{(+,-)}}{G_{(+,+)} + G_{(+,-)}} = g$$

with the relative probability for a diffusive evolution to change into a contracting one. This allows us to say that among the states coming from a condition farther from equilibrium, almost all persist on the "right way" toward equilibrium, sharing a common evolution with the great majority of the microstates in the reached macrostate. In the example, if the gas is contained in $V/2$ and expanded to $V$, then it will not come back into $V/2$ but will continue toward $2V$, just as if it had started from $V$. Finally, notice that, since $g(N_R = N/2) = 1/2$, a non-changing value for $N_R = 1/2$ is assured.

### *Qualitative considerations: the Boltzmann entropy*

The above result is at the heart of a possible explanation, by means of statistical reasoning, of the compatibility between observed macroscopic irreversibility and microscopic time-symmetric dynamical equations. Time-symmetry invariance remains fully active at the microscopic level, i.e. all the required states are there, but the macroscopic evolutions are almost insensible to it. In fact the overwhelming majority of microscopic evolutions lead to a sequence of values of global observables, with an ordering that their time-inverted evolutions would also generate. That is, microscopic time invariance is hidden to the global variables we usually experience.

This implies that the parameter ordering the history of macrovariables is not the microscopic time. In the approach we are discussing, the evolution of a global variable $F$ (e.g. number of particles in a subvolume $V$), when a constraint defining the initial equilibrium state (e.g. all $N$ particles in $V$) is relaxed, points toward a new equilibrium state, characterized by an $F$ value (e.g. $N/2$ particles in $V$) common to an exaggerated majority of microscopic states; this makes it a nearly non-changing (or equilibrium) value. Thus it is apparent that a suitable macroscopic evolution parameter is the number of microscopic states sharing the same values of global variables that determine the macrostate of the system. If the logarithm of the number of states is considered, then one sees that a macroscopic quantity (proportional to $N$, see Eq. (5.7)) may be conceived to give an ordering to the observed evolution of macroscopic systems toward equilibrium. At this point one can identify this quantity with entropy, just as Boltzmann did, thus obtaining its possible microscopic interpretation.

According to Boltzmann's ideas, the statistical theory of the approach to equilibrium can be presented as follows (Lebowitz 1993a, 1993b). An isolated thermodynamic system out of equilibrium is in a microstate, $\mathbf{X}$, belonging to the class of all microstates giving rise to the same value of the non-equilibrium macrostate $M(\mathbf{X})$, that describes the actual macroscopic state of the system. The (Boltzmann) entropy of this system, in this state, is:

$$S_{\mathrm{B}}(M) = k_{\mathrm{B}} \ln \|M\| \,, \tag{5.10}$$

where $k_{\mathrm{B}}$ is the Boltzmann constant and $\|M\|$ is the number of states in the class (as given by a suitable integral in the system phase space) measuring the extension of the macrostate. Letting the system evolve, $\|M\|$ typically increases, and so does $S_{\mathrm{B}}(M)$, until the system reaches the macrostate of maximal extension in phase space, and the growth ends:

$$S_{\mathrm{B}}(M(t')) \geq S_{\mathrm{B}}(M(t)) \quad \text{for} \quad t' \geq t \,. \tag{5.11}$$

It is useful to remember that, in the proposed scenario, the end of macroscopic evolution, i.e. reaching equilibrium values, would be due only to the fact that in thermodynamic systems (for which $S_{\mathrm{B}} \propto N$) a class of maximal extension can exist that dominates, exponentially in $N$, all the other classes. Starting from the microscopic dynamics, one would like to prove Eq. (5.11) for the systems of interest or, even better, to write out explicit evolution equations for the relevant macrovariables.

An important point to be stressed (again) is that, by experience, we expect to see the appropriate unidirectional evolution of macrovariables in each actual realization of a thermodynamic experiment *on a single system*. Thus one has to show that the evolution of Boltzmann's entropy, or of the macrovariables, is the same for (almost) all initial microstates: the thermodynamic behavior must be shown to be a "typical" behavior of the involved microstates. Indeed in the quantitative considerations discussed above for the ideal gas in a box, the statistics of the events shows that, in a suitable limit, anti-thermodynamic behavior can appear only in a going-to-zero fraction of cases.

### 5.3.2 *The Lorentz gas*

To continue along this path we consider a more complex system, the Lorentz gas. This has been analyzed with mathematical rigor, demonstrating the existence of irreversible macroscopic behavior. We limit ourselves to recalling briefly the interesting results. In the Lorentz system a macroscopic number of classical non-interacting particles move in a fixed periodic array of hard convex scatterers. In contrast to the ideal gas considered above, each particle now undergoes a chaotic motion and, if it can travel only a bounded distance between collisions (what is called a *finite horizon*), then the motion has a diffusive character (Bunimovich and Sinai 1981a, 1981b). One considers an ensemble of particles, whose initial spatial distribution is given by a smooth density $n_0(\mathbf{r})$. Since the collisions with the scatterers are elastic, single-particle kinetic energy is a conserved quantity and one considers all particles as having unit speed, with uniform orientations. These conditions define a macrostate, and its class of microstates. In a suitably defined limit that cleanly separates the macroscopic and microscopic scales (Lebowitz and Spohn 1982a, 1982b), one can prove that the evolution of the density, $n(\mathbf{r}, t)$, is governed by the standard diffusion equation, for each microstate typical of the initial macrostate.

Since in the Lorentz gas molecules do not interact, the model is not very satisfactory. This means that not all the dynamical variables can evolve toward equilibrium. The relevant macroscopic variable for this model, $n(\mathbf{r}, t)$, has irreversible behavior but, for example, the particle kinetic energies are frozen and so no thermalization is allowed. Notice, however, that if one initializes the system with an arbitrary distribution of velocity orientation, then the chaotic single-particle dynamics drives the particles toward an isotropic distribution of the velocities.

### 5.3.3 The dilute gas with collisions

*The Boltzmann equation and the $\mathcal{H}$-theorem*

The next important system to be considered is a dilute gas, a large collection of particles, interacting by short-range forces, in a state of very low density. The diluteness condition allows us to neglect the contribution of interaction terms to the numerical value of the total energy of the system, even if interactions have a fundamental role in the evolution process. In this condition a meaningful macrostate description of the gas is given by the occupation number function, or one-particle distribution function, $f(\mathbf{x}, \mathbf{p})$, defined so that

$$f(\mathbf{x}, \mathbf{p})\, \mathrm{d}\mathbf{x}\, \mathrm{d}\mathbf{p}$$

is the number of particles whose position and momentum is found in a volume $\mathrm{d}\mathbf{x}\mathrm{d}\mathbf{p}$ around the one-particle phase space point $(\mathbf{x}, \mathbf{p})$. One has

$$\int f(\mathbf{x}, \mathbf{p})\, \mathrm{d}\mathbf{x}\, \mathrm{d}\mathbf{p} = N,$$

where, as before, $N \gg 1$ is the number of particles of the gas. This function is a generalization of the $n(\mathbf{r}, t)$, defined above, taking into account that in this system exchanges of momenta between particles are possible (and necessary, to reach a complete equilibrium). The distribution function is defined in the six-dimensional one-particle phase space, and it selects a class of microstates in the full $6N$-dimensional phase space of the gas: those states whose particles are (approximately) distributed as specified by $f$.

For an ideal gas, $f(\mathbf{x}, \mathbf{p})$ evolves in time according to a conservation law that, in the absence of external forces, is (exactly):

$$\frac{\mathrm{d}}{\mathrm{d}t} f(\mathbf{x}, \mathbf{p}, t) = \left( \frac{\partial}{\partial t} + \frac{\mathbf{p}}{m} \cdot \nabla_{\mathbf{x}} \right) f(\mathbf{x}, \mathbf{p}, t) = 0. \tag{5.12}$$

Equation (5.12) says that, since the particles evolve independently, the one-body dynamics is volume conserving and, in the single-particle phase space, the density cannot change along the motion. Note that a similar conservation law could also be written for the Lorentz gas, for the full one-particle distribution, but not for the density $n(\mathbf{r}, t)$, obtained from the latter by integration on the momenta.

The interactions among particles cause transitions of particles between volume elements, and then the density can change:

$$\left( \frac{\partial}{\partial t} + \frac{\mathbf{p}}{m} \cdot \nabla_{\mathbf{x}} \right) f(\mathbf{x}, \mathbf{p}, t) \equiv \widetilde{(\partial_t f)}. \tag{5.13}$$

In the dilute gas approximation, it is possible to take into account binary collisions alone, and one can write (see e.g. Huang 1987):

$$\widetilde{(\partial_t f)} = \int \mathrm{d}\mathbf{p}_1 \mathrm{d}\Omega \; |\mathbf{v} - \mathbf{v}_1| \left( \frac{\mathrm{d}\sigma}{\mathrm{d}\Omega} \right) (F_2(\mathbf{x}, \mathbf{p}', \mathbf{p}_1', t) - F_2(\mathbf{x}, \mathbf{p}, \mathbf{p}_1, t)), \tag{5.14}$$

where $(\mathrm{d}\sigma/\mathrm{d}\Omega)$ is the collision differential cross-section, depending on the details of the interparticle potential, and the two-particle correlation function $F_2(\mathbf{x}, \mathbf{p}, \mathbf{p}_1, t)$ has been introduced such that

$$F_2(\mathbf{x}, \mathbf{p}, \mathbf{p}_1, t) \; \mathrm{d}\mathbf{x} \, \mathrm{d}\mathbf{p} \, \mathrm{d}\mathbf{p}_1$$

is the number of couples of particles one finds, at time $t$, in a volume $\mathrm{d}\mathbf{x}$ about $\mathbf{x}$, with momenta in the volume elements $\mathrm{d}\mathbf{p}$ and $\mathrm{d}\mathbf{p}_1$ about $\mathbf{p}$ and $\mathbf{p}_1$, respectively. Of course, in Eq. (5.14) $\mathbf{p}'$ and $\mathbf{p}_1'$ are related to $\mathbf{p}$ and $\mathbf{p}_1$ by conservation of energy and momentum. Boltzmann made the crucial assumption of "molecular chaos," according to which the momenta of the particles, interacting around $\mathbf{x}$, are independent, so that one can write

$$F_2(\mathbf{x}, \mathbf{p}, \mathbf{p}_1, t) = f(\mathbf{x}, \mathbf{p}, t) \cdot f(\mathbf{x}, \mathbf{p}_1, t) \; . \tag{5.15}$$

This factorization condition, based on physical common sense, allows us to turn Eq. (5.13) into a closed equation, giving rise to the Boltzmann transport equation for the one-particle distribution function:

$$\left(\frac{\partial}{\partial t}+\frac{\mathbf{p}}{m}\cdot\nabla_{\mathbf{x}}\right) f(\mathbf{x},\mathbf{p},t)=J[f,f]\ , \tag{5.16}$$

where $J[f, f]$ is the quantity in (5.14) with the approximation (5.15). In contrast to Newton's equation (for the microstates), Boltzmann's equation (for a particular macrostate) is not invariant under time reversal. Indeed, if one lets $t \to -t$ and $\mathbf{p} \to -\mathbf{p}$, one does not recover (5.16), but rather the *anti-Boltzmann* equation

$$\left(\frac{\partial}{\partial t}+\frac{\mathbf{p}}{m}\cdot\nabla_{\mathbf{x}}\right) f(\mathbf{x},\mathbf{p},t)=-\,J[f,f]\ , \tag{5.17}$$

with a changed sign in front of the collision term $J$. This non-invariance has a striking manifestation. In the case of the macrostate defined by $f(\mathbf{x}, \mathbf{p}, t)$ the entropy, according to (5.10), can be written as follows:

$$S_{\mathrm{B}}(f)\equiv k_{\mathrm{B}}\ln\|f\|=-k_{\mathrm{B}}\int \mathrm{d}\mathbf{x}\,\mathrm{d}\mathbf{p}\ f(\mathbf{x},\mathbf{p},t)\ln f(\mathbf{x},\mathbf{p},t)\equiv-\mathcal{H}(f)\ . \tag{5.18}$$

The remarkable fact is that if Eq. (5.16) holds (i.e. if condition (5.15) is applicable) then the celebrated Boltzmann $\mathcal{H}$-theorem follows:

$$\frac{\mathrm{d}\mathcal{H}(t)}{\mathrm{d}t}\le 0 \quad \text{or} \quad \frac{\mathrm{d}S_{\mathrm{B}}(t)}{\mathrm{d}t}\ge 0\ , \tag{5.19}$$

where the equality holds if, and only if, the involved $f$ is the Maxwell–Boltzmann distribution, having the property of the equilibrium distribution. The theorem reflects the irreversible (or time-asymmetric) character of the Boltzmann equation; and it is strictly dependent on the assumption of molecular chaos. Indeed the factorization (5.15) introduces (by hand) a substantial difference in the relevance of events before and after a collision, because interactions create correlations between particles. Moreover, because of definition (5.10), this irreversibility shows itself in an increase in the number of macroscopically similar microstates.

### *The objections of Loschmidt and of Zermelo*

When Boltzmann put forward his microscopic interpretation of entropy, equipped with the transport equation and $\mathcal{H}$-theorem, two major criticisms were raised by his opponents, against his explanation of the Second Principle. One is the "reversibility objection" or "Loschmidt paradox," named after the nineteenth century Bohemian scientist who formulated it. The other is the "recurrence objection" or "Zermelo paradox," named after the German mathematician who formulated it.

The essence of the Loschmidt criticism is that Boltzmann's $\mathcal{H}$-theorem cannot be a general theorem of mechanics, valid for all microscopic evolutions. This is due to the fact that if in a system $\mathcal{H}$ becomes smaller, then by reversing the velocities of all atoms an evolution with an increasing $\mathcal{H}$ is obtained, violating the "theorem."

Indeed this is the conflict between the time invariance of Newton's equations and the time-asymmetric thermodynamic behavior, the subject of this chapter. At the end, the aim of this chapter is to show that Loschmidt's assertion, though undoubtedly right, is not relevant in the explanation of macroscopic behaviors, that characterize practically all, although not rigorously, the microstates. In systems with a small number of degrees of freedom the time-symmetric dynamics would certainly show all its effectiveness. However, when the number of degrees of freedom is comparable with Avogadro's number then macroscopic observables can be defined. In this case, in suitable (non-equilibrium) conditions a non-symmetric behavior emerges, with respect to macroscopic properties, leading almost all the involved microstates toward a unique macrostate, without violating the microscopic time symmetry. Recalling the discussion on the quantity $N_{\mathrm{R}}$ for the ideal gas in a box, we can see that, in the present case, an irreversible evolution of $\mathcal{H}$ would derive from the fact that, in the overwhelming majority of microstates, $\mathcal{H}(t)$ decreases in either time direction. So, as for the increase in $N_{\mathrm{R}}$, the $\mathcal{H}$-theorem surely is not valid for all the microstates, nonetheless it rules their typical macroscopic behavior.

The criticism of Zermelo stems from Poincaré's recurrence theorem. The latter states that a mechanical system, evolving in a limited region of the phase space, within a finite time $T_{\mathrm{R}}$, will come again near to its initial condition. So, if the initial condition is a non-equilibrium microstate (i.e. corresponding to a non-equilibrium macrostate), and $\mathcal{H}$ begins decreasing, after a time $T_{\mathrm{R}}$ the $\mathcal{H}$ function will also be back near to its initial value, and this implies a non-monotonic behavior. Also in this case the objection is formally correct, but has no practical consequences. Boltzmann himself noted that $T_{\mathrm{R}}$, for a macroscopic system, is extremely long and, in fact, unobservable. For instance, he estimated that, for a cubic centimeter of gas in normal conditions, the return time to an initial state, with a tolerance of $10^{-9}$ m on atomic positions and 1 m/s on velocities, would be $10^{10^{19}}$ yr (to be compared with the age of the Universe, about $10^{10}$ yr). More generally, an order of magnitude estimate of the recurrence time, appropriate to the relevant macrovariables, is given by $\|M\|_{\mathrm{eq}}/\|M\|_0$, the ratio between the number of final equilibrium states to the number of initial out of equilibrium states. For macrovariables this ratio is exponentially large in $N$: in any usual unit of time it results in a non-human time and can be considered as physically irrelevant. In the mathematical world the return is certain, however in the physical world we are not able to verify it and, surely, the system itself would not keep its properties unaltered over such a long time.

### *The Boltzmann–Grad limit and the Lanford theorem*

Coming back to Eq. (5.16) and its consequence (5.19), it can be shown with all mathematical rigor (Lanford 1981) that in a well-defined limit, the so-called Boltzmann–Grad limit, at least for short times, the Boltzmann equation exactly rules the evolution of a given distribution function for almost all the microstates representing it. It is important to underline that, in this limit, the result of $f(\mathbf{x}, \mathbf{p}, t)$ being driven by Eq. (5.16) follows directly from Newton's equation. The limit can be described as follows. Consider a system consisting of $N$ particles interacting by short-range forces, in the simplest case elastic spheres with diameter $\epsilon$. The limiting regime is obtained by letting $N \to \infty$ and $\epsilon \to 0$, such that $N\epsilon^2$ remains finite: this allows the mean free path to be finite (i.e. there are collisions) while the volume occupied by the molecules vanishes (since $N\epsilon^3 \to 0$). By depressing the granularity of the gas, the Boltzmann–Grad limit is able to reduce the macroscopic fluctuations of a macrovariable such as $f(\mathbf{x}, \mathbf{p})$, while it is still evolving, because of collisions. The theorem ensures that the absence of macroscopic fluctuations in $f(\mathbf{x}, \mathbf{p})$, a property of the initial set of microstates by construction, propagates to later times, with the expected behavior of the macrovariable. The good behavior (5.16) pertains to a fraction of the initial microstates that is as close to 1 as desired, if $N$ is large enough. This means that almost all the microstates $\{\mathbf{X}\}_0$, representing initially $f(\mathbf{x}, \mathbf{p}, 0)$, at time $t$ represent the macrostate $f(\mathbf{x}, \mathbf{p}, t)$, as evolved according to (5.16). However, because of (5.19), $f(\mathbf{x}, \mathbf{p}, t)$ has a class of representative points $\{\mathbf{X}\}_t$ that is much greater than $\{\mathbf{X}\}_0$: (almost all) the set $\{\mathbf{X}\}_0$ becomes, by evolution, a tiny fraction of $\{\mathbf{X}\}_t$, but the future behavior of its points conforms to the typical behavior of the points in the extended set $\{\mathbf{X}\}_t$.

It is to be stressed that the above results do not concern simply an average behavior of the considered system, on the contrary they predict a practically certain macroscopic behavior for it, i.e. (almost) independently of the microscopic state realizing the initial macrostate. This expected feature is clearly linked to the large value of the number $N$ ($\to \infty$) of the constituent subsystems.

All this gives support to the idea that the irreversibility scenario outlined in points (a) and (b) of Section 5.2.3 can also hold for real macroscopic systems, i.e. systems with $N$ very large but finite.

## 5.4 About ensembles, the number of degrees of freedom and chaos

We conclude this chapter with some remarks on the usual ensemble point of view of statistical mechanics, the relevance of the large number of degrees of freedom and the small role of chaos in the irreversibility problem.

### 5.4.1 Boltzmann and Gibbs entropies: their evolutions

If $\rho_0$ describes a situation of initial equilibrium, its Gibbs entropy is defined by:[1]

$$S_{\mathrm{G}}(\rho_0) = -k_{\mathrm{B}} \int \mathrm{d}^{6N}\mathbf{X}\ \rho_0(\mathbf{X}) \ln \rho_0(\mathbf{X}). \tag{5.20}$$

[1] We can consider all the phase space volumes as adimensional quantities, obtained by dividing them by $h^{3N}$ (see Section 5.3.1); e.g. in Eq. (5.20) $\mathrm{d}^{6N}\mathbf{X}$ replaces $\mathrm{d}^{6N}\mathbf{X}/h^{3N}$. In this way the normalized $\rho(\mathbf{X})$ is also an adimensional quantity.

The microstates that are "reasonably probable" with respect to $\rho_0$ (those that are "typical" of the given macroscopic equilibrium) can be found in a phase space region whose (adimensional) volume is $W_{\rho_0}$ (Jaynes 1965); so the equilibrium entropy, according to Boltzmann, is $S_B(\rho_0) = k_B \ln W_{\rho_0}$. $S_G(\rho_0)$ turns out to be the same as $S_B(\rho_0)$ and the two expressions for the entropy agree with the thermodynamic Clausius equilibrium entropy.

After the relaxing of a constraint, the equilibrium is broken. If we follow these microstates (they are contained essentially in a region of volume $W_{\rho_0}$), in almost all cases, the relevant macrovariables change their values according to the rule of maximizing the extension in phase space, (5.11). This means that, with respect to the considered observables, as time goes on, practically all the initial microstates have the same behavior, that, moreover, is common to an ever larger number of microstates (that were not in the initial set). From the point of view we are embracing, this is the sign of the irreversible going toward equilibrium. It is worth remembering that, because of the known results, here we are taking for granted a behavior that nonetheless should be derived, from the appropriate dynamical laws, in each specific case.

If one lets $\rho_0(\mathbf{X})$ evolve, by the Liouville equation, $\rho(\mathbf{X}, t)$ will be an adequate density to predict the behavior of a macrosystem, independently (almost) of its initial microstate and at least concerning the macrostate under investigation. However, after the elapsing time has taken the macrovariables to the new equilibrium values, $\rho(\mathbf{X}, t)$ will be spread over the phase space on a domain that will be, typically, very convoluted. This is because the stretching of the initial distribution into finer and finer filaments develops structures on smaller and smaller scales. Then the evolved density will not look at all like a standard equilibrium ensemble, by which one would describe the new equilibrium state (although they share the same values for the macroscopic observables). Moreover, if one computes the Gibbs entropy of the evolving ensemble

$$S_G(\rho(t)) = -k_B \int d^{6N}\mathbf{X}\; \rho(\mathbf{X}, t) \ln \rho(\mathbf{X}, t)\ , \tag{5.21}$$

then, as is known, it does not change. In fact, $S_G(\rho(t))$ measures the extension in phase space of the set supporting $\rho(t)$ that, because of the Liouville theorem, is constant. The Gibbs entropy of the density evolved by the Liouville equation misses the point, that is the enlargement of the phase space extension of the "common property" associated with the current macrostate. Now it does not seem difficult to remedy this deficiency. One has to coarse grain the phase space: in this way the increase in $S_G(\rho)$ is assured (Tolman 1980) and the distribution itself will appear uniformly smeared out, as the standard equilibrium ensembles. The coarse graining can be considered to be a consequence of the unavoidable inaccuracy in the measurement, and/or as a technical device to arrive at a meaningful definition of entropy.

One also follows this path for the Kolmogorov–Sinai entropy (see Chapter 2): in a deterministic dynamical system, there is no room to ask how many trajectories of a given length start from a given point of phase space. On the contrary, one can ask (to obtain an unknown answer) how many trajectories of a given length start from a given cell of the coarse-grained phase space (eventually, one lets the graining go to zero). In this case, when a sufficiently long time has elapsed the coarse-grained density, typically, has smoothly invaded all the available phase space, where the macrovariable keeps almost a (macroscopically) unique value. For a long time, i.e. at equilibrium, the relevant volumes, and the associated entropies, are again equal and the two descriptions are again equivalent. However, the Gibbs and Boltzmann entropies, in the time-dependent stage of the going to equilibrium process, continue measuring different quantities. The coarse-grained Gibbs entropy, including at each time step all the microstates belonging to the current grain, now keeps track of the "effective" expansion in phase space of the initial set of microstates. It does not (cannot) yet take care of the expansion properties with respect to the relevant macroscopic observables, as the Boltzmann entropy does, by definition, including all the microstates corresponding to the current macrostate. Indeed, in a high-dimensional phase space, the time an initial set of points needs to reach a substantial part of the allowed region of motion (so that $S_G$ stops growing), can be very long, also for a chaotic system. In contrast, the time a macrovariable needs to enter its region of stability (so that $S_B$ stops growing) can be very short. Anyhow, there is no evident reason for the two times to be equal. The straightforward way of implementing the time evolution in the Gibbs ensemble approach does not agree with the Boltzmann point of view.

It seems one can reasonably think that, for macroscopic systems, and at least in the non-transient regime, the Boltzmann (single system) and Gibbs (ensemble) descriptions agree. Moreover, the Boltzmann point of view provides a justification of the Gibbs ensemble description.

### 5.4.2 About $N$

It is worth repeating that all the foregoing discussion concerns macroscopic bodies, composed of a huge number $N$ of elementary constituents. The rigorous results on irreversible behavior of single systems pertain to the limit $N \to \infty$ and are thought to be meaningful also for $N$ large but finite.

An interesting question one can put at last is what about systems with small $N$ values. As noted at the beginning, we do not expect to see any kind of irreversibility in a single system with few degrees of freedom, e.g. a two billiard ball system. On the contrary, we see "irreversible behavior" with 16 billiard balls, which is not so large a number!

As we tried to illustrate, irreversibility of thermodynamic systems is tightly linked to the properties of their high-dimensional phase spaces and, in particular, to the very peculiar partitioning induced by classification of microstates according to macrovariables: the domination of one class allows both the existence of equilibrium values and their description by (equilibrium) ensembles. A set of states, representing given macroscopic variables at $t = 0$, can be described consistently by an initial (equilibrium) distribution $\rho(0)$, where the macrovariables have no macroscopic fluctuations. Since $N \gg 1$, this can be done in a non-unique way, i.e. different (but equivalent) equilibrium ensembles exist. After a constraint has been relaxed, eventually the set is dispersed over the new accessible region of phase space where, however, the macrovariables attain again almost constant new values, so that a final (equilibrium) distribution can be introduced. The system has irreversibly moved toward the new equilibrium state. The initial and final equilibrium ensembles provide good descriptions of the properties of almost all the single macroscopic systems, with respect to the concerned macrovariables.

If $N$ is small, then the classification by observables, in general, is no longer able to generate dominating classes in phase space. This means that, in general, the (non-conserved) observables of one system, which is wandering about its phase space, will undergo large relative fluctuations around a mean value, and there will no longer be a notion of equilibrium, and the consequent irreversibility, as for a macrovariable of a macroscopic system.

In this regard the calculations performed by Hobson and Loomis (1968) are very instructive. A system consisting of $N$ identical non-interacting point particles, with mass $m$, in a rectangular box is considered. The initial set of microstates is selected by assigning values for the center of mass, the total energy and the total momentum of the system. These values are considered as the averages of the observables with respect to an initial distribution. One way to select an appropriate generalized Gibbs density $\rho(\mathbf{X}, 0)$ is via the "maximum entropy principle" (Jaynes 1963). Given the Hamiltonian for this Knudsen gas:

$$H = \sum_{i=1}^{N} \frac{\mathbf{p}_i^2}{2m} + V(\mathbf{x}_i),$$

where the potential energy $V(\mathbf{x})$ takes into account the rigid walls of the box, i.e. it is zero when $\mathbf{x}$ is inside the box and infinite for $\mathbf{x}$ outside; then the Liouville equation is solved exactly to obtain $\rho(\mathbf{X}, t)$. Even if, as we stressed above, the transient behavior of $\rho$ is not able to give a completely correct description of the physical process, nonetheless its long time limit is expected to provide, somehow, a meaningful description of the new equilibrium. From $\rho(\mathbf{X}, t)$ one can derive expressions for various reduced distributions and moments. One obtains the following. The full exact distribution itself does not have a time-independent limit: $\rho(\mathbf{X}, t)$ undergoes oscillatory behavior at any fixed point $\mathbf{X}$, i.e. it does not relax to equilibrium. Indeed, we know that some kind of coarse graining is needed for this to happen, by overcoming the Liouville theorem. For instance, the reduced distribution for the configuration space, $\rho(\mathbf{x}_1, \mathbf{x}_2, \ldots, \mathbf{x}_N, t)$, obtained by integrating away the impulses, relaxes, as it must do, to the uniform distribution inside the box. However, the reduced distribution for the momenta, $\rho(\mathbf{p}_1, \mathbf{p}_2, \ldots, \mathbf{p}_N, t)$, in this case cannot relax, since the modulus of the momentum of every particle is conserved in the elastic bouncing from the walls. Another way to extract coarse information from the ensemble is by considering averages of phase space functions. One has:

$$\langle x^\alpha p^\gamma \rangle_t = \langle x^\alpha p^\gamma \rangle_{\text{eq}} + F_{\alpha\gamma}(t)$$

where $x$ and $p$ are any two out of the $6N$ phase space variables, $F_{\alpha\gamma}(t)$ is a function that, for $\alpha \neq 0$, decays to zero exponentially in $t$ and the time-independent part of the expected value,

$$\langle x^\alpha p^\gamma \rangle_{\text{eq}} = \int x^\alpha p^\gamma P_{\text{eq}}(x, p)\, \mathrm{d}x\, \mathrm{d}p,$$

is given by an "equilibrium" distribution, $P_{\text{eq}}$, satisfying the time-independent Liouville equation. In the general case, $P_{\text{eq}}$ differs from the canonical distribution:

$$P_{\text{eq}}(x, p) = \frac{\sqrt{\beta/2\pi m}}{2L}\left[\exp[-\beta(p - p_0)^2/2m] + \exp[-\beta(p + p_0)^2/2m]\right],$$

where $\beta = 1/k_{\text{B}}T$, $L$ is the length of the container in the (real space) direction of the coordinate $x$ and $p_0$ is the initial mean momentum per particle in the (momentum space) direction of the coordinate $p$. But this is because, in this system, any even power of the single-particle momentum is conserved. However, if $p_0 = 0$ then $P_{\text{eq}}$ is just the canonical distribution.

These results imply that, although the full $N$-particle distribution does not have a long time limit, the expected value of any analytic phase function does have a time-independent limit, and the system is said to relax "weakly" to equilibrium. It is to be stressed that in the discussion above, $N$ is an arbitrary parameter: it can have a small as well as a large value. In either case the weak relaxation exists, as a *property of the average quantities* that does not imply, by itself, irreversible behavior of a single system. The latter feature, called predictability by Hobson and Loomis (1968), is only present in sum type observables when $N$ is very large, as a consequence of suppression of the relative fluctuations, rendering the observables macroscopically well defined. The former, namely the "statistical relaxation" is a general property of an ensemble that is not automatically related to the single system relaxation.

### 5.4.3 On chaos and irreversibility

Let us now discuss why chaos in dynamical systems, although very important from many points of view, has a rather marginal role in the irreversibility problem. Since, it seems to us, a certain confusion exists concerning this point, we believe it is important to try to clarify the situation by considering the problem from both technical and conceptual points of view.

We saw in Chapter 1 that in mixing systems one has a relaxation to the invariant measure, i.e. independently of the initial density distribution $\rho(\mathbf{x}, 0)$, for large $t$ one has

$$\rho(\mathbf{x}, t) \to \rho^{\mathrm{inv}}(\mathbf{x}). \tag{5.22}$$

The above property at first glance appears analogous to the irreversibility,[2] but it is not so.

(a) The relaxation to the invariant measure is a property of the "ensemble of the initial conditions," and it just reflects the fact that different points in the support of $\rho(\mathbf{x}, 0)$, even if they are very close to each other at $t = 0$, will be separated as a consequence of chaos. However this kind of "irreversible" behavior is rather different from the irreversibility of real life, i.e. concerning a unique (large) system.
(b) The property (5.22) refers to a generic dynamical system, even a low-dimensional one. In such systems, there is no sense of a micro/macro distinction, since it is not possible to define macroscopic variables.

We know that there is not complete consensus on this. For example, Driebe (1994) in a criticism of Lebowitz (1993b) writes that irreversible processes are well observed in systems with few degrees of freedom, such as the baker or multibaker transformation. Let us present a simple argument against this point of view. Consider one particle of the (non-interacting) Lorentz gas or, equivalently, any chaotic low-dimensional symplectic system (e.g. the baker or the standard map). Of course the particle, as a consequence of chaos, shows irregular behavior. At a certain time we invert its momentum and look. We do not observe anything surprising. It is easy to realize that this is not in disagreement with the "irreversibility" in (5.22) which is a property of the ensemble of the initial conditions.

The point is that, as already previously discussed in the example of the billiard balls in Section 5.2.1, the irreversible phenomena deal somehow with macroscopic variables. If one looks at a single molecule, or at a small number of molecules, there is no sense in which one can speak of irreversibility which is a "global" property of a large system. Only looking at time-reversed behavior of a large number of particles do we observe something "strange."

[2] Some authors claim this explicitly. For a detailed discussion of this controversy see Bricmont (1995).

Let us now discuss again the two objections of Zermelo and Loschmidt. The main reason, as already realized by Boltzmann, for the non-relevance of the Poincaré recurrence theorem in macroscopic systems, is that the recurrence time increases exponentially with $N$. This is true in both chaotic and regular systems. In order to understand this point it is enough to consider a periodic system with periods $T_1, \ldots, T_N$; even if $T_k/T_j$ are rational numbers the recurrence time $\mathcal{T}$ increases exponentially with $N$.

Consider now the Loschmidt paradox regarding the consequences of velocity inversion for the time evolution of a system in computer simulation, e.g. a two-dimensional diluted gas of 100 hard disks in a periodic box (Orban and Bellemans 1967). One defines the quantity $\mathcal{H}$, according to Eq. (5.18), via the function $f(\mathbf{x}, \mathbf{p}, t)$ that is obtained by making histograms of the positions and velocities of the disks. Starting from a particular non-equilibrium configuration, one follows the time behavior of $\mathcal{H}$ in direct evolution and in evolutions with velocity inversion at certain times. If very small errors are introduced into the initial data after the velocity inversion, in a chaotic system one does not observe the exact antikinetic behavior. If the errors increase, the antikinetic behavior is suppressed more or less completely, as shown in Figure 5.1.

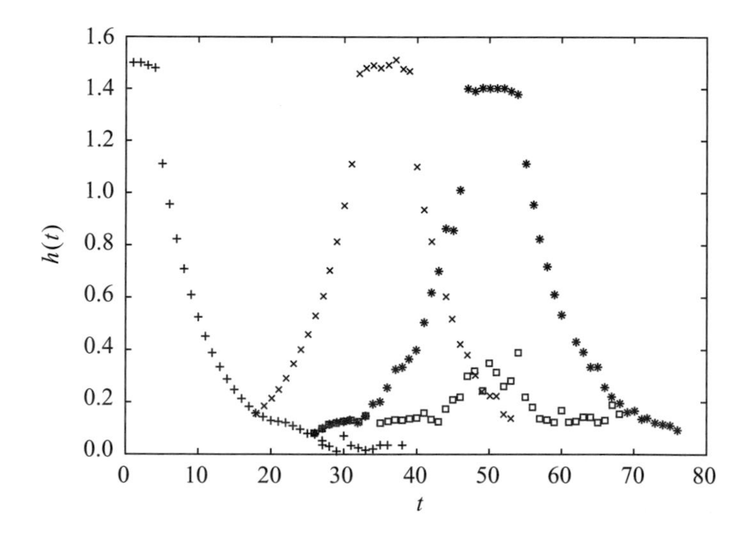


Figure 5.1 The effect of velocity inversion on $h(t) = \mathcal{H}(t) - \mathcal{H}_{\min}$. The pluses correspond to direct evolution; the crosses, the stars and the squares are obtained by reversing the velocities after 18 collisions with very small errors, after 26 collisions with small errors and after 26 collisions with (relatively) large errors, respectively. This picture shows in a schematic way the computer experiment of Orban and Bellemans (1967).

Such a phenomenon is interesting in itself, for example in the context of quantum chaos and quantum information (Veble and Prosen 2004, 2005). If $N$ is not large the deterministic chaos is, of course, important: the difference between the "actual" initial condition after the velocity inversion and the "true" one is quickly amplified both by the chaos and the rounding-off errors of the computers. Therefore the antikinetic behavior is easily reduced (or erased). On the other hand, if $N \gg 1$ the solution of the paradox does not involve chaos. In the velocity reversion operation one unavoidably introduces some errors. According to Lanford (1981) for the "actual" reversed initial condition (which is different from the "true" one), with probability close to one, molecular chaos holds and therefore $\mathcal{H}(t)$ must decrease.

### *5.4.4 Some general conclusions*

It seems to us that Boltzmann was basically able to understand the essence of the mechanism of the second law. The process of clarification, which started with Boltzmann himself, has been rather slow and culminated with the work of Grad (1949) and the precise formulation of Lanford (1981), and other mathematicians, see Cercignani *et al.* (1994). The technical "tricks" are essentially the following:

(a) assume $N \to \infty$ and rescale the size of the molecules in a proper way (i.e. the Grad–Boltzmann limit);
(b) select "good" initial conditions in such a way that the molecular chaos hypothesis is satisfied, and prove that in the limit $N \to \infty$ such initial conditions have measure approaching to one.

With the above assumptions one can eliminate the fluctuations in the time behavior of $\mathcal{H}(t)$ versus $t$ and therefore the objections of Loschmidt and Zermelo are overcome. It is thus very surprising that there still exists a large literature, including books aimed at a wide audience (Prigogine and Stengers 1979), which claims that the explanation *à la* Boltzmann of the irreversibility needs to be revised according to modern approaches (mainly using arguments from deterministic chaos).[3] We recall a famous Latin adage:

Contra facta non valent argumenta
(It is not possible to contrast facts with words).

[3] For example Hoover (1999) claims that "Our exploration of time reversibility from the perspective of computer simulation and chaos has provided us with insights into the breaking of time symmetry which were not available to Bolzmann or Gibbs ... Simulations have clarified the formation and significance of time-reversible ergodic multifractal phase-space structures."

## References


Berger, T. (1971). *Rate Distortion Theory*, Englewood Cliffs, NJ: Prentice-Hall.

Bricmont, J. (1995). Science of chaos or chaos in science?, *Physicalia Mag.* **17**, 159.

Bunimovich, L. and Sinai, Y. (1981a). Markov partitions for dispersed billiards, *Commun. Math. Phys.* **78**, 247.

Bunimovich, L. and Sinai, Y. (1981b). Statistical properties of Lorentz gas with periodic configuration of scatterers, *Commun. Math. Phys.* **78**, 479.

Cercignani, C., Illner, R. and Pulvirenti, M. (1994). *The Mathematical Theory of Dilute Gases*, Berlin: Springer.

Cover, T. M. and Thomas, J. A. (1991). *Elements of Information Theory*, New York: Wiley.

Cronin, J. W. (1981). CP symmetry violation – the search for its origin, *Rev. Mod. Phys.* **53**, 373.

Driebe, D. J. (1994). Is Boltzmann entropy time's arrow's archer? *Phys. Today*, November 1994, p. 13.

Grad, H. (1949). On the kinetic theory of rarified gases, *Commun Pure Appl. Math.* **2**, 325.

Hahn, E. L. (1950). Spin echoes, *Phys. Rev.* **80**, 580.

Hobson, A. and Loomis, D. N. (1968). Exact classical nonequilibrium statistical-mechanical analysis of the finite ideal gas, *Phys. Rev.* **173**, 285.

Hoover, W. G. (1999). *Time Reversibility, Computer Simulation, and Chaos*, Singapore: World Scientific.

Huang, K. (1987). *Statistical Mechanics*, New York: Wiley.

Hurley, J. (1980). Resolution of the time-asymmetry paradox, *Phys. Rev. A* **22**, 1205.

Hurley, J. (1981). Time-asymmetry paradox, *Phys. Rev. A* **23**, 268.

Jaynes, E. T. (1963). *Lectures in Theoretical Physics, Brandeis, 1962, Statistical Physics*, Vol. 3, p. 181, New York: Benjamin.

Jaynes, E. T. (1965). Gibbs vs Boltzmann entropies, *Am. J. Phys.* **33**, 391.

Khinchin, A. I. (1949). *Mathematical Foundations of Statistical Mechanics*, New York: Dover.

Kolmogorov, A. N. (1956). On the Shannon theory of information transmission in the case of continuous signals, *IRE Trans. Inf. Theory* **1**, 102.

Lanford III, O. E. (1981). The hard sphere gas in the Boltzmann–Grad limit, *Physica A* **106**, 70.

Lebowitz, J. L. (1993a). Macroscopic laws, microscopic dynamics, time's arrow and Boltzmann's entropy, *Physica A* **194**, 1.

Lebowitz, J. L. (1993b). Boltzmann's entropy and time's arrow, *Physics Today*, **46** September 1993, p. 32.

Lebowitz, J. L. and Spohn, H. (1982a). Microscopic basis for Fick's law for self-diffusion, *J. Stat. Phys.* **28**, 539.

Lebowitz, J. L. and Spohn, H. (1982b). Steady state self-diffusion at low density, *J. Stat. Phys.* **29**, 39.

Ma, S.-K. (1985). *Statistical Mechanics*, Singapore: World Scientific.

Mazur, P. and Montroll, E. (1960). Poincaré cycles, ergodicity, and irreversibility in assemblies of coupled harmonic oscillators *J. Math. Phys.* **1**, 70.

Mazur, P. and van der Linden, J. (1963). Asymptotic form of the structure function for real systems, *J. Math. Phys.* **4**, 271.

Mehra, J. (2001). *The Golden Age of Theoretical Physics*, Singapore: World Scientific.

Orban, J. and Bellemans, A. (1967). Velocity-inversion and irreversibility in a dilute gas of hard disks, *Phys. Lett. A* **24**, 620.

Prigogine, I. (1999). Laws of nature, probability and time symmetry breaking, *Physica A* **263**, 528.

Prigogine, I. and Stengers, I. (1979). *Entre le Temps et l'Éternité*, Paris: Fayard.

Rhim, W.-K., Pines, A. and Waugh, J. S. (1971). Time-reversal experiments in dipolar-coupled spin systems, *Phys. Rev. B* **3**, 684.

Shannon, C. E. (1948). A mathematical theory of communication, *Bell Syst. Tech. J.* **27**, 379; **27**, 623. This work can be downloaded from: http://cm.bell-labs.com/cm/ms/what/shannonday/paper.html.

Tolman, R. C. (1980). *The Principles of Statistical Mechanics*, New York: Dover.

Veble, G. and Prosen, T. (2004). Faster than Lyapunov decays of the classical Loschmidt echo, *Phys. Rev. Lett.* **92**, 034101.

Veble, G. and Prosen, T. (2005). Classical Loschmidt echo in chaotic many-body systems, *Phys. Rev. E* **72**, 025202.

# 6

# The role of chaos in non-equilibrium statistical mechanics

> I am conscious of being only an individual struggling weakly against the stream of time. But still remains in my power to contribute ...
>
> *Ludwig Boltzmann*

In Chapter 4 we discussed the connection between chaos and equilibrium statistical mechanics, in particular with respect to the ergodic hypothesis. We saw that in systems with many degrees of freedom, chaos (in the sense of at least one positive Lyapunov exponent) is not strictly necessary (nor sufficient) to obtain good agreement between the time average and phase average. This is due, as Boltzmann himself thought and Khinchin proved for an ideal gas, to the fact that in systems with many components, for a large class of observables, the validity of the ergodic hypothesis is basically a consequence of the law of large numbers, and it has a rather weak connection with the underlying dynamics.

From a conceptual point of view the ergodic approach (possibly in a "weak" variant, only pertaining to some interesting macroscopic variables) can be seen as a natural way to introduce probabilistic concepts in a deterministic context. In addition, since one deals with a unique system (although with many degrees of freedom) the ergodicity is a possible (unique?) way to found the equilibrium statistical mechanics on a physical ground, i.e. by exploiting the frequentistic interpretation of probability to extract a statistical description from the analysis of a single experimental trajectory. Finally, on the basis of the results in Chapter 4, and not forgetting that thermodynamics, as a physical theory, was developed to describe the properties of *single* systems made of many microscopic interacting parts, it seems to us that it is quite fair to conclude that the ensemble viewpoint is just a useful mathematical tool.

On the other hand the situation is much more intricate for the non-equilibrium[1] problems than for equilibrium problems, and even after many years of debate, there is no general agreement about the fundamental ingredients needed for the validity of statistical laws. The discovery of chaos, implying that even deterministic systems with a few degrees of freedom may present some statistical features typical of probabilistic evolutions, forced physicists to reconsider the foundations of statistical mechanics from a new perspective (Bricmont 1995). These issues are the source of heated debates, of which an example is the debate on irreversibility originated by Lebowitz's paper (Lebowitz 1993), and the partly contrasting views on irreversible entropy production reported in Gaspard (1998), Tél and Vollmer (2000), and Rondoni and Cohen (2000).

The aim of this chapter is to discuss the role of deterministic chaos, of coarse-graining procedures and of the many degrees of freedom in non-equilibrium statistical mechanics. The following issues will be treated.

(a) *The connection between chaos and entropy production rate.* If one adopts the Gibbs point of view for the ensembles, this problem can also be treated in systems with few degrees of freedom. We will see that in general there is just a weak (if any) connection between the Kolmogorov–Sinai entropy and the production rate of the coarse-grained Gibbs entropy.
(b) *The relevance of chaos and of the coarse-graining procedure for the time evolution of the Boltzmann entropy $S_B(t)$ of a large assembly of weakly interacting systems.* As a consequence of the interaction, a characteristic scale emerges, so that the behavior of $S_B(t)$ versus $t$, at variance with that of the Gibbs entropy, does not depend on the coarse-graining resolution, as long as it is finer than this dynamical scale. Therefore the growth of $S_B(t)$ is an intrinsic property of the system.
(c) *The linear response theory and the dissipation-fluctuation theorem.* It is possible to see that, in any system with good (but quite weak and general) chaotic properties, i.e. with a mixing invariant measure, a fluctuation-response relation holds which links the response to a finite perturbation with a suitable correlation function computed for the unperturbed system. This relation is rather general and also holds in non-equilibrium systems under the assumption that a stationary invariant measure exists.
(d) *The role of chaos for the validity of statistical mechanics laws*, e.g. *diffusion and conduction.* We will see that the main ingredient is not chaos in the technical sense (i.e. a positive Lyapunov exponent); the presence of pseudochaos (i.e. $\lambda_1 = 0$) and some "large-scale instability" is enough to give a proper diffusive (or conductive) feature.

[1] By the term "non-equilibrium" we indicate (a) cases where, at $t = 0$, the probability density function is not the microcanonical (or canonical) one or, if one is dealing with a unique large system, the initial condition is not "typical" (in the sense of the results of Khinchin, see Section 4.2), (b) systems whose invariant measure is not the microcanonical (or canonical) distribution.

## 6.1 On the connection between the Kolmogorov–Sinai entropy and production rate of the coarse-grained Gibbs entropy

In statistical mechanics, the physical entropy of a (macroscopic) system in equilibrium is identified with the Gibbs entropy, that is the quantity[2]

$$S(\rho_{\text{eq}}) = -\int \rho_{\text{eq}}(\mathbf{x}) \ln[\rho_{\text{eq}}(\mathbf{x})] \mathrm{d}\mathbf{x}, \tag{6.1}$$

where $\rho_{\text{eq}}$ is the appropriate equilibrium density. In this section and in the following one, for convenience, we set Boltzmann's constant equal to 1. Almost immediately, this formulation of equilibrium entropy leads to the idea of defining an entropy for non-equilibrium situations, by simply extending definition (6.1) to the cases of a time-dependent density. In such a way one obtains a time-dependent entropy functional, $S(\rho_t)$, that, in the long time, hopefully should reduce to $S(\rho_{\text{eq}})$. However, from a general point of view, setting aside the question of its physical meaning and relation with the Second Principle, $S(\rho_t)$ may be considered merely as a well-defined quantity of a given dynamical system whose evolution properties deserve investigation. From such an aseptic perspective, the system involved does not need to be either a Hamiltonian system or a high-dimensional system. The behavior of $\mathrm{d}S(\rho_t)/\mathrm{d}t$ is among the interesting problems about $S(\rho_t)$. On the basis of reasonable arguments, see the following, and numerical computations (Latora and Baranger 1999), a simple relation has been suggested between $r_{\text{G}}$, the variation rate of a suitably average coarse-grained Gibbs entropy, and $h_{\text{KS}}$, the Kolmogorov–Sinai entropy of the dynamical system: $r_{\text{G}} = h_{\text{KS}}$, at least in a certain range of time. However, the situation is not so clear and, as we are going to show in the following, such a relationship between $r_{\text{G}}$ and $h_{\text{KS}}$ does not hold in the generic case.

### *6.1.1 A brief overview of basic facts about the Gibbs entropy*

Consider a deterministic dynamical law

$$\mathbf{x} \to T^t \mathbf{x} \tag{6.2}$$

(where $\mathbf{x}$ is a $D$-dimensional vector) and a probability density $\rho(\mathbf{x}, t)$ for the state of the system throughout its phase space, at time $t$. The time-dependent Gibbs entropy of the system is defined as follows

$$S(\rho_t) = -\int \rho(\mathbf{x}, t) \ln[\rho(\mathbf{x}, t)] \mathrm{d}\mathbf{x}. \tag{6.3}$$

[2] As noted in Chapter 5, all volumes of the phase space may be considered as adimensional quantities, when they are related to an elementary volume $h^{3N}$. In this way a normalized $\rho(\mathbf{x})$ can also be considered as an adimensional quantity.

For chaotic dissipative systems, where $\rho(\mathbf{x}, t)$ tends to a singular (fractal) measure, definition (6.3) becomes meaningless. Nevertheless, following a nice idea of Kolmogorov and Ruelle (1989), one can avoid this difficulty by adding (or considering unavoidably present) a small noise term in the evolution law. In such a way one obtains a $\rho(\mathbf{x}, t)$ continuous with respect to the Lebesgue measure. Denoting by $J(\mathbf{x}, t)$ the Jacobian of transformation (6.2), a straightforward computation gives

$$S(\rho_t) = S(\rho_0) + \int \rho(\mathbf{x}, t) \ln |J(\mathbf{x}, t)| \mathrm{d}\mathbf{x}. \tag{6.4}$$

In the case of volume-conserving evolutions, where $|J(\mathbf{x}, t)| = 1$, one has $S(\rho_t) = S(\rho_0)$; therefore in order to allow entropy variation a coarse graining is necessary (Tolman 1980). Let us consider a hypercubic partition, with cells of linear size $\Delta$, and let us define the probability $P^{\Delta}(i, t)$ of finding the state of the system in the cell $i$ at time $t$:

$$P^{\Delta}(i, t) = \int_{\Lambda_i^{\Delta}} \rho(\mathbf{x}, t) \mathrm{d}\mathbf{x} \tag{6.5}$$

where $\Lambda_i^{\Delta}$ is the region singled out by the $i$th cell. We can now introduce the $\Delta$-grained Gibbs entropy

$$S^{\Delta}(P_t) = -\sum_i P^{\Delta}(i, t) \ln P^{\Delta}(i, t). \tag{6.6}$$

If one considers a distribution of initial conditions different from zero over a small number of cells, and $\Delta$ small enough (with respect to the typical linear size of the phase space region where the motion evolves), then one can argue qualitatively, as follows, about the behavior of $S^{\Delta}(P_t)$. Assume that $\rho(\mathbf{x}, 0)$ is localized around $\mathbf{x}^{\mathrm{c}}(0)$ and that it has a Gaussian shape. For the sake of simplicity also assume that the variance in any direction is $\sigma_0^2$. In a suitable reference system (with the axes along the eigendirections of the Lyapunov exponents), for some times $\rho(\mathbf{x}, t)$ is well approximated by a product of $D$ one-dimensional Gaussian distributions with variances

$$\sigma_j^2(t) = \sigma_0^2 \mathrm{e}^{2\lambda_j t}, \tag{6.7}$$

that is:

$$\rho(\mathbf{x}, t) \simeq \prod_{j=1}^{D} \frac{1}{\sqrt{2\pi\sigma_j^2(t)}} \exp\left\{ -\frac{(x_j - x_j^{\mathrm{c}}(t))^2}{2\sigma_j^2(t)} \right\} \tag{6.8}$$

where $\mathbf{x}^c(t)$ is the state evolved from $\mathbf{x}^c(0)$. Therefore in the non-grained case one has

$$S(\rho_t) \simeq S(\rho_0) + \sum_j \ln \frac{\sigma_j(t)}{\sigma_0} = S(\rho_0) + \sum_{j=1}^{D} \lambda_j t. \tag{6.9}$$

If the phase space volume is conserved by the dynamics, then $\sum_{j=1}^{D} \lambda_j = 0$, and so $S(\rho_t) = S(\rho_0)$. In the coarse-grained case, supposing that the system possesses $m$ positive Lyapunov exponents, for a long enough $t$, one has:

$$\sigma_k(t) \approx \sigma_0 \mathrm{e}^{-|\lambda_k| t} < \Delta, \tag{6.10}$$

for $k > m$, i.e. along the $D - m$ directions of the negative Lyapunov exponents. The previous inequality is satisfied for all $k$ when $t \gtrsim (1/\ell) \ln(\sigma_0/\Delta)$, where $\ell$ is the smaller among the absolute values of the negative Lyapunov exponents. In the qualitative argument, the differences among Lyapunov exponents can be disregarded, and as the characterizing time one can consider

$$t_\lambda = \frac{1}{\lambda_1} \ln \frac{\sigma_0}{\Delta},$$

where $\lambda_1$ is the maximal exponent. Thus for $t > t_\lambda$

$$\rho(\mathbf{x}, t) \simeq \prod_{j=1}^{m} \frac{1}{\sqrt{2\pi \sigma_j^2(t)}} \exp\left\{ - \frac{(x_j^{(i)} - x_j^c(t))^2}{2\sigma_j^2(t)} \right\}$$

and

$$S^\Delta(P_t) \simeq S^\Delta(P_0) + \sum_{j=1}^{m} \lambda_j t. \tag{6.11}$$

Therefore using Pesin's formula one can write

$$S^\Delta(P_t) \simeq S^\Delta(P_0) + h_{\mathrm{KS}} t. \tag{6.12}$$

However, at this point, one observes that this linear regime cannot last indefinitely. It must change when the considered states, that were initially well concentrated in a small region of the phase space, begin to be spread all over the accessible region and the entropy begins to saturate, that is, when the width of $\rho(\mathbf{x}, t)$ becomes of the same order as the system size. According to Eq. (6.7), this happens when the time is greater than

$$t_{\mathrm{sat}} = \frac{1}{\lambda_1} \ln \frac{L}{\sigma_0},$$

where $L$ is the typical linear size of the system. Of course, the phase space is filled faster along the more expanding directions.

Summing up, we see that in a conservative system a non-constant entropy is allowed by the fact that, in the presence of a coarse graining, contraction below the coarse-graining resolution $\Delta$ is no longer "efficient," i.e. the contracting eigendirections (corresponding to the negative values of the Lyapunov exponents) cannot balance the effects of the expanding eigendirections.

One can roughly identify three different stages of the evolution. In the first stage no entropy variation is present. This lasts until a time $t_\lambda$, when the scales along the contracting directions have been shrunk down to the graining size, and the second stage begins. This is the linear regime of changing entropy, ending when the saturation effects begin, around a time $t_{\text{sat}}$.

Finally, a behavior like (6.12) may be expected to show up only during the time interval

$$t_\lambda = \frac{1}{\lambda_1} \ln \frac{\sigma_0}{\Delta} \lesssim t \lesssim \frac{1}{\lambda_1} \ln \frac{L}{\sigma_0} = t_{\text{sat}}. \tag{6.13}$$

### 6.1.2 Remarks and numerical results

We believe that in general there is no room for relation (6.12) to hold, and that, on the contrary, its realization is just a "lucky" coincidence (Falcioni *et al.* 2005). With regard to this, we begin by stressing the following points.

(a) The usual physical justification of a coarse graining is the limited accuracy available to the experimenter. From such a perspective, the graining must be realized by cells of strictly finite size. Since we want to study the rate of change of $S^\Delta(P_t)$, in order to speed up the appearance of the interesting time behavior, the numerical computations described below are performed with initial distributions concentrated in single cells, i.e. $\sigma_0 = \Delta$. In this way, according to (6.13), the useful time interval is extended to

$$t \lesssim t_{\text{sat}} = \frac{1}{\lambda_1} \ln \frac{L}{\Delta}, \tag{6.14}$$

because the first time stage, $t < t_\lambda$, during which there is no entropy variation, is completely suppressed.

(b) The Kolmogorov invariant $h_{\text{KS}}$ is a global characterizing property of a dynamical system. In contrast, by definition, the Gibbs entropy (6.3) depends explicitly on the particular chosen initial density. In the discussion above this dependence may be labeled by a point, $\mathbf{x}^c$, around which the distribution is different from zero at the starting time. A generic system possesses a certain degree of intermittency, so that, for instance, the expanding and contracting properties may depend strongly on the phase space region the trajectory is visiting. So, one expects that density-independent behavior, as in (6.12), can be found only in "friendly" dynamical systems, i.e. systems with no fluctuations.

In all the other cases, the remark we made calls for averaging over the initial condition $\mathbf{x}^c$, weighted, say, by the natural invariant measure of the system:

$$S(\rho_t) \to S(t) = \int S(t|\mathbf{x}^c)\rho_{eq}(\mathbf{x}^c)d\mathbf{x}^c, \tag{6.15}$$

where $S(t|\mathbf{x}^c)$ is $S(\rho_t)$ with $\rho_0(\mathbf{x})$ localized around $\mathbf{x}^c$. The same averaging procedure leads to $S^\Delta(t)$ from $S^\Delta(P_t)$. This operation yields intrinsic quantities, and can produce a dependence of $dS^\Delta(t)/dt$ on the Kolmogorov–Sinai entropy of the system as simple as

$$\frac{dS^\Delta(t)}{dt} \simeq h_{KS}t. \tag{6.16}$$

(c) However, both the Kolmogorov–Sinai entropy and Lyapunov exponents are quantities defined in the limits of long time ($t \to \infty$) and high resolution ($\Delta \to 0$). In the present situation, where a coarse-grained Gibbs entropy is involved and a finite $\Delta$ is mandatory, condition (6.14) makes it clear that the two limits are both out of reach, and so may be a growing rate like (6.16).

Let us show by numerical evidence that, notwithstanding the astute averaging discussed in point (b), a certain degree of intermittency combined with a finite resolution prevent Eq. (6.16) from holding.

We study the following simple one-dimensional system, which is a modified tent map:

$$x_{t+1} = \begin{cases} x_t/p & \text{if} \quad 0 \le x_t \le p \\ (1-x_t)/(1-p) & \text{if} \quad p \le x_t \le 1. \end{cases} \tag{6.17}$$

This map reduces to the usual (symmetric) tent map if $p = 1/2$, and for $p \neq 1/2$ it has an intermittent behavior characterized by two different local Lyapunov exponents, namely:

$$\lambda_+ = -\log p, \quad \lambda_- = -\log(1-p). \tag{6.18}$$

The stationary distribution is constant between 0 and 1 and consequently we have

$$h_{KS} = \lambda_1 = p\lambda_+ + (1-p)\lambda_- = -p\log p - (1-p)\log(1-p), \tag{6.19}$$

assuming its maximum value for $p = 1/2$.

The intermittent behavior of this system is rather weak and it is not as critical as in strongly intermittent maps, for example Manneville's map (see the following), and the distribution of the time-length of the "laminar" zones (the permanence in the interval $[p, 1]$) is simply exponential without a power-law tail. Nevertheless one can observe that in this system Eq. (6.12) already does not hold. We recall that, as explained in point (a), we consider ensembles that, at the beginning, are concentrated in a single cell, so that the evolution is already inside the changing stage. The numerical analysis is performed as follows:

(1) we select several starting conditions $x_1^j(0), x_2^j(0), \ldots, x_N^j(0)$ all located in the $j$th interval of size $\Delta$, $\Lambda_j^\Delta$;
(2) we let evolve all the $N \gg 1$ starting conditions up to a time $t$, obtaining $x_1^j(t), x_2^j(t), \ldots, x_N^j(t)$;
(3) we calculate

$$P^{\Delta,j}(k,t) = \frac{1}{N}\sum_{i-1}^{N} \delta(x_i^j(t), k)$$

with

$$\delta(x,k) = \begin{cases} 1 & \text{if } x \in \Lambda_k^\Delta \\ 0 & \text{otherwise;} \end{cases}$$

(4) we compute the entropy of $P^{\Delta,j}(k,t)$, according to (6.6)

$$S^\Delta(j,t) - -\sum_k P^{\Delta,j}(k,t)\log P^{\Delta,j}(k,t)$$

(we recall that $j$ refers to the interval where the initial conditions are chosen);
(5) we average this quantity on the coarse-grained invariant distribution $P_{\text{eq}}(j)$, obtained from $\rho_{\text{eq}}(x)$ using the procedure of Eq. (6.5); recall that, in this case, $\rho_{\text{eq}}(x) = 1$, so we have

$$S^\Delta(t) = \sum_j P_{\text{eq}}(j) S^\Delta(j,t). \tag{6.20}$$

The results are shown in Figure 6.1. As one can see, the agreement with Eq. (6.12) is good only for $p = 1/2$ (the case with no intermittency) and becomes worse and worse for decreasing $p$. The main point is that, according to the heuristic arguments leading to Eq. (6.13), the linear behavior of $S^\Delta(t)$ should hold, until a time about

$$t_{\text{sat}} = \frac{1}{\lambda_1}\log\left(\frac{1}{\Delta}\right) \tag{6.21}$$

corresponding, on the scaled time of Figure 6.1, to the value $-\log\Delta \simeq 7$. This is observed for $p = 1/2$, but for small values of $p$, $S^\Delta(t)$ increases in time with a "wrong" slope (i.e. different from $h_{\text{KS}}$) and later it exhibits a rather long crossover. The origin of this effect is the intermittent behavior of the system, that is, strong inhomogeneities in the system evolution, depending on its initial condition. Indeed if $p$ is small then the realizations of $S^\Delta(j,t)$ with $x^j(0)$ starting in the zone $[0, p]$ are quickly spread over the whole interval $[0, 1]$, almost reaching the asymptotic value of $-\log\Delta$ after a few steps; the realizations starting, for example, near the unstable equilibrium point $x = 1/(2-p)$ take several time steps to reach saturation, giving

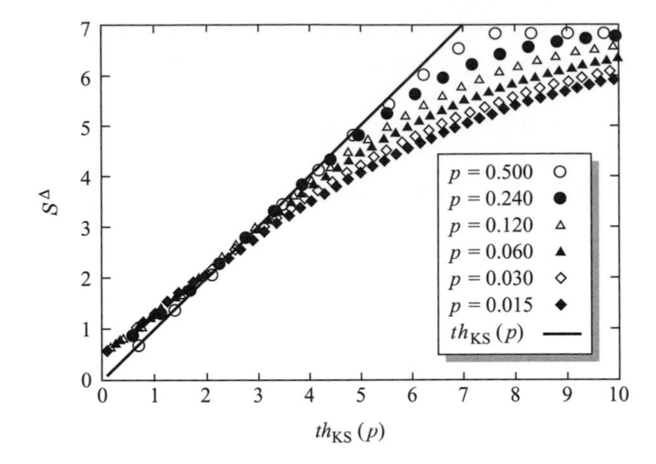


Figure 6.1 $S^\Delta(t)$ as a function of $t\, h_{KS}(p)$ for the modified tent map of Eq. (6.17) with $\Delta = 10^{-3}$, for different values of $p$. Note that Eq. (6.12) should give the straight line $t\, h_{KS}(p)$ for all values of $p$.

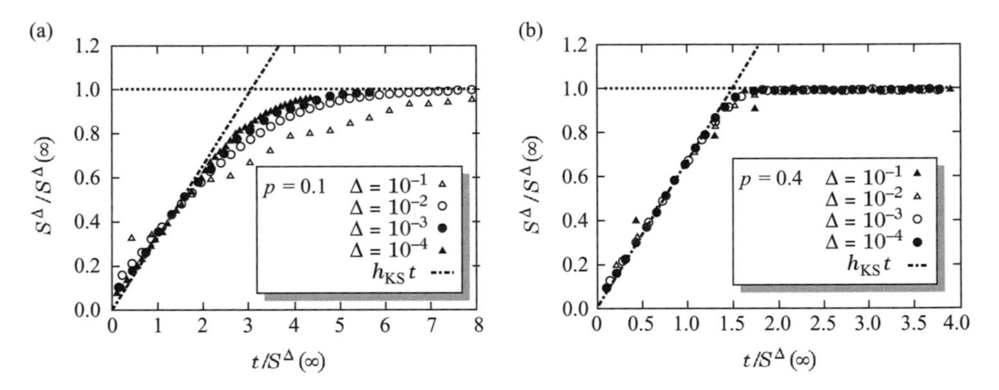


Figure 6.2 $S^\Delta(t)/S^\Delta(\infty)$ as a function of $t/S^\Delta(\infty)$ for the modified tent map of Eq. (6.17) with different values of $\Delta$. (a) $p = 0.1$, (b) $p = 0.4$.

a dominant contribution to the rate of increase of $S^\Delta(t)$. In this way the $S^\Delta(t)$ computed with Eq. (6.20) does not increase in time following the naive argument yielding Eq. (6.12).

Figure 6.2 shows other interesting properties of the behavior of $S^\Delta(t)$ for $p = 0.1$ and $p = 0.4$ with varying value of $\Delta$. As one can see, in the slightly intermittent case $p = 0.4$, the rescaled curves collapse together, confirming the assumption of Eq. (6.12); in the intermittent case $p = 0.1$, the curves do not collapse and only for very low values of $\Delta$ is a linear growth of $S^\Delta(t)$ present. Let us stress the fact that, at variance with the case shown in Figure 6.2(b), in the intermittent case

(Figure 6.2(a)) the crossover regime (after the linear regime and before saturation) is very long and is comparable with the duration of the linear regime.

The above results are generic (Falcioni *et al.* 2005): for example, the same qualitative behavior has been observed in systems with different time scales, in automata with zero $h_{\mathrm{KS}}$, as well as for the Manneville map

$$x_{t+1} = \begin{cases} x_t + kx_t^z & \text{if} \quad x_t < d \\ (1 - x_t)/(1 - d) & \text{if} \quad d < x_t < 1, \end{cases} \tag{6.22}$$

where $d$ fulfills the relation $d + kd^z = 1$, and an invariant distribution exists for $3/2 < z < 2$ (Gaspard and Wang 1993).

## 6.2 Gibbs and Boltzmann entropies: the role of chaos, interaction and coarse graining

In the previous section we introduced, and partially faced, the problem of the evolution of the time-dependent Gibbs entropy (6.3). This problem, besides its abstract interest in dynamical systems theory, can be regarded as relevant in characterizing the entropic behavior of a system out of equilibrium. This can be called the Gibbs approach to irreversibility. In Chapter 5 we discussed irreversibility according to Boltzmann and his definition of entropy. In this section we compare the two points of view, with the numerical analysis of a simple model. We want to show that an interaction among the very many elementary constituents of a single (macroscopic) system is crucial to distinguish between the two entropies and to recognize Boltzmann's entropy as more appropriate in a physical context.

### *6.2.1 A brief overview of basic facts*

For the sake of self-consistency we briefly recall some well-known facts. Because of the physical perspective, in this section we are only concerned with either Hamiltonian or symplectic systems.

Consider a Hamiltonian system of $N$ particles, and let the vector $\mathbf{X}(t) = (\mathbf{q}_1(t), \mathbf{p}_1(t), \ldots, \mathbf{q}_N(t), \mathbf{p}_N(t))$ define its microscopic state in the phase space $\Gamma$. If one supposes that the system, at a certain time, can be found in a variety of states constituting an ensemble, then, denoting by $\rho(\mathbf{X}, t)\mathrm{d}\mathbf{X}$ the probability that the microscopic state is in a phase space volume $\mathrm{d}\mathbf{X}$ around $\mathbf{X}$, at time $t$, the Gibbs entropy is defined as in (6.3):

$$S_{\mathrm{G}}(\rho_t) = -\int \rho(\mathbf{X}, t) \ln[\rho(\mathbf{X}, t)]\mathrm{d}\mathbf{X}, \tag{6.23}$$

where we again let Boltzmann's constant be 1. We also add the subscript $G$ to distinguish this from Boltzmann's entropy to be introduced later. As previously discussed, a coarse graining of $\Gamma$, by cells of size $\Delta$, leads to a coarse-grained density and to a corresponding coarse-grained entropy $S_{\mathrm{G}}^{\Delta}(t)$ that changes with time. When the system is chaotic, and the initial probability distribution is supported over a small region of linear size $\sigma$, simple arguments (see Section 6.1.1) suggest that $S_{\mathrm{G}}^{\Delta}(t)$ increases linearly in time, after a short transient of length $t_\lambda$:

$$S_{\mathrm{G}}^{\Delta}(t) - S_{\mathrm{G}}^{\Delta}(0) \simeq \begin{cases} 0 & t < t_\lambda \\ h_{\mathrm{KS}}(t - t_\lambda) & t \geq t_\lambda \end{cases} \tag{6.24}$$

where

$$t_\lambda \sim \frac{1}{\lambda} \ln\left(\frac{\sigma}{\Delta}\right) \tag{6.25}$$

and $\lambda$ is the maximal Lyapunov exponent of the Hamiltonian equations. As discussed in the previous section, prediction (6.24) is valid only when intermittency effects are negligible. But this is now a marginal aspect, therefore in the following, for simplicity, we will assume its validity.

Some authors consider Eq. (6.24) to be an indicator of the deep connection between chaos and non-equilibrium statistical mechanics. However, the kind of behavior considered may have no relation at all to thermodynamics, essentially because $S_{\mathrm{G}}^{\Delta}(t)$ may have no relation to thermodynamics (in a non-equilibrium situation). This is suggested by the following remarks:

(a) the quantity $S_{\mathrm{G}}^{\Delta}(t)$ describes properties of the $\Gamma$-space that cannot be computed from a single-system measurement (at least this is practically impossible in any realistic situation);
(b) the time increase of $S_{\mathrm{G}}^{\Delta}(t)$ depends on an arbitrary coarse-graining procedure and this appears as a non-ontological result.

Consider now Boltzmann's viewpoint, applied to a macroscopic system of weakly interacting identical particles. A one-body probability distribution function $f(\mathbf{q}, \mathbf{p}, t)$ (the probability density of finding a particle in a given volume of the single-particle $\mu$-space) can be defined without any reference to an ensemble of different states of the given system and the corresponding distribution in $\Gamma$-space. This can be done by introducing cells of linear size $\Delta$ in the $\mu$-space of the system. If the system lives in a $d$-dimensional physical space, let $N \gg \Delta^{-2d}$, so that each cell contains a statistically relevant number of particles. Then, given one state $\mathbf{X}(t)$, evolving in $\Gamma$, the one-particle time-dependent coarse-grained distribution is defined as:

$$f_\Delta(\mathbf{q}^{(j)}, \mathbf{p}^{(k)}, t) = \frac{1}{N}\sum_{i=0}^{N} \Theta\left(1 - \frac{2|\mathbf{q}^{(j)} - \mathbf{q}_i(t)|}{\Delta}\right) \Theta\left(1 - \frac{2|\mathbf{p}^{(k)} - \mathbf{p}_i(t)|}{\Delta}\right) \tag{6.26}$$

where $\Theta(z)$ is the Heaviside step function and $\mathbf{q}^{(j)}$, $\mathbf{p}^{(k)}$ are the coordinates of the center of each cell $C_{jk}$ having volume $\Delta^{2d}$, in the appropriate units. For a dilute gas, this $\mu$-space function is able to identify a meaningful macrostate, whose volume $\Delta\Gamma$, according to Boltzmann's relation, gives the entropy as follows:

$$S_\mathrm{B} = \log \Delta\Gamma. \tag{6.27}$$

Thus, for a dilute gas, $\log \Delta\Gamma$ and $S_\mathrm{B}$ can be well approximated by

$$S_\mathrm{B} = \log \Delta\Gamma \approx -N \sum_{j,k} f_\Delta(\mathbf{q}^{(j)}, \mathbf{p}^{(k)}) \ln f_\Delta(\mathbf{q}^{(j)}, \mathbf{p}^{(k)}), \tag{6.28}$$

where terms dependent on $\Delta$ and $N$ have been disregarded. The Boltzmann entropy for these systems can also be written, as usual, as

$$S_\mathrm{B}(t) = -N \int f(\mathbf{q}, \mathbf{p}, t) \ln f(\mathbf{q}, \mathbf{p}, t) \mathrm{d}\mathbf{q}\, \mathrm{d}\mathbf{p} \tag{6.29}$$

where $f$ is the regular $\mu$-space probability distribution, obtained in the $N \to \infty$, $\Delta \to 0$ limit of Eq. (6.26).[3] For dilute systems according to the hypothesis of molecular chaos, the celebrated Boltzmann $\mathcal{H}$-theorem holds (Huang 1987):

$$\frac{\mathrm{d}S_\mathrm{B}}{\mathrm{d}t} \geq 0. \tag{6.30}$$

The validity of the molecular chaos hypothesis has been demonstrated for the class of dilute systems in the Grad limit, where $N \to \infty$ and the interaction range between particles goes to zero, in order to keep the total cross-section constant (Lanford 1981, Illner and Pulvirenti 1986).

The two main approaches are equivalent when describing equilibrium conditions, both coinciding with plain thermodynamic entropy. Some textbooks try to connect them out of equilibrium too, noticing that in dilute systems:

$$\rho(\mathbf{X}, t) \simeq \prod_{j=1}^{N} f(\mathbf{q}_j, \mathbf{p}_j, t), \tag{6.31}$$

which implies $S_\mathrm{G} \simeq S_\mathrm{B}$. However, relation (6.31) is a very delicate assumption and must be interpreted *cum grano salis* otherwise, for instance, because of (6.30), one would infer that the Gibbs entropy (6.23) increases.

[3] Since, for small $\Delta$, $f_\Delta(\mathbf{q}^{(j)}, \mathbf{p}^{(k)}) \approx f(\mathbf{q}^{(j)}, \mathbf{p}^{(k)})\Delta^{2d}$, the quantities in (6.28) and (6.29) differ just by a term $O(\ln \Delta)$ which is constant in time and is not relevant.

With respect to the relation between $S_G$ and $S_B$, some important conceptual differences should be stressed.

- The Gibbs point of view is based on the ensemble, i.e. an abstract collection of macroscopically identical systems, and does not depend on the number of particles of which each system is made. In contrast, Boltzmann's approach does not require an ensemble of copies of the same system, but needs $N \gg 1$, in order to define a meaningful $f(\mathbf{q}, \mathbf{p}, t)$ for the single system.
- The Gibbs entropy deals with the $\Gamma$-space, and necessitates a coarse-graining procedure in order to escape the consequences of the Liouville theorem, so as to grow during an irreversible evolution. The Boltzmann entropy has no no-grow theorem and the graining of the $\mu$-space is only introduced to deal with a smooth distribution.
- The two entropies capture different features of a given system. The Gibbs entropy is concerned with the spreading in phase space of the initial ensemble of microstates, as it is selected by the macroscopic observables of interest. In this picture, the time evolution of these observables plays no role. In contrast, the Boltzmann entropy is concerned with the evolution of phase space volumes, as determined by the evolution of the interesting observables along a single history.

At this point one can wonder whether there is a means to reveal, in a quantitative way, the possible differences in the time behavior of the two entropies.

### 6.2.2 *Looking for the differences: the Boltzmann entropy of a chaotic system*

We study here a system made of many chaotic elements and we compare the behavior of quantity (6.28), as a function of the graining size $\Delta$, in two different settings: either no interaction among elements is present, or the elements interact.

A widely studied system consists of $N$ particles that do not interact with each other, and move in a periodic array of fixed convex scatterers, with which they collide elastically. It is well established that such a system, commonly known as the Lorentz gas,[4] is chaotic and displays asymptotic diffusion.

The study of this system, for a large number of particles, is expensive from a computational point of view. A reasonable substitute, which shares its main features, can be given in terms of symplectic maps. For instance, one can consider a two-dimensional map, with one "coordinate" and one "momentum," in place of each particle of the Lorentz gas. At a second stage one can introduce a form of interaction among these "particles." Thus we require the following:

- in the absence of interactions among the "particles," the single-particle dynamics in the corresponding $\mu$-space is chaotic and volume preserving;

[4] In Lorentz's original model (Lorentz 1905), the moving particles were considered in thermal equilibrium with the scatterers, which is impossible to achieve without energy exchanges between scatterers and particles, as in the present model.

- in the presence of interactions among particles, the dynamics of the whole system, described by the vector $\mathbf{X} = (\mathbf{Q}, \mathbf{P})$, $\mathbf{Q} = (q_1 \dots q_N)$, $\mathbf{P} = (p_1 \dots p_N)$, is symplectic and volume preserving in the $\Gamma$-space.

The resulting model will be easier to handle numerically, and still have some important properties of the particle system. To this end, we introduce the symplectic map:

$$\begin{cases} q_i = \partial G(\mathbf{Q}', \mathbf{P})/\partial p_i \mod 1 \\ p'_i = \partial G(\mathbf{Q}', \mathbf{P})/\partial q'_i \mod 1 \end{cases} \tag{6.32}$$

with generating function

$$G(\mathbf{Q}', \mathbf{P}) = \sum_{i=1}^{N} q'_i p_i - \frac{k}{2\pi} \sum_{i=1}^{N} \sum_{j=1}^{N_S} \cos[2\pi(q'_i - Y_j)] - \frac{\epsilon}{4\pi} \sum_{i=1}^{N} \sum_{n=-M/2}^{M/2} \cos[2\pi(q'_i - q'_{i+n})], \tag{6.33}$$

where $q_i, p_i \in [0, 1]$. $N_S$ is the number of fixed "obstacles" having positions $Y_j$, which play the role of the convex scatterers in the Lorentz gas, and $N$ is the number of "particles." In order to make the numerical simulations faster, we assume that each particle interacts only with a limited number $M$ of other particles. The parameters $k$ and $\epsilon$ represent the interaction strengths between particles and obstacles and among particles respectively. If $k = \epsilon = 0$ one has free particles. The functional form of the generating function is reminiscent of the standard map, which is a paragon of symplectic dynamics.

Substituting Eq. (6.33) into (6.32), one finds:

$$\begin{cases} q'_i = q_i + p_i & \mod 1 \\ p'_i = p_i + k \sum_{j=0}^{N_S} \sin[2\pi(q'_i - Y_j)] + \epsilon \sum_{n=-M/2}^{M/2} \sin\left[2\pi(q'_i - q'_{i+n})\right] & \mod 1. \end{cases} \tag{6.34}$$

We are going to discuss the growth in time of the Boltzmann entropy per particle,

$$\frac{S_{\mathrm{B}}(t, \Delta)}{N} \equiv \eta(t, \Delta) = -\sum_{j,k} f_\Delta(q^{(j)}, p^{(k)}, t) \log f_\Delta(q^{(j)}, p^{(k)}, t), \tag{6.35}$$

by varying the cell size $\Delta$.

We stress again that Eq. (6.35) (times $N$) is an acceptable expression for the Boltzmann entropy (6.27) only in the case of dilute systems. In such a case the function $f(q, p, t)$ is able to describe properly the macrostates of the system. Since we are considering the case of weakly interacting subsystems we use Eq. (6.35).

Simulations of system (6.34) have been performed with $N_S = 10^3$, $k = 0.017$, $N = 10^7$. The positions of the $N_S$ obstacles are selected at random with uniform distribution. As initial non-equilibrium condition, we take a cloud of points distributed according to a Gaussian of variance $\sigma^2 = (0.01)^2$, in a fixed region of phase space. Then we compute the time evolution of differences between the entropy per particle and its initial value at several resolutions $\Delta$:

$$\delta S(t, \Delta) = \eta(t, \Delta) - \eta(0, \Delta). \tag{6.36}$$

We begin with the case $\epsilon = 0$, about which an important general remark should be made. A one-particle distribution, $f_\Delta(q^{(j)}, p^{(k)}, t)$, of one system consisting of $N \gg 1$ non-interacting identical particles, can also be seen as a phase space distribution, $P^\Delta(q^{(j)}, p^{(k)}, t)$, of an ensemble of $N$ identical (and independent) systems composed of one particle. The coordinates and velocities of the $N$ particles represent one state of the macrosystem, in its $6N$-dimensional $\Gamma$-space, by which one can build a one-particle distribution in a six-dimensional $\mu$-space. However, they can also represent $N$ independent states in a six-dimensional $\Gamma$-space of a single microsystem, by which one builds an ensemble distribution. The functional forms of $f_\Delta$ and $P^\Delta$ are the same. Moreover, since the particles do not interact with each other, $f_\Delta$ evolves in the same manner as $P^\Delta$. Thus in the non-interacting case, the quantity $\delta S(t, \Delta)$ represents either a $\delta S_{\mathrm{B}}(t, \Delta)$ or a $\delta S_{\mathrm{G}}^\Delta(t)$, according to the adopted point of view. We stress that the two entropies refer to different systems, but they share identical behavior.

The increase in entropy shown in Figure 6.3 must originate exclusively from the discretization procedure since, when $\epsilon = 0$, the evolution of $f(q, p, t)$ obeys the

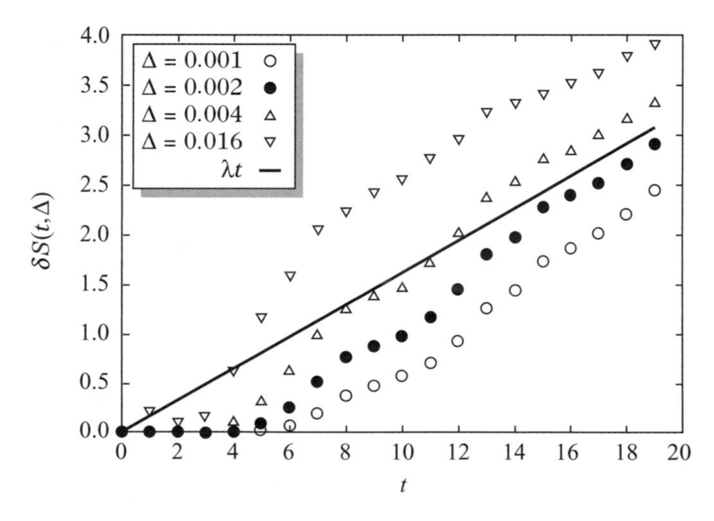


Figure 6.3 $\delta S(t, \Delta)$ with $\epsilon = 0$ (non-interacting particles) as a function of $t$ for different values of $\Delta$. The slope of the straight line equals the Lyapunov exponent.

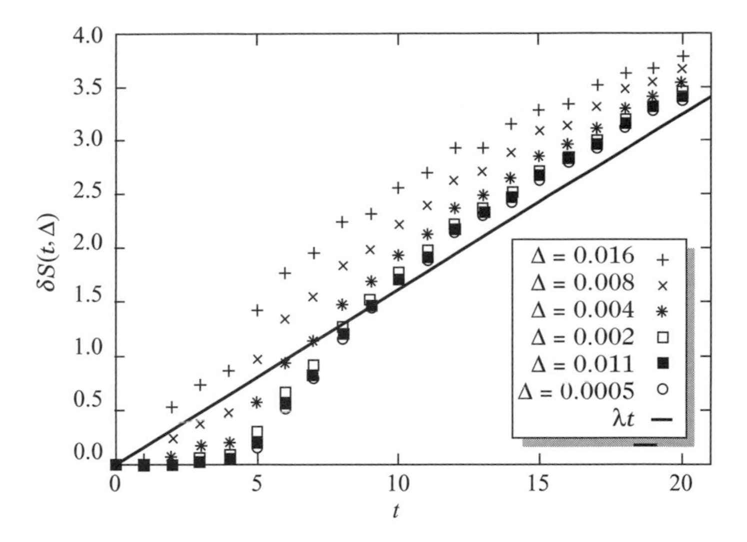


Figure 6.4 $\delta S(t, \Delta)$ with $\epsilon = 10^{-4}$ as a function of $t$ for different values of $\Delta$. The straight line slope equals the Lyapunov exponent.

Liouville theorem. Indeed, as one can see in the figure, the curves of the entropy differences as functions of time stay constant up to a time $t_\lambda$, depending on $\Delta$. After this transient, the slope of $\delta S(t, \Delta)$ is practically the same for all the curves and is given approximately by $h_{\mathrm{KS}} = \lambda$ (the numerically computed Lyapunov exponent is $\lambda = 0.162$). One has $\delta S \simeq 0$ for $t \lesssim t_\lambda(\Delta)$ and $\delta S \simeq \lambda(t - t_\lambda(\Delta))$ for $t \gtrsim t_\lambda(\Delta)$. This is in agreement with Eq. (6.24) and with the indistinguishability between $f_\Delta$ and $P^\Delta$ in the non-interacting case. Since $t_\lambda$ increases as $\Delta$ decreases, the "true" Boltzmann entropy, together with the Gibbs entropy, for $\Delta \to 0$ is constant in time.

Note that, when considering $\delta S(t, \Delta)$ as a coarse-grained Gibbs entropy, we are analyzing its behavior during the first two stages of the evolution, as described in Section 6.1.1. This is obtained by keeping the extension of the density fixed ($\sigma = 0.01$), when varying $\Delta$. We choose this because here, in contrast to the preceding section, we are also interested in the transient behavior before $\delta S(t, \Delta)$ begins to increase.

Consider now the interacting case, with $\epsilon > 0$. We underline that, in this situation, $\delta S(t, \Delta)$ can only represent the evolution of the Boltzmann entropy (of a dilute system), and has nothing to do with the Gibbs entropy. Figure 6.4 shows the curves of $\delta S(t, \Delta)$ as a function of $t$ for different values of $\Delta$. In this case, the entropy curves as functions of $\Delta$ no longer extrapolate to zero. After a characteristic time depending on $\epsilon$, $t_*(\epsilon)$, the entropy shows just a weak dependence on $\Delta$ and correctly extrapolates to a finite value when $\Delta \to 0$.

### 6.2.3 Comments and interpretations

Let us summarize and comment on the previous results.

(a) For non-interacting systems ($\epsilon = 0$), $\delta S(t, \Delta)$ represents the evolution of the Boltzmann entropy and of the Gibbs entropy at the same time. Its growth behavior reflects the properties of the observation tools, and has a kind of "subjective" character, i.e. it does not depend on the system alone. Since the time evolution of the coarse-grained $S$ can be delayed further and further, by taking finer and finer partitions, not only does the value of the entropy depend on the coarse graining, but also the increase in entropy depends (for "small" $t$) on the resolution scale.

(b) For weakly interacting systems, there is an effective cell size $\Delta_*(\epsilon, \lambda)$, such that if $\Delta < \Delta_*(\epsilon, \lambda)$ the value of $\delta S(t, \Delta)$ does not depend on $\Delta$. In such a case, the growth in Boltzmann entropy is an objective property, meaning that the limit for $\Delta \to 0$ of $\delta S(t, \Delta)$ exists, is finite, and is an intrinsic property of the system.

(c) The role of chaos in the limit of vanishing coupling is relevant, i.e. the slope of $\delta S(t, \Delta)$, for $t$ large enough, is given by the Lyapunov exponent, but the existence of an effective cell size $\Delta_*(\epsilon, \lambda)$ and the corresponding $t_*(\epsilon, \lambda)$ depends on the coupling strength $\epsilon$, and on $\lambda$.

(d) In the above procedure, the quantities defined by Eqs. (6.26) and (6.35), are evaluated with no assumptions such as the hypothesis of molecular chaos or dilution of the system. From a mathematical point of view, we can define $\eta(t, \Delta)$ in Eq. (6.35) in full generality. However, we consider only the weakly interacting limit of small $\epsilon$ for the physical reason that only in such a case is $f(q, p, t)$ endowed with an appropriate thermodynamic meaning.

(e) The time evolution of $f(q, p, t)$ for small values of $\epsilon$ is different from the case $\epsilon = 0$ only on very small graining scales. In other words, the coupling is necessary for the "genuine" growth of the entropy, but it does not have any dramatic effect on $f(q, p, t)$ at scales $\Delta \gtrsim \Delta_*$. Indeed, the non-interacting and the weakly interacting cases do not appear to be very different, in terms of the single-particle phase space distribution.

The previous results suggest the following interpretation: since the number of particles is large, one can expect that the effect on each particle of the interactions can be reasonably described by some kind of thermal bath. The single particle dynamics can then be mimicked by the chaotic dynamics (corresponding to the symplectic map of Eq. (6.34) with $\epsilon = 0$) coupled to a noise term whose strength is $O(\epsilon)$:

$$\begin{cases} q_i(t+1) = q_i(t) + p_i(t) & \mod 1 \\ p_i(t+1) = p_i(t) + k \sum_j \sin[2\pi(q_i(t+1) - Y_j)] + \sqrt{2D}\eta_i(t) & \mod 1 \end{cases} \tag{6.37}$$

where the $\eta_i(t)$ are i.i.d. Gaussian variables with zero mean and unitary variance, i.e.

$$\langle \eta_i(t) \rangle = 0, \quad \langle \eta_i(t)\eta_j(t') \rangle = \delta_{t,t'}\delta_{i,j}. \tag{6.38}$$

With this approximation, one basically assumes that $f(q, p, t)$ evolves according to a discrete time Fokker–Planck equation. Supposing that each particle gives an uncorrelated contribution to the noise term, one can roughly estimate the diffusion coefficient $D$ as $M\epsilon^2/4$. This heuristic estimate is well supported by numerical simulations of (6.34): the quantity $\delta S(t, \Delta)$ practically does not change for varying $M$ and $\epsilon$, keeping $M\epsilon^2$ constant.

In this stochastic framework, one can introduce a characteristic time $t_c$, defined as the time over which the scale of the noise-induced diffusion reaches the smallest scale originated by the deterministic chaotic dynamics (Pattanayak 2001). This definition of $t_c$ would correspond to $t_*(\epsilon, \lambda)$ introduced above. Noting that the typical length due to noise behaves as $\sqrt{M\epsilon^2 t/2}$ and the smallest chaotic scale behaves as $\sigma \exp(-\lambda t)$,[5] the time $t_c$ may be estimated as the solution of the following transcendent equation:

$$\epsilon\sqrt{Mt_c/2} = \sigma \exp(-\lambda t_c). \tag{6.39}$$

Consequently, one can introduce a characteristic scale:

$$\Lambda_c = \epsilon\sqrt{Mt_c/2}\,, \tag{6.40}$$

beyond which the value of the entropy still depends on the size of $\Delta$. Numerical checks confirm the consistency of this approach.

A similar reasoning leads to the decoherence mechanism proposed by Zurek and Paz (1994, 1995), see also Casati and Chirikov (1995) for the semiclassical limit of quantum mechanics.

## 6.3 Fluctuation-response relation and chaos

There exist physical situations where a system, on which a weak perturbation is applied, produces a reaction, or response, that is proportional to the perturbation. A simple example is an electric resistor, in which (in the Ohm regime) a density of electric current is generated which is proportional to the applied electric field. Another example is a particle in a fluid on which a constant force **F** is acting. The particle will attain an asymptotic velocity, along the direction of the force, and

[5] A rough explanation of this kind of behavior is obtained by considering that during a chaotic evolution, in a two-dimensional volume-conserving case, an initial region $\sigma$ is stretched by a factor $\exp(\lambda t)$ along the unstable direction while it shrinks along the stable direction by the factor $\exp(-\lambda t)$.

proportional to it: $v_d = |\mathbf{F}|/\gamma$, where $\gamma$ is the friction constant of the particle in the given fluid. The latter phenomenon, concerning a Brownian particle, and analyzed from a molecular point of view, led Einstein to a physical remark with important consequences. The impacts of the fluid molecules with the Brownian particle are at the origin both of the viscous resistance (when the external force is present) and of the wandering behavior (when there is no force). Thus it is sensible to expect some relation between the parameters characterizing the two different manifestations of the same physical process, i.e. a relation between the friction constant $\gamma$ and the diffusion coefficient $D$. Indeed (for simplicity restricting to the one-dimensional case along the $x$ direction) Einstein was able to prove that $D = kT/\gamma$. Moreover he linked $D$ in a quantitative way to the fluctuations of the particle displacement per unit time:

$$\frac{kT}{\gamma} = D = \frac{\langle(\Delta x)^2\rangle}{2\Delta t}. \tag{6.41}$$

This equation, connecting the dissipative properties of the system, as embodied in $\gamma$, with the equilibrium fluctuations of the particle position, is the first example of the so-called *fluctuation-dissipation* or *fluctuation-response* relations. If one rewrites the diffusion coefficient in the form

$$D = \lim_{t\to\infty} \frac{1}{2t}\big\langle(x(t) - x(0))^2\big\rangle = \int_0^\infty \big\langle u(t_0)u(t_0 + t)\big\rangle \mathrm{d}t,$$

with $u = \mathrm{d}x/\mathrm{d}t$, then Eq. (6.41) can also be read as a relation between a response coefficient and a suitable time correlation function of the unperturbed system.

The connection between "non-equilibrium" features (e.g. response to an external perturbation) and "equilibrium" properties (e.g. time correlations computed according to an invariant measure) is a rather important issue in statistical mechanics. Starting with Callen and Welton (1951), a systematic exploration of the subject has been pursued, for both classical and quantum Hamiltonian systems, making use of first-order time-dependent perturbation theory, and giving rise to the *linear response theory* (Kubo 1966, 1986). The fluctuation-response theory was originally developed in the context of (equilibrium) statistical mechanics of Hamiltonian systems; this can generate a misleading idea about the fluctuation-response relations. For instance, some authors claimed, using qualitative arguments, that in fully developed turbulence (which is a non-Hamiltonian and non-equilibrium system) there is no relation between spontaneous fluctuations and relaxation to the statistical steady state (Rose and Sulem 1978). We will see that a generalized fluctuation-response relation holds under rather general hypotheses, independently from the Hamiltonian nature of the systems (Falcioni *et al.* 1990). As a matter of fact, fluctuation-response relations can also be important in non-Hamiltonian systems, for instance, systems with chaotic dynamics (Ruelle 1989), in particular in hydrodynamics (Kraichnan 1959). This issue has an obvious relevance in geophysics and climate research (Leith 1975) where one essential point is the possibility that the recovery of the climate system from a perturbation (e.g. a volcanic eruption) can be estimated from its time history (correlation time of the unperturbed system).

The fluctuation-response problem has often been studied for infinitesimal perturbations; in statistical mechanics this is not a serious limitation. In a similar way this problem has an importance in analytical approaches to the statistical description of hydrodynamics, where Green functions are naturally involved both in perturbative theories and in closure schemes (Kraichnan 1959, McComb and Kiyani 2005). On the other hand, in different contexts, for example in geophysical or climate problems, the interest in infinitesimal perturbations is rather academic, while the interesting problem is the relaxation of large fluctuations in the system due to rapid changes of the parameters (Boffetta *et al.* 2003).

### 6.3.1 A derivation of the fluctuation-response relation

Consider a dynamical system with states $\mathbf{x}$ belonging to an $N$-dimensional vector space. For the sake of generality, we will consider the case in which the time evolution may also not be completely deterministic (e.g. stochastic differential equations). We assume the existence of an invariant probability distribution $\rho(\mathbf{x})$ and the mixing character of the system (from which its ergodicity follows). Note that no assumption is made on $N$.

Our aim is to express the average response of a generic observable $A$ to a perturbation, in terms of suitable correlation functions, computed according to the invariant measure of the unperturbed system. As the first step we study the behavior of one component of $\mathbf{x}$, say $x_i$, when the system, described by $\rho(\mathbf{x})$, is subjected to an initial (non-random) perturbation such that $\mathbf{x}(0) \to \mathbf{x}(0) + \Delta\mathbf{x}_0$.[6] This instantaneous kick modifies the density of the system to $\rho'(\mathbf{x})$, related to the invariant distribution by $\rho'(\mathbf{x}) = \rho(\mathbf{x} - \Delta\mathbf{x}_0)$. We introduce the probability of transition from $\mathbf{x}_0$ at time 0 to $\mathbf{x}$ at time $t$, $W(\mathbf{x}_0, 0 \to \mathbf{x}, t)$. For a deterministic system, with evolution law $\mathbf{x}(t) = U^t\mathbf{x}(0)$, the probability of transition reduces to $W(\mathbf{x}_0, 0 \to \mathbf{x}, t) = \delta(\mathbf{x} - U^t\mathbf{x}_0)$, where $\delta(\cdot)$ is the Dirac delta. Then we can write an expression for the mean value of the variable $x_i$, computed with the density of the perturbed system:

$$\langle x_i(t)\rangle' = \int\int x_i \rho'(\mathbf{x}_0) W(\mathbf{x}_0, 0 \to \mathbf{x}, t)\, d\mathbf{x}\, d\mathbf{x}_0. \tag{6.42}$$

[6] The study of an "impulsive" perturbation is not a serious limitation because in the linear regime from the (differential) linear response one understands the effect of a generic perturbation.

The mean value of $x_i$ during the unperturbed evolution can be written in a similar way:

$$\langle x_i(t) \rangle = \int\int x_i \rho(\mathbf{x}_0) W(\mathbf{x}_0, 0 \to \mathbf{x}, t) \, d\mathbf{x} \, d\mathbf{x}_0. \tag{6.43}$$

Therefore, defining $\overline{\delta x_i} = \langle x_i \rangle' - \langle x_i \rangle$, we have:

$$\begin{aligned} \overline{\delta x_i}(t) &= \int\int x_i \frac{\rho(\mathbf{x}_0 - \Delta\mathbf{x}_0) - \rho(\mathbf{x}_0)}{\rho(\mathbf{x}_0)} \rho(\mathbf{x}_0) W(\mathbf{x}_0, 0 \to \mathbf{x}, t) \, d\mathbf{x} \, d\mathbf{x}_0 \\ &= \langle x_i(t) F(\mathbf{x}_0, \Delta\mathbf{x}_0) \rangle \end{aligned} \tag{6.44}$$

where

$$F(\mathbf{x}_0, \Delta\mathbf{x}_0) = \left[ \frac{\rho(\mathbf{x}_0 - \Delta\mathbf{x}_0) - \rho(\mathbf{x}_0)}{\rho(\mathbf{x}_0)} \right]. \tag{6.45}$$

Let us note here that the mixing property of the system is required so that the decay to zero of the time-correlation functions ensures the switching off of the deviations from equilibrium.

For an infinitesimal perturbation $\delta\mathbf{x}(0) = (\delta x_1(0) \cdots \delta x_N(0))$, if $\rho(\mathbf{x})$ is non-vanishing and differentiable, the function in (6.45) can be expanded to first order and one obtains:

$$\begin{aligned} \overline{\delta x_i}\,(t) &= -\sum_j \left\langle x_i(t) \left. \frac{\partial \ln \rho(\mathbf{x})}{\partial x_j} \right|_{t=0} \right\rangle \delta x_j(0) \\ &\equiv \sum_j R_{i,j}(t) \delta x_j(0) \end{aligned} \tag{6.46}$$

which defines the linear response

$$R_{i,j}(t) = -\left\langle x_i(t) \left. \frac{\partial \ln \rho(\mathbf{x})}{\partial x_j} \right|_{t=0} \right\rangle \tag{6.47}$$

of the variable $x_i$ with respect to a perturbation of $x_j$. One can repeat the computation for a generic observable $A(\mathbf{x})$:

$$\overline{\delta A}\,(t) = -\sum_j \left\langle A(\mathbf{x}(t)) \left. \frac{\partial \ln \rho(\mathbf{x})}{\partial x_j} \right|_{t=0} \right\rangle \delta x_j(0). \tag{6.48}$$

At this point one could object that in a chaotic deterministic dissipative system the above machinery cannot be applied, because the invariant measure is not smooth at all. Nevertheless, a small amount of noise, that is always present in a physical system, smoothes the $\rho(\mathbf{x})$ and the fluctuation-response relation can be derived. We recall that this "beneficial" noise has the important role of selecting the natural measure, see Section 1.4, and, in the numerical experiments, it is provided by the roundoff errors of the computer. We stress that the assumption on the smoothness of the invariant measure allows us to avoid subtle technical difficulties. In chaotic dissipative systems, where the invariant measure is singular, the fluctuation-response relation is valid only along the expanding directions. For a mathematically oriented presentation see Ruelle (1998).

In Hamiltonian systems, taking the canonical ensemble as the equilibrium distribution, one has $\ln \rho = -\beta H(\mathbf{Q}, \mathbf{P}) +$ constant. If $x_i$ denotes the component $q_k$ of the position vector $\mathbf{Q}$ and $x_j$ the corresponding component $p_k$ of the momentum $\mathbf{P}$, from Hamilton's equations ($\mathrm{d}q_k/\mathrm{d}t = \partial H/\partial p_k$) one has

$$\frac{\overline{\delta q_k(t)}}{\delta p_k(0)} = \beta \left\langle q_k(t) \frac{\mathrm{d}q_k(0)}{\mathrm{d}t} \right\rangle = -\beta \frac{\mathrm{d}}{\mathrm{d}t} \langle q_k(t) q_k(0) \rangle \tag{6.49}$$

which is just the differential form of the usual fluctuation-response relation (Kubo 1966, 1986).

In non-Hamiltonian systems, where usually the shape of $\rho(\mathbf{x})$ is not known, relation (6.47) does not give very detailed information. It only guarantees the existence of a connection between the mean response function $R_{i,j}$ and a suitable correlation function, computed in the non-perturbed systems:

$$\langle x_i(t) f_j(\mathbf{x}(0)) \rangle, \quad \text{with} \quad f_j(\mathbf{x}) = -\frac{\partial \ln \rho}{\partial x_j}, \tag{6.50}$$

where, in the general case, the function $f_j$ is unknown. Let us stress that in spite of the technical difficulty in the determination of the function $f_j$, which depends on the invariant measure, a fluctuation-response relation always holds in mixing systems whose invariant measure is "smooth enough." We note that the nature of the statistical steady state (either equilibrium or non-equilibrium) has no effect at all on the validity of the fluctuation-response relation.

In the case of Gaussian distribution, $\rho(\mathbf{x})$ can be factorized and the elements of the linear response matrix recover the normalized correlation functions:

$$R_{i,j}(t) = \frac{\langle x_i(t) x_i(0) \rangle}{\langle x_i^2 \rangle} \delta_{ij}. \tag{6.51}$$

One important non-trivial class of systems with a Gaussian invariant measure is inviscid hydrodynamics, where the Liouville theorem holds, and a quadratic invariant exists

$$\sum_{i,j}^{N} \alpha_{i,j} x_i x_j = \text{constant}, \tag{6.52}$$

with $\alpha_{i,j}$ a positive matrix (Bohr *et al.* 1998). In such a case the $\{x_j\}$ are the coefficients of the Fourier series (or other similar expansions) of the velocity field. Under the hypothesis of ergodicity, the invariant measure is the microcanonical measure on the hypersurface $\sum_{i,j} \alpha_{i,j} x_i x_j = \text{constant}$, and therefore for large $N$ the $\rho(\mathbf{x})$ is close to the multivariate Gaussian distribution (Kraichnan and Montgomery 1980, Bohr *et al.* 1998).

In the case of finite perturbations, the fluctuation-response relation (6.44) is typically non-linear in the perturbation $\Delta\mathbf{x}_0$, and thus no simple relations analogous to (6.47) exist. Nevertheless Eq. (6.44) guarantees the existence of a link between equilibrium properties of the system and the response to finite perturbations. This fact has a straightforward consequence for systems with one single characteristic time, for example a low-dimensional system such as the Lorenz model. A generic correlation function in principle gives information on the relaxation time of finite size perturbations, even when the invariant measure $\rho$ is not known (Boffetta *et al.* 2003).

In systems with many different characteristic times, such as fully developed turbulence, one has a more complicated scenario: different correlation functions can show different behaviors (i.e. different ranges and scales), which depend on the observables. In addition, the amplitude of the perturbation can play a major role in determining the response, because different amplitudes may trigger different response mechanisms with different time properties (Boffetta *et al.* 2003).

### *6.3.2 Remarks on van Kampen's objection and the connections between the fluctuation-response relation and chaos*

Since the fluctuation-response relation involves the evolution of differences between variables computed on different realizations of the system, it seems natural to expect that this issue is related to the predictability problem and, more in general, to chaotic behavior. Actually it is possible to show that the two topics, i.e. the fluctuation-response relation and predictability, have a subtle connection, making the linear-response theory heavily indebted to chaos.

The standard derivation of the linear-response results, for Hamiltonian systems, relies on a first-order truncation of the time-dependent perturbation theory, for the evolution of probability density (or density matrix) (Kubo 1966). This procedure was severely criticized by van Kampen (1971). In a nutshell, using the dynamical system terminology, van Kampen's argument is as follows. Given an initial perturbation $\delta\mathbf{x}(0)$, one can write a Taylor expansion for $\delta\mathbf{x}(t)$, the difference between the perturbed trajectory and the unperturbed trajectory:

$$\delta x_i(t) = \sum_j \frac{\partial x_i(t)}{\partial x_j(0)} \delta x_j(0) + O(|\delta\mathbf{x}(0)|^2). \tag{6.53}$$

Averaging over the initial condition one has the mean response function:

$$R_{i,j}(t) = \left\langle \frac{\partial x_i(t)}{\partial x_j(0)} \right\rangle = \int \frac{\partial x_i(t)}{\partial x_j(0)} \rho(\mathbf{x}(0))\mathrm{d}\mathbf{x}(0). \tag{6.54}$$

After an integration by parts the previous formula becomes Eq. (6.47). In the presence of chaos the terms $\partial x_i(t)/\partial x_j(0)$ grow exponentially as $\mathrm{e}^{\lambda t}$, therefore it is not possible to use the linear expansion (6.53) for a time larger than $(1/\lambda)\ln(L/|\delta\mathbf{x}(0)|)$, where $L$ is the typical fluctuation of the variable $\mathbf{x}$. On account of that estimate, the linear-response theory is expected to be valid only for extremely small and unphysical perturbations (or times). For instance, if one wants the fluctuation-response relation to hold up to 1 s when applied to the electrons in a conductor then, according to this reasoning, a perturbing electric field should be smaller than $10^{-20}$ V/m, in clear disagreement with experience. However, as shown in the previous subsection, Eq. (6.47) can be derived without making any approximation on the evolution of $\delta\mathbf{x}(t)$. Starting with the correct and complete expression (6.44) for the response, only a linearization on the initial time-perturbed density is needed, and this implies nothing but the smallness of the initial perturbation.

With regard to this point we have to observe that, from the evolution of the difference in trajectories, one can define the leading Lyapunov exponent $\lambda$ by considering the absolute values of $\delta\mathbf{x}(t)$: at small $|\delta\mathbf{x}(0)|$ and large enough $t$ one has

$$\langle \ln|\delta\mathbf{x}(t)|\rangle \simeq \ln|\delta\mathbf{x}(0)| + \lambda t. \tag{6.55}$$

On the other hand, in the fluctuation-response issue one deals with averages of quantities with sign, such as $\langle\delta\mathbf{x}(t)\rangle$. This apparently marginal difference is very important and is the basis of an answer to van Kampen's objection.

Indeed the reason for the seemingly inexplicable effectiveness of the linear-response theory may reside in the "constructive role of chaos" because, as Kubo suggested, "*instability [of the trajectories] instead favors the stability of distribution functions, working as the cause of the mixing*" (Kubo 1986). An analysis performed in the general case of perturbations taking place over finite times, makes clear the positive role of mixing in restoring the validity of linear-response theory, and shows, for instance, that in a conductor, with typical values for temperature and effective electron mass, the fluctuation-response relation holds for an electric field of up to $O(10^3$ V/m), instead of $10^{-20}$ V/m (Falcioni and Vulpiani 1995).

Nevertheless the objection that van Kampen raised remains relevant for numerical computations. In numerical simulations, $R_{i,j}(t)$ is obtained by perturbing the variable $x_i$ at time $t = t_0$ with a small perturbation of amplitude $\delta x_i(0)$ and then evaluating the separation $\delta x_i(t)$ between the two trajectories $\mathbf{x}(t_1)$ and $\mathbf{x}'(t_1)$ which are integrated up to a prescribed time $t_1 = t_0 + t$. At time $t = t_1$ the variable $x_i$ of the reference trajectory is again perturbed with the same $\delta x_i(0)$, and a new sample $\delta \mathbf{x}(t)$ is computed and so forth. The procedure is repeated $M \gg 1$ times and the mean response is then evaluated. In the presence of chaos, the two trajectories $\mathbf{x}(t)$ and $\mathbf{x}'(t)$ typically separate exponentially in time, therefore the mean response is the result of a delicate balance of terms which grow in time in different directions. The average error in the computation of $R_{i,j}(t)$ typically increases in time as $e^{L(2)t/2}/\sqrt{M}$, where $L(2)$ is the generalized Lyapunov exponent. Thus very high statistics (that is, very large $M$) is needed in order to capture this balance properly and to compute $R_{i,j}(t)$ for large $t$.

## 6.4 Chaos and pseudochaos for diffusion and conduction

Several simulations and theoretical works have suggested that, in systems with very strong chaos (namely hyperbolic systems), a close relationship exists between transport coefficients (e.g. viscosity, diffusivity, thermal and electrical conductivity) and indicators of chaos (Lyapunov exponents, Kolmogorov–Sinai entropy, escape rates) (Gaspard 1998, Dorfman 1999). At first glance, the existence of such relations seems to support the point of view according to which chaos is the basic ingredient for the validity of statistical mechanics laws. However, some counterexamples show that chaos is not a necessary condition for the emergence of robust statistical behaviors (Dettmann and Cohen 2001, Cecconi *et al.* 2003). The present section aims to clarify this point.

### *6.4.1 Diffusion in deterministic non-chaotic systems*

Soon after the (re)discovery of dynamical chaos (Lorenz 1963), it was realized that simple low-dimensional deterministic systems may also exhibit diffusive behavior. In this framework, the two-dimensional Lorentz gas (Lorentz 1905), describing the motion of a free particle through a lattice of hard convex obstacles, provided the paradigmatic example. Particle trajectories are chaotic as a consequence of the convexity of the obstacles, and diffusive behavior is observed, i.e. the mean square displacement from the initial position of the particle grows linearly in time (at large time). The Lorentz system is closely related to the Sinai billiard (Sinai 1979, Bunimovich and Sinai 1981) which can be obtained from the Lorentz gas by folding the trajectories into the unitary lattice cell. On the other hand, even non-chaotic deterministic systems, such as a bouncing particle on a two-dimensional billiard table with polygonal, randomly distributed, obstacles (wind-tree Ehrenfest model), may exhibit diffusion-like properties (Dettmann and Cohen 2001).

The relevant question can be recast as follows: what are the necessary microscopic conditions for observing large-scale diffusion? We will see that the study of chaotic models exhibiting diffusion and their non-chaotic counterparts is important for a better understanding of the role of microscopic chaos in macroscopic diffusion.

As argued by Dettmann and Cohen (2001), even an accurate numerical analysis based on $\epsilon$-entropy has no chance of detecting differences in the diffusive behavior between a chaotic Lorentz gas and its non-chaotic counterpart, such as the wind-tree Ehrenfest model. In this model, the maximal Lyapunov exponent is zero, since the reflection by the flat edges of the obstacles cannot produce exponential separation of trajectories. Let us note that the presence of corners can produce a stringent separation of trajectories; even if these unprobable events give a vanishing contribution to the Lyapunov exponents they can have non-negligible effects on the overall dynamics and its diffusive features. Thus, the disorder in the distribution of the obstacles may happen to be crucial. In particular, one may conjecture that a finite spatial (Shannon-like) entropy density, $h_S$, is necessary for observing diffusion. Indeed it is generally believed that deterministic diffusion might be a consequence either of a non-zero "dynamical" entropy ($h_{KS} > 0$) in chaotic systems or of a non-zero "static" entropy ($h_S > 0$), due to the random (quenched) positions of the obstacles, in non-chaotic systems, as in the present case. This is a key point, because one can argue that a deterministic infinite system with spatial randomness can be interpreted as an effective stochastic system (this is probably a "matter of taste"). With the aim of clarifying this point, we consider here a spatially disordered non-chaotic model (Cecconi *et al.* 2003), which is the one-dimensional analog of a two-dimensional non-chaotic Lorentz system with polygonal obstacles. Let us start with the map already discussed in Section 3.1:

$$x(t+1) = [x(t)] + F(x(t) - [x(t)]), \tag{6.56}$$

where [ ] indicates the integer part, and $F$ is

$$F(u) = \begin{cases} 2(1+a)u & \text{if } u \in [0, 1/2[ \\ 2(1+a)(u-1)+1 & \text{if } u \in [1/2, 1] \end{cases} \tag{6.57}$$

with $a > 0$. Now we introduce some modifications to make the above map non-chaotic. One can proceed as exemplified in Figure 6.5, that is by replacing the function (6.57) on each unit cell $C_\ell \equiv [\ell, \ell+1[$ by its step-wise approximation generated as follows. The first half of the interval $[\ell, \ell+1[$, is partitioned in $N$ micro-intervals $[\ell+\xi_{n-1}, \ell+\xi_n[$ , $n = 1, \ldots, N$, with $\xi_0 = 0 < \xi_1 < \xi_2 < \ldots < \xi_{N-1} < \xi_N = 1/2$. In each interval the map is defined by its linear approximation

$$F_\Delta(u) = u - \xi_n + F(\xi_n) \quad \text{if } u \in [\xi_{n-1}, \xi_n[. \tag{6.58}$$

The map in the second half of the unit cell is then determined by the anti-symmetry condition with respect to the middle of the cell. The quenched random

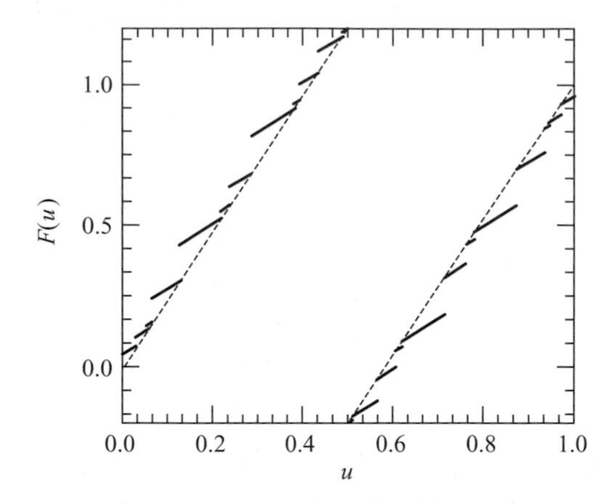


Figure 6.5 Sketch of the random staircase map in the unitary cell. The parameter $a$ is set to 0.23. The half-domain [0, 1/2] is divided into $N = 12$ micro-intervals of random size. The map on [1/2, 1] is obtained by applying the antisymmetric transformation with respect to the center of the cell (1/2, 1/2).

variables $\{\xi_k\}_{k=1}^{N-1}$ are uniformly distributed in the interval [0, 1/2], i.e. the micro-intervals have a *random* extension. Furthermore they are chosen independently in each cell $C_\ell$ (so one should more properly write $\xi_n^{(\ell)}$). All cells are partitioned into the same number $N$ of randomly chosen micro-intervals (of mean size $\Delta = 1/2N$). This modification of the continuous chaotic system is conceptually equivalent to replacing circular obstacles with polygonal obstacles in the Lorentz system (Dettmann and Cohen 2001). Since $F_\Delta$ has slope 1 almost everywhere, the map is no longer chaotic; for $\Delta \to 0$ (i.e. $N \to \infty$) the continuous chaotic map (6.57) is recovered. However, this limit is singular and as soon as the number of intervals is finite, even if arbitrarily large, chaos is absent.

It has been found (Cecconi *et al.* 2003) that this model still exhibits diffusion in the presence of both quenched disorder and a quasi-periodic external perturbation

$$x(t+1) = [x(t)] + F_\Delta(x(t) - [x(t)]) + \gamma \cos(\alpha t). \tag{6.59}$$

The strength of the external forcing is controlled by $\gamma$ and $\alpha$ defines its frequency, while $\Delta$ indicates a specific quenched disorder realization, i.e. the random positions of $\xi_n^{(\ell)}$. The results, summarized in Figure 6.6, show that $D$ is significantly different from zero only for $\gamma > \gamma_c = O(1/N)$, where $D$ exhibits saturation to a value close to that obtained with the chaotic system defined by Eqs. (6.56) and (6.57). The existence of a threshold $\gamma_c$ can be understood as follows. Owing to the staircase nature of the system, the perturbation has to exceed the typical discontinuity $O(1/N)$ of $F_\Delta$ to activate the "macroscopic" instability, which is the first step toward the diffusion. Data collapsing, obtained by plotting $D$ versus $\gamma N$ in Figure 6.6, confirms this argument. These findings are robust and do not depend

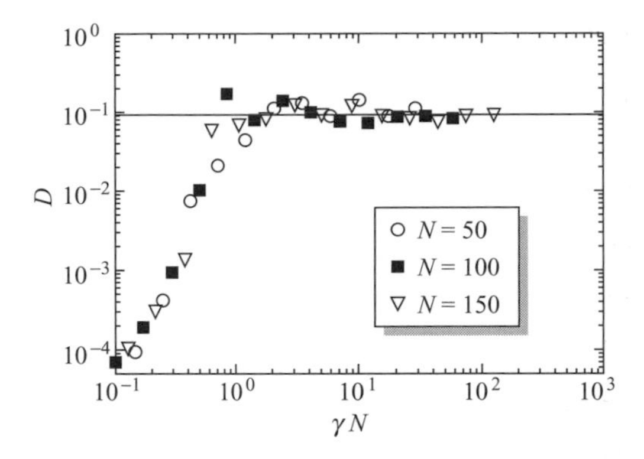


Figure 6.6 Log–log plot of the dependence of the diffusion coefficient $D$ on the external forcing strength $\gamma$. Different data relative to a number of cell micro-intervals $N = 50$, 100 and 150 are plotted *versus* the natural scaling variable $\gamma N$ to obtain a collapse of the curves. The horizontal line represents the result for the chaotic system (6.56), (6.57).

on the details of forcing. Therefore, we have an example of a model that is, by construction, non-chaotic (in the sense that the Lyapunov exponent is not positive) and which exhibits diffusion.

A tentative explanation of the diffusive behavior could be the presence of a quenched randomness with non-zero spatial entropy per unit length. To clarify this point, similar to Dettmann and Cohen (2001), the model can be modified in such a way that the spatial entropy per unit cell is forced to be zero. This can be obtained by repeating the same disorder configuration every $M$ cells, i.e. $\xi_n^{(\ell)} = \xi_n^{(\ell+M)}$. Looking at an ensemble of walkers it was observed that diffusion is still present, with $D$ very close to the value obtained with the chaotic system. A careful analysis (see Cecconi *et al.* 2003), showed that the system displays genuine diffusion for very long times even with a vanishing (spatial) entropy density, at least for sufficiently large $M$.

The above results are in perfect agreement with those of Dettmann and Cohen, and allow us to draw some conclusions on the fundamental ingredients for observing deterministic diffusion, in both chaotic and non-chaotic systems.

- An instability mechanism is necessary to ensure particle dispersion at small scales (in the model above, small scale means that the distance between two particles is smaller than $\Delta$). In chaotic systems this is realized by the sensitivity to the initial conditions. In non-chaotic systems this may be induced by a finite-size instability mechanism. Also, with zero maximal Lyapunov exponent one can have a rapid increase in the distance between two trajectories which are initially close (Torcini *et al.* 1995). In the wind-tree Ehrenfest model this stems from the edges of the obstacles, in the "stepwise" system of Figure 6.5 it stems from the jumps.
- A mechanism able to suppress periodic orbits and therefore to allow for diffusion at large scale.

The first requirement is not particularly strong while the second is more subtle. It is certainly fulfilled in systems with "strong chaos," where all periodic orbits are unstable. In non-chaotic systems, such as the map (6.59) and the polygonal billiards studied by Dettmann and Cohen, the stable periodic orbits seem to be suppressed or, at least, strongly depressed, by the quenched randomness (also in the limit of zero spatial entropy). However, unlike the two-dimensional non-chaotic billiards, in the one-dimensional system, the periodic orbits may survive in the presence of disorder, so a quasi-periodic perturbation is needed, to produce their destruction and the consequent diffusion.

The above results give a clear indication of the importance of pseudochaos, i.e. non-trivial behavior of systems with zero Lyapunov exponent. The relevance of this issue was already understood in the work of the late Ford (Vega *et al.* 1993, 1996), who stressed both the possible interesting behavior of some non-chaotic systems, and the connection between pseudochaos and the semiclassical limit; see also Mantica (2000) and Zaslavsky and Edelman (2004).

### 6.4.2 Heat transport in chaotic and non-chaotic systems

We saw that chaotic dynamics is not so important in the statistical mechanics description of some equilibrium and out-of-equilibrium phenomena; in particular, transport may also occur in the absence of deterministic chaos. This is so because a chaotic motion has the statistical properties of a "random walk," when observed at finite resolution. However this "random walking behavior" on a large scale can originate not only in a deterministic system with exponential instability, but also in a system with intrinsic disorder and "finite size instabilities." Heat conduction is another important example of such an issue.

In the context of the conduction problem, FPU chains (see Section 4.4) have recently played an important role in clarifying the transport properties of low spatial dimension systems. FPU models represent simple but non-trivial candidates to study heat transport by phonons in solids, whenever their boundaries are kept at different temperatures. This issue becomes even more interesting at low spatial dimension, e.g. $d = 1$, where the constraints set by the geometry may induce anomalous transport properties, characterized by the presence of divergent transport coefficients in the thermodynamic limit (Lepri *et al.* 2003). Thermal conductivity $\chi$, defined via Fourier's law,

$$J = -\chi \nabla T,$$

relates the heat (energy) flux $J$ to a temperature gradient. When a small temperature difference $\delta T = T_1 - T_2$ is applied to the ends of a system of linear size $L$, the heat current across the system is expected to be

$$J = -\frac{\chi \delta T}{L}.$$

For some one- and even two-dimensional systems, theoretical arguments, confirmed by several simulations, predict a scaling behavior $J \sim L^{\alpha-1}$ implying a size-dependent conductivity (Lepri *et al.* 2003):

$$\chi(L) \sim L^{\alpha}. \tag{6.60}$$

As a consequence, $\chi$ diverges in the limit $L \to \infty$ with a power law whose exponent $\alpha > 0$ depends on the specific system considered. The presence of this divergence is referred to as *anomalous heat conduction*, in contrast with normal conduction which, according to dimensional analysis of Fourier's law, prescribes a finite limit for the conductivity $\chi$.

FPU chains are systems where the anomaly in the heat transport is clearly observed. Its origin can be traced back to the existence of low-energy modes which survive long enough to propagate freely before scattering with other modes. Such modes can carry much energy and since their motion is mainly ballistic rather than diffusive, the overall heat transport happens to be anomalous. Models other than FPU indeed reveal such peculiar conduction, as shown widely in the literature (Grassberger *et al.* 2002, Eckmann and Young 2004, Gruber and Lesne 2005), therefore the issue is the general understanding of the conditions leading to this phenomenon.

A chaotic system, such as the Lorentz gas in a channel configuration, provides an example of normal heat conduction (Alonso *et al.* 1999). This model consists of a series of semicircular obstacles with radius $R$ arranged in a lattice along a slab of size $L \times h$ ($h \ll L$) see Figure 6.7. Two thermostats, at temperatures $T_1$ and $T_2$ respectively, are placed at each end of the slab to induce a non-equilibrium situation and presumably transport. They reinject into the system those particles reaching the ends with a velocity drawn from a Gaussian velocity distribution with variance

Figure 6.7 Example of channel geometry used to study heat transport in low-dimensional chaotic (upper panel) and non-chaotic billiards (lower panel).

proportional to the temperatures $T_1$ and $T_2$. In the case of semicircular obstacles the system is chaotic and one observes a standard Fourier's law (Alonso *et al.* 1999). Li *et al.* (2003) proposed some non-chaotic variants of the Lorentz channel, in order to unravel the role of exponential instabilities in heat conduction. In these models, called the Ehrenfest channel, the semicircular obstacles are replaced by triangular obstacles, so that the system is trivially non-chaotic, since collisions with flat edges of the obstacles cannot separate close trajectories more than algebraically. Numerical results (Grassberger *et al.* 2002, Li *et al.* 2003) show that when two angles ($\phi$ and $\psi$ in Figure 6.7) of the triangles are irrational multiples of $\pi$, the system exhibits normal heat conduction. However, for rational ratios, such as isosceles right triangles, the conduction becomes anomalous. The single-particle heat flux across $N$ cells, $J_1(N)$, scales as $J_1(N) \sim N^{\alpha}$, while the temperature gradient behaves as $1/N$ implying that $\chi(N)$ diverges as $N \to \infty$. The explanation of such a divergence can be found in the single-particle diffusivity along the channel direction which occurs in an anomalous way. Indeed, the evolution of an ensemble of particles has a mean square displacement from initial conditions which grows in time with a power law behavior:

$$\langle [x(t) - x(0)]^2 \rangle \sim t^b$$

with an exponent $1 < b < 2$. Such super-diffusion is uniquely responsible for divergent thermal conductivity, independently of Lyapunov instabilities, since the model has zero Lyapunov exponents.

When an Ehrenfest channel with anomalous thermal conductivity is disordered, for instance by randomly modulating the height of triangular obstacles or their positions along the channel, the conduction follows Fourier's law, becoming normal (Li *et al.* 2002).

Lepri *et al.* (2003) and Li and Wang (2003) suggested that the anomalous conduction is associated with the presence of a mean free path of energy carriers that can behave abnormally in the thermodynamic limit. For FPU the long mean free path is due to soliton-like ballistic modes. In the channels the long free flights, between consecutive particle collisions, become relevant.

The considerations above, and the results in the previous subsection on the diffusion of non-chaotic maps, suggest a very weak role of chaos in heat transport, and for transport in general, since systems without exponential instability may also show transport, even an anomalous kind.

## 6.5 Remarks and perspectives

Let us sum up and conclude with some remarks and perspectives on the issues discussed in this chapter.

(a) The relation between the Kolmogorov–Sinai entropy and the growth of the coarse-grained (Gibbs-like) entropy is rather weak. The main reason for this is the asymptotic nature of $h_{\mathrm{KS}}$ (and of the $\epsilon$-entropy), i.e. its value is determined by the behavior of the system at very large time intervals. In contrast, the growth of the coarse-grained entropy involves short time intervals. In addition, during the early time evolution of the coarse-grained Gibbs entropy one can have entanglement of behaviors at different characteristic space scales. This phenomenon is similar to that observed in the spreading of passive tracers in closed basins (Artale *et al.* 1997). In such a case, if the characteristic length scale of the Eulerian velocities is not very small, compared with the size of the basin, neither the diffusion coefficient nor the Lyapunov exponent gives relevant information about the mechanism of large-scale spreading.

(b) The study of a system made of weakly coupled subsystems gives evidence of the objective nature of the growth with time of the Boltzmann entropy. An intrinsic characteristic scale $\Delta_*$ emerges and the Boltzmann entropy does not depend on the coarse-graining resolution as long as it is smaller than this scale. Let us summarize the main conceptual results.
   - In spite of their apparent resemblance, systems composed of non-interacting subsystems and systems composed of weakly interacting subsystems are rather different, for some aspects. This is shown by an analysis of their Boltzmann and Gibbs entropies.
   - The chaotic properties of the single subsystem rule the rate of the increase in Boltzmann entropy, but the coupling is a fundamental ingredient which allows for the emergence of the intrinsic characteristic scale $\Delta_*$.
   - The Boltzmann entropy, which is a property of the $\mu$-space, does not require the introduction of an ensemble.

(c) A generalized fluctuation-dissipation theorem holds under rather general hypotheses. As a result one has that, if the system is mixing and the invariant measure is "smooth" enough, a connection exists between the "non-equilibrium" properties (response to external perturbations of the state or evolution law) and the "equilibrium" properties (correlation functions computed according to the invariant measure of the unperturbed system). The validity of the fluctuation-response relation does not depend on the deterministic or stochastic nature of the system or on the "equilibrium" or "non-equilibrium" character of its statistical steady state.

(d) Chaos (in the technical sense of positive Lyapunov exponent) is not a necessary ingredient for the validity of statistical mechanics laws such as diffusion and conduction. Detailed numerical results show that the basic elements are the following.
   - An instability mechanism able to induce particle dispersion at small scales. In chaotic systems this is realized by the sensitivity to the initial conditions; in systems with zero maximal Lyapunov exponent this mechanism may be induced by a finite-size instability mechanism.
   - A mechanism able to suppress periodic orbits and therefore to allow diffusion at large scale.

## References


Alonso, D., Artuso, R., Casati, G. and Guarneri, I. (1999). Heat conductivity and dynamical instability, *Phys. Rev. Lett.* **82**, 1859.

Artale, V., Boffetta, G., Celani, A., Cencini, M. and Vulpiani, A. (1997). Dispersion of passive tracers in closed basins: beyond the diffusion coefficient, *Phys. Fluids* **9**, 3162.

Boffetta, G., Lacorata, G., Musacchio, S. and Vulpiani, A. (2003). Relaxation of finite perturbations: beyond the fluctuation-response relation, *Chaos* **13**, 806.

Bohr, T., Jensen, M. H., Paladin, G. and Vulpiani, A. (1998). *Dynamical Systems Approach to Turbulence*, Cambridge: Cambridge University Press.

Bricmont, J. (1995). Science of chaos or chaos in science?, *Physicalia Mag.* **17**, 159.

Bunimovich, L. A. and Sinai, Ya. G. (1981). Statistical properties of Lorentz gas with periodic configuration of scatterers, *Commun. Math. Phys.* **78**, 479.

Callen, H. B. and Welton, T. A. (1951). Irreversibility and generalized noise, *Phys. Rev.* **83**, 34.

Casati, G. and Chirikov, B. V. (1995). Comment on decoherence, chaos, and the second law, *Phys. Rev. Lett.* **75**, 350.

Cecconi, F., del-Castillo-Negrete, D., Falcioni, M. and Vulpiani, A. (2003). The origin of diffusion: the case of non-chaotic systems, *Physica D* **180**, 129.

Dettmann, C. P. and Cohen, E. G. D. (2001). Note on chaos and diffusion, *J. Stat. Phys.* **103**, 589.

Dorfman, J. R. (1999). *An Introduction to Chaos in Nonequilibrium Statistical Mechanics*, Cambridge: Cambridge University Press.

Eckmann, J. P. and Young, L. S. (2004). Temperature profiles in Hamiltonian heat conduction, *Europhys. Lett.* **68**, 790.

Falcioni, M. and Vulpiani, A. (1995). The relevance of chaos for the linear response theory, *Physica A* **215**, 481.

Falcioni, M. Isola, S. and Vulpiani, A. (1990). Correlation functions and relaxation properties in chaotic dynamics and statistical mechanics, *Phys. Lett. A* **144**, 341.

Falcioni, M., Palatella, L. and Vulpiani, A. (2005). Production rate of the coarse-grained Gibbs entropy and the Kolmogorov–Sinai entropy: a real connection?, *Phys. Rev. E* **71**, 016118.

Gaspard, P. (1998). *Chaos, Scattering and Statistical Mechanics*, Cambridge: Cambridge University Press.

Gaspard, P. and Wang, X.-J. (1993). Noise, chaos, and ($\epsilon$, $\tau$)-entropy per unit time, *Phys. Rep.* **235**, 291.

Grassberger, P., Nadler, W. and Yang, L. (2002). Heat conduction and entropy production in a one-dimensional hard-particle gas, *Phys. Rev. Lett.* **89**, 180601.

Gruber, C. and Lesne, A. (2005). Hamiltonian model of heat conductivity and Fourier law, *Physica A* **351**, 358.

Huang, K. (1987). *Statistical Mechanics*, New York: Wiley.

Illner, R. and Pulvirenti, M. (1986). Global validity of the Boltzmann equation for a two-dimensional rare gas in a vacuum, *Commun. Math. Phys.* **105**, 189.

Kraichnan, R. H. (1959). Classical fluctuation-relaxation theorem, *Phys. Rev.* **113**, 1181.

Kraichnan, R. H. and Montgomery, D. (1980). Two-dimensional turbulence, *Rep. Prog. Phys.* **43**, 547.

Kubo, R. (1966). The fluctuation-dissipation theorem, *Rep. Prog. Phys.* **29**, 255.

Kubo, R. (1986). Brownian motion and nonequilibrium statistical mechanics, *Science* **233**, 330.

Lanford III, O. E. (1981). The hard sphere gas in the Boltzmann–Grad limit, *Physica A* **106**, 70.
Latora, V. and Baranger, M. (1999). Kolmogorov–Sinai entropy rate versus physical entropy, *Phys. Rev. Lett.* **82**, 520.
Lebowitz, J. (1993). Boltzmann's entropy and time's arrow, *Phys. Today* **46**, 32. Several comments on the paper were published on the same journal, together with Lebowitz's replies: Barnum, H., Caves, C. M., Fuchs, C. and Schack, R. **47**, 11 (1994). Driebe, J. **47**, 13 (1994). Hoover, W. G., Posch, H. and Holian, B. L. **47**, 15 (1994). Peierls, R. **47**, 115 (1994).
Leith, C. E. (1975). Climate response and fluctuation dissipation, *J. Atmos. Sci.* **32**, 2022.
Lepri, S., Livi, R. and Politi, A. (2003). Thermal conduction in classical low-dimensional lattices, *Phys. Rep.* **377**, 1.
Li, B. and Wang, J. (2003). Anomalous heat conduction and anomalous diffusion in one-dimensional systems, *Phys. Rev. Lett.* **91**, 044301.
Li, B., Wang, L. and Hu, B. (2002). Finite thermal conductivity in 1D models having zero Lyapunov exponents, *Phys. Rev. Lett.* **88**, 223901.
Li, B., Casati, G. and Wang, J. (2003). Heat conductivity in linear mixing systems, *Phys. Rev. E* **67**, 021204.
Lorentz, H. A. (1905). The motion of electrons in metallic bodies, *Proc. Amst. Acad.* **7**, 438, 585, 684.
Lorenz, E. N. (1963). Deterministic nonperiodic flow, *J. Atmos. Sci.* **20**, 130.
Mantica, G. (2000). Quantum algorithmic integrability: the metaphor of classical polygonal billiards, *Phys. Rev. E* **61**, 6434.
McComb, W. D. and Kiyani, K. (2005). Eulerian spectral closures for isotropic turbulence using a time-ordered fluctuation-dissipation relation, *Phys. Rev. E* **72**, 016309.
Pattanayak, A. K. (2001). Characterizing the metastable balance between chaos and diffusion, *Physica D* **148**, 1.
Rondoni, L. and Cohen, E. G. D. (2000). Gibbs entropy and irreversible thermodynamics, *Nonlinearity* **13**, 1905.
Rose, R. H. and Sulem, P. L. (1978). Fully developed turbulence and statistical mechanics, *J. Phys (Paris)* **39**, 441.
Ruelle, D. (1989). *Chaotic Evolution and Strange Attractors*, Cambridge: Cambridge University Press.
Ruelle, D. (1998). General linear response formula in statistical mechanics, and the fluctuation-dissipation theorem far from equilibrium, *Phys. Lett. A* **245**, 220.
Sinai, Ya. G. (1979). Ergodic properties of the Lorentz gas, *Funct. Anal. Appl.* **13**, 192.
Tél, T. and Vollmer, J. (2000). Entropy balance, multibaker maps, and the dynamics of the Lorentz gas, in *Hard Ball Systems and the Lorentz Gas*, ed. D. Szasz, Springer Encyclopaedia of Mathematical Sciences.
Tolman, R. C. (1980). *The Principles of Statistical Mechanics*, New York: Dover.
Torcini, A., Grassberger, P. and Politi, A. (1995). Error propagation in extended chaotic systems, *J. Phys. A* **28**, 4533.
van Kampen, N. G. (1971). The case against linear response theory, *Phys. Norv.* **5**, 279.
Vega, J. L., Uzer, T. and Ford, J. (1993). Chaotic billiards with neutral boundaries, *Phys. Rev. E* **48**, 3414.
Vega, J. L., Uzer, T., Borondo, F. and Ford, J. (1996). Deterministic diffusion in almost integrable systems, *Chaos* **6**, 519.

Zaslavsky, G. M. and Edelman, M. A. (2004). Fractional kinetics: from pseudochaotic dynamics to Maxwell's demon, *Physica D* **193**, 128.
Zurek, W. H. and Paz, J. P. (1994). Decoherence, chaos, and the second law, *Phys. Rev. Lett.* **72**, 2508.
Zurek, W. H. and Paz, J. P. (1995). A reply to the comment by Giulio Casati and B. V. Chirikov, *Phys. Rev. Lett.* **75**, 351.

# 7

# Coarse-graining equations in complex systems

> To develop the skill of correct thinking is in the first place to learn what you have to *disregard*. In order to go on, you have to know what to leave out: this is the essence of effective thinking.
>
> *Kurt Gödel*

Almost all the interesting dynamic problems in science and engineering are characterized by the presence of more than one significant scale, i.e. there is a variety of degrees of freedom with very different time scales. Among numerous important examples we can mention protein folding and climate. While the time scale of vibration of covalent bonds is $O(10^{-15}\ \mathrm{s})$, the folding time for proteins may be of the order of seconds. Also in the case of climate, the characteristic times of the involved processes vary from days (for the atmosphere) to $O(10^3\ \mathrm{yr})$ (for the deep ocean and ice shields). In such a situation one says that the system has a *multiscale*[1] character (E and Engquist 2003).

The necessity of treating the "slow dynamics" in terms of effective equations is both practical (even modern supercomputers are not able to simulate all the relevant scales involved in certain difficult problems) and conceptual: effective equations are able to catch some general features and to reveal dominant ingredients which can remain hidden in the detailed description. The study of multiscale problems has a long history in science (in particular in mathematics): an early important example is the averaging method in mechanics (Arnold 1976). Consider the Hamiltonian equations written in the action-angle variables:

$$
\begin{aligned}
\frac{\mathrm{d}\phi}{\mathrm{d}t} &= \frac{1}{\epsilon}\big[\omega(I) + f(\phi, I)\big] \\
\frac{\mathrm{d}I}{\mathrm{d}t} &= g(\phi, I)
\end{aligned}
\tag{7.1}
$$

[1] In the literature one also finds multi-scale and multiple-scale.

where the functions $f$ and $g$ are $2\pi$-periodic in $\phi$. Assuming $\epsilon \ll 1$, $\phi$ and $I$ are the fast (i.e. of time scale $O(\epsilon)$) and slow (of time scale $O(1)$) variables respectively. In the averaging method one introduces a smoothed action $J$, which describes the "slow motion," as obtained by the filtering of the fast variable oscillations. The dynamics of $J$ is ruled by the force acting on $I$ averaged over the fast variable $\phi$:

$$\frac{\mathrm{d}J}{\mathrm{d}t} = G(J) = \frac{1}{2\pi}\int_0^{2\pi} g(\phi, J)\mathrm{d}\phi. \tag{7.2}$$

The evolution of $J$ gives the leading order behavior of $I$, see Arnold (1976).

Perhaps, at a fundamental level, the most important multiscale problem is the rigorous derivation of the hydrodynamic equations from the microscopic level (i.e. the deterministic Newton equations). This issue has an old and noble story, starting with the seminal work by Boltzmann (Cercignani 1988, 1998, Cercignani *et al.* 1994) and continuing up to the recent attempts to model fluidodynamics with a coarse-graining procedure of cellular automaton systems (Frisch *et al.* 1986). Here we do not treat this intriguing and difficult topic, but we believe it is important to recall the basic facts that allow us to derive macroscopic equations from the microscopic dynamics.

The particle configurations on a molecular scale are quickly randomized by collisions, therefore a local equilibrium[2] is attained, described by a few macroscopic quantities (mass density, temperature, momentum density, pressure and so on). The key ingredient is the molecular chaos, discussed in Chapter 5. The partial differential equations of hydrodynamics rule the large-scale evolution of such macroscopic variables. The interesting fact is that the macroscopic equations are determined essentially by very general features (such as symmetry properties and conservation laws) and not by the precise details of the underlying microscopic dynamics. Therefore one has the remarkable result that the same macroscopic equation can be obtained for systems with completely different microscopic dynamics, for example

(a) deterministic systems evolving in a continuous way (i.e. molecules interacting with a short-range potential), and
(b) cellular automata with discrete character, in both time and configuration states, whose evolution is driven by probabilistic rules.

This is basically due to the large-scale separation between the microscopic characteristic time $\tau_{\text{micro}} = \ell/v_t$ ($\ell$ is a characteristic microscopic length, e.g. the mean

[2] There is seemingly a paradox here, ubiquitous in all multiscale approaches: we speak at the same time of "local equilibrium" and "evolution" of the macroscopic variables. The solution lies in the very notion of equilibrium, meaning equilibrium at short time scales, large enough compared to molecular scales and the mixing time $\tau_{\text{micro}}$ associated with collisions and molecular chaos, but small compared to macroscopic (observation) scales. As in "quasistatic" evolutions encountered in thermodynamics, the large-scale evolution will be composed of a continued succession of local equilibrium states.

free path, and $v_t$ is the typical speed of the molecules) and the macroscopic time $\tau_{\text{macro}} = L/U$ ($L$ is a macroscopic length, for instance the size of the vessel containing the fluid, and $U$ is the typical velocity at hydrodynamic level).

We can sketch briefly the passage from the microscopic level to the macroscopic level with the following scheme (Español 2003):

I microscopic level, $\Gamma$-space description (Liouville equation);
II microscopic level, $\mu$-space description (Boltzmann equation);
III mesoscopic level, $\mu$-space description but at "large scale" (Fokker–Planck equation);
IV macroscopic level, fluidodynamics description (Navier–Stokes equation, Fourier law, ...).

The crossing from one level of description to another is determined by a coarse-graining and/or a projection procedure with a "loss of information"; and rather delicate mathematical singular limits are involved.

## 7.1 A short parenthesis: secular terms and multiscale analysis

Let us start with a brief discussion of multiscale analysis in the context of the perturbation theory techniques for ordinary differential equations (Bender and Orszag 1978). We can see at work, in a rather simple example, many of the basic tools necessary for the study of important phenomena, such as diffusion and mesoscopic description of non-equilibrium statistical mechanics.

Consider a non-linear oscillator ruled by Duffing's equation

$$\frac{d^2x}{dt^2} + x + \epsilon x^3 = 0, \tag{7.3}$$

with the initial conditions $x(0) = 1$ and $dx(0)/dt = 0$. It is easy to show that a naive perturbation approach is plagued by the appearence of secular terms, that are unbounded in time, typically increasing as $t$, and that imply a non-uniform validity of the perturbation expansion at large $t$. Let us expand $x(t)$ as a power series in $\epsilon$:

$$x(t) = x_0(t) + \epsilon x_1(t) + \epsilon^2 x_2(t) + \cdots \tag{7.4}$$

where $x_0(0) = 1, dx_0(0)/dt = 0$ and $x_n(0) = dx_n(0)/dt = 0$ for $n \geq 1$. Substituting (7.4) into (7.3) and equating to zero each coefficient of the resulting power series in $\epsilon$ one obtains a sequence of ordinary differential equations:

$$\frac{d^2x_0}{dt^2} + x_0 = 0 \tag{7.5}$$

$$\frac{d^2x_1}{dt^2} + x_1 = -x_0^3 \tag{7.6}$$

and so on. Elementary manipulations give the solution

$$x_0(t) = \cos(t), \quad x_1(t) = \frac{1}{32}\cos(3t) - \frac{1}{32}\cos(t) - \frac{3t}{8}\sin(t) \tag{7.7}$$

showing that $x_1(t)$ contains a secular term $O(t)$. Because of the conservation of energy,

$$\frac{1}{2}\left(\frac{\mathrm{d}x}{\mathrm{d}t}\right)^2 + \frac{1}{2}x^2 + \frac{\epsilon}{4}x^4, \tag{7.8}$$

the exact solution $x(t)$ must be bounded, therefore $x_0 + \epsilon x_1$ can be a fair approximation only for $t < t_*(\epsilon) = O(1/\epsilon)$.

The secular term appears because of the forcing term in Eq. (7.6) $x_0^3(t) = \cos^3(t) = \cos(3t)/4 + 3\cos(t)/4$ which contains a component with the same frequency as the unperturbed equation for $x_0$, i.e. $3\cos(t)/4$. A straightforward, but lengthy (and inelegant), way to eliminate the most diverging secular contributions to the perturbation theory is to sum all orders in powers of $\epsilon$ to recover the bounded exact behavior. This computation can be avoided with multiscale analysis. Let us introduce a new variable $\tau = \epsilon t$; since $\tau$ is appreciable only when $t > t_*(\epsilon) = O(1/\epsilon)$, it corresponds to a long time scale. The solution $x(t)$ is a function of $t$ alone, but in the multiscale analysis one treats $t$ and $\tau$ as independent variables. The additional freedom thus introduced will be compensated for in the course of the computation by imposing a *solvability condition* ensuring the vanishing of the secular terms and the consistency of the perturbation method. The procedure consists in assuming a perturbation expansion:

$$x(t) = x_0(t,\tau) + \epsilon x_1(t,\tau) + \epsilon^2 x_2(t,\tau) + \cdots \tag{7.9}$$

and replacing $\mathrm{d}/\mathrm{d}t$ with $\partial/\partial t + \epsilon\partial/\partial\tau$. This last recipe follows from the chain rule for partial differentiation

$$\frac{\mathrm{d}f}{\mathrm{d}t} = \frac{\partial f}{\partial t} + \frac{\partial f}{\partial \tau}\frac{\partial \tau}{\partial t} = \left(\frac{\partial}{\partial t} + \epsilon\frac{\partial}{\partial \tau}\right) f. \tag{7.10}$$

Simple computations give

$$\frac{\partial^2 x_0}{\partial t^2} + x_0 = 0 \tag{7.11}$$

$$\frac{\partial^2 x_1}{\partial t^2} + x_1 = -x_0^3 - 2\frac{\partial^2 x_0}{\partial t \partial \tau}. \tag{7.12}$$

The general solution of (7.11) is

$$x_0(t,\tau) = A(\tau)\mathrm{e}^{\mathrm{i}t} + A^*(\tau)\mathrm{e}^{-\mathrm{i}t}, \tag{7.13}$$

so the $t$-dependence contributes only to fast, bounded evolution. The function $A(\tau)$ must be determined by the condition that secular terms do not appear in $x_1(t,\tau)$. The right-hand side of Eq. (7.12) is

$$\left(-3A^2A^* - 2\mathrm{i}\frac{\mathrm{d}A}{\mathrm{d}\tau}\right)\mathrm{e}^{\mathrm{i}t} + \left(-3(A^*)^2A + 2\mathrm{i}\frac{\mathrm{d}A^*}{\mathrm{d}\tau}\right)\mathrm{e}^{-\mathrm{i}t} - A^3\mathrm{e}^{3\mathrm{i}t} - (A^*)^3\mathrm{e}^{-3\mathrm{i}t}. \tag{7.14}$$

Since $\mathrm{e}^{\mathrm{i}t}$ and $\mathrm{e}^{-\mathrm{i}t}$ are solutions of (7.11), in order to preclude the appearance of secular terms, the coefficients of $\mathrm{e}^{\mathrm{i}t}$ and $\mathrm{e}^{-\mathrm{i}t}$ in (7.14) must be equal to zero. Therefore $A(\tau)$ must satisfy the equation

$$-3A^2A^* - 2\mathrm{i}\frac{\mathrm{d}A}{\mathrm{d}\tau} = 0. \tag{7.15}$$

The solution of the above equation can be obtained by writing $A(\tau) = R(\tau)\mathrm{e}^{\mathrm{i}\theta(\tau)}$, where $R$ and $\theta$ are real functions. A simple computation gives

$$x_0(t,\tau) = 2R(0)\cos\left[\theta(0) + \frac{3}{2}R^2(0)\tau + t\right], \tag{7.16}$$

where $R(0)$ and $\theta(0)$ are determined by the initial conditions; if $x(0) = 1$ and $\mathrm{d}x(0)/\mathrm{d}t = 0$ one has

$$x(t) = \cos\left[t\left(1 + \frac{3}{8}\epsilon\right)\right] + O(\epsilon). \tag{7.17}$$

The above approximation is now in good agreement with the actual solution also for large values of $t$.

## 7.2 From molecular level to Brownian motion

Brownian motion played a central role in the development of physics and mathematics, see Section 1.2.1.

A rigorous general derivation of Brownian motion from first principles is a formidable task. Here we want to discuss the steps (via coarse-graining procedures) from molecular dynamics to Brownian motion, stressing mainly the conceptual aspects (Español 2003).

Consider a system of colloidal particles suspended in a liquid. At the microscopic level we introduce the canonical coordinates $(\mathbf{Q}_i, \mathbf{P}_i)$ and $(\mathbf{q}_n, \mathbf{p}_n)$ of colloidal particles and solvent molecules respectively. Omitting external potentials, the complete

Hamiltonian is

$$H = \sum_i \frac{\mathbf{P}_i^2}{2M} + \sum_n \frac{\mathbf{p}_n^2}{2m} + \sum_{n,l,i,j} (V^{ss}(\mathbf{q}_j - \mathbf{q}_i) + V^{cc}(\mathbf{Q}_n - \mathbf{Q}_l) + V^{sc}(\mathbf{Q}_n - \mathbf{q}_i)), \tag{7.18}$$

where $m$ is the mass of a solvent molecule, $M$ is the mass of a colloidal particle (we assume $M \gg m$), $V^{ss}$, $V^{sc}$ and $V^{cc}$ are the potentials of the forces between solvent molecules, solvent and colloidal particles, colloidal particles, respectively.

The evolution of such a system is ruled by the Hamiltonian equations:

$$\begin{aligned} \frac{d\mathbf{Q}_i}{dt} &= \frac{\partial H}{\partial \mathbf{P}_i}, & \frac{d\mathbf{q}_n}{dt} &= \frac{\partial H}{\partial \mathbf{p}_n}, \\ \frac{d\mathbf{P}_i}{dt} &= -\frac{\partial H}{\partial \mathbf{Q}_i}, & \frac{d\mathbf{p}_n}{dt} &= -\frac{\partial H}{\partial \mathbf{q}_n}. \end{aligned} \tag{7.19}$$

The solutions of these equations give the most detailed description of the system. If we are interested in the colloidal subsystem alone, the next, less accurate, level of description is obtained by integrating over the degrees of freedom of the solvent particles. In this case, the future state of the suspended particles is not determined solely by a given $\{\mathbf{Q}_i, \mathbf{P}_i\}$ configuration, but also depends on the past history of the subsystem (a unique evolution is obtained only if one knows the complete microscopic state of the system at a given time). This means that the dynamical equations for the variables $(\mathbf{Q}_i, \mathbf{P}_i)$ must contain memory effects and, in general, cannot be first order in time. However, since in comparison with the solvent molecules the colloidal particles have a much larger mass, they have a much slower evolution. Then, because of this time-scale separation between the two subsystems, and because of the huge number of solvent particles, we can suppose that the fast solvent dynamics can be consistently decoupled from the slow colloid dynamics, by approximating its effects on the large suspended particles by means of an effective force. This force can be decomposed into a systematic part, of viscous type, and a truly stochastic fluctuating part, almost like white noise. In such a limit of very different masses, we recover a practically Markovian evolution (i.e. first order in time) for the colloidal subsystem, that is driven by a stochastic equation:

$$\begin{aligned} \frac{d\mathbf{Q}_i}{dt} &= \mathbf{V}_i \\ \frac{d\mathbf{P}_i}{dt} &= \mathbf{F}_i - \sum_j \widetilde{\zeta}_{ij} \mathbf{V}_j + \mathbf{G}_i \end{aligned} \tag{7.20}$$

where $\mathbf{V}_i = \mathbf{P}_i/M$, $\mathbf{F}_i$ is the force on the $i$th particle due to the interactions with other colloidal particles (and possibly external potentials), and $\widetilde{\zeta}_{ij}$ is the friction tensor (originating from the interaction of the solvent with the colloidal particles) that can depend on the **Q** variables. The stochastic component of the force, $\mathbf{G}_i$, is a Gaussian process with $\langle G_i^k(t)\rangle = 0$ and $\langle G_i^k(t)\, G_j^l(t')\rangle = \alpha_{ij}^{kl}\,\delta(t-t')$, where $G_i^k$ $(k = 1, 2, 3)$ indicates the $k$th spatial component of $\mathbf{G}_i$. Here $\widetilde{\alpha}_{ij}$ is a tensorial quantity in the spatial indices, like the friction tensor $\widetilde{\zeta}_{ij}$. The fluctuation-dissipation theorem requires that $\alpha_{ij}^{kl} = 2k_\mathrm{B}T\zeta_{ij}^{kl}$, where $T$ is the temperature of the solvent, and $k_\mathrm{B}$ is the Boltzmann constant (van Kampen 1990).

If the colloidal suspension is dilute then we expect that the mutual influence among the colloidal particles is negligible, so $\mathbf{F}_i$ is due only to possible external potentials and the friction tensor reduces to a scalar quantity: $\zeta_{ij}^{kl} = \zeta\delta_{ij}\delta^{kl}$, $\zeta$ being the friction coefficient. In this case Eq. (7.20) becomes the well-known Langevin equation, for the independent evolution of each colloidal particle:

$$\begin{aligned}\frac{\mathrm{d}\mathbf{Q}}{\mathrm{d}t} &= \mathbf{V}\\ \frac{\mathrm{d}\mathbf{P}}{\mathrm{d}t} &= \mathbf{F}(\mathbf{Q}) - \zeta\mathbf{V} + \sqrt{2k_\mathrm{B}T\zeta}\;\mathbf{g}\end{aligned} \tag{7.21}$$

where, for the components of the random vector **g**, one has $\langle g^k(t)\rangle = 0$ and $\langle g^k(t)g^l(t')\rangle = \delta^{kl}\delta(t-t')$.

However, at this point we observe that, for a typical colloidal suspension, the time scale over which the variables **P** evolve is very short, $O(10^{-6}$ s), compared with the time scale, $O(10^{-3}$ s), of the evolution of **Q**. This suggests that, if one is interested only in phenomena occurring on time scales above $O(10^{-3}$ s), the possibility exists of another level of description, looking only at the **Q** variables. In this case, a suitable equation for the colloidal particle position variables is:

$$\frac{\mathrm{d}\mathbf{Q}_i}{\mathrm{d}t} = \frac{1}{k_\mathrm{B}T}\sum_j \widetilde{D}_{ij}\mathbf{F}_j + \sum_j \frac{\partial}{\partial\mathbf{Q}_j}\widetilde{D}_{ij} + \mathbf{W}_i, \tag{7.22}$$

where the force $\mathbf{F}_i$ is the same as in Eq. (7.20), $\widetilde{D}_{ij}$ is the diffusion tensor (which, in general, depends on **Q**) and $\mathbf{W}_i$ is a Gaussian stochastic contribution to the velocity of the particle. We require that $\langle\mathbf{W}_i\rangle = 0$ and, by the fluctuation-dissipation theorem, $\langle W_i^k(t)W_j^l(t')\rangle = 2D_{ij}^{kl}\,\delta(t-t')$. Also in this case, for a dilute suspension the equations simplify, since the diffusion tensor becomes a scalar quantity: $D_{ij}^{kl} = \delta_{ij}\delta^{kl}D$, where $D$ is the self-diffusion coefficient of the colloidal particles. Equation (7.22) becomes

$$\frac{\mathrm{d}\mathbf{Q}}{\mathrm{d}t} = \frac{D}{k_\mathrm{B}T}\mathbf{F} + \frac{\partial D}{\partial\mathbf{Q}} + \sqrt{2D}\,\mathbf{w}, \tag{7.23}$$

where the random vector **w** enjoys the same properties as the vector **g** defined above.

Finally we see that, owing to the large difference of masses, and to the large number of solvent molecules, the memory effects, that arise from disregarding some details, can be well approximated by a stochastic action of Markovian type on the colloidal particles. That is, the unpredictability remains, but the future state of the subsystem depends only on its present state.

Once the original deterministic dynamics, because of the graining, has been transformed into a stochastic evolution, it is unavoidable to reason in terms of probabilities for the state of the studied (sub)system. This means that, for instance, in the first level of graining, the problem to be posed is: given the colloidal system in the state $\{\mathbf{Q}_i, \mathbf{P}_i\}_0$ at time $t = 0$, what is the probability density function $\rho_c(\{\mathbf{Q}_i, \mathbf{P}_i\}, t)$ of finding it in the state $\{\mathbf{Q}_i, \mathbf{P}_i\}$ at a later time $t$ (we do not write explicitly the dependence of $\rho_c$ on the initial state).

Moreover, usually one does not know the initial microscopic state of a system, so that another source of uncertainty has to be considered, in the form of a distribution on the possible initial states. This is not an uncertainty springing from the randomness of the dynamics: indeed it is also present when the detailed dynamical equations (7.19) drive the system. It is at the base of the ensemble point of view, *à la* Gibbs. Thus, we begin from the description at the microscopic level. In this case the probability density at time $t = 0$ is defined in the whole $\Gamma$-space $\rho_L(t = 0) = \rho_L(\{\mathbf{Q}_i, \mathbf{P}_i\}_0, \{\mathbf{q}_n, \mathbf{p}_n\}_0)$ and evolves according to the Liouville equation

$$\frac{\partial \rho_L}{\partial t} = -\{\rho_L, H\}, \tag{7.24}$$

where $\{\rho_L, H\}$ is the Poisson bracket between $\rho_L$ and $H$, with respect to the full set of canonical variables. If one is able to find the solution $\rho_L(t)$, then one also has the density involving only the variables of the colloidal particles. For the first level of approximation one gets

$$\rho_c(\{\mathbf{Q}_i, \mathbf{P}_i\}, t) = \int \rho_L(\{\mathbf{Q}_i, \mathbf{P}_i\}, \{\mathbf{q}_n, \mathbf{p}_n\}, t)\ \Pi_n \mathrm{d}\mathbf{q}_n \mathrm{d}\mathbf{p}_n, \tag{7.25}$$

and for the second level

$$\rho(\{\mathbf{Q}_i\}, t) = \int \rho_c(\{\mathbf{Q}_i, \mathbf{P}_i\}, t)\ \Pi_i \mathrm{d}\mathbf{P}_i. \tag{7.26}$$

Of course, when the exact solution is not attainable, one can resort to an approximate one.

In the first graining level one has to solve the evolution equation for the probability density function of systems evolving according to the random dynamics (7.20), that is the Fokker–Planck equation:[3]

$$\frac{\partial \rho_c}{\partial t} = -\sum_i \left( \mathbf{V}_i \frac{\partial}{\partial \mathbf{Q}_i} + \mathbf{F}_i \frac{\partial}{\partial \mathbf{P}_i} \right) \rho_c + k_B T \sum_{i,j} \frac{\partial}{\partial \mathbf{P}_i} \tilde{\zeta}_{ij} \left( \frac{\partial}{\partial \mathbf{P}_j} + \frac{\mathbf{P}_j}{M k_B T} \right) \rho_c. \tag{7.27}$$

When system (7.20) gives a satisfying approximation of the dynamics, the solution of Eq. (7.27), with initial condition obtained by $\rho_L(t = 0)$, must be a good approximation of the density (7.25).

At the next level of graining, where the stochastic dynamics is given by Eq. (7.22), the probability density function $\rho(\{\mathbf{Q}_i\}, t)$ of the colloidal particle positions evolves according to the so-called Smoluchowski equation:

$$\frac{\partial}{\partial t} \rho = -\sum_{ij} \frac{\partial}{\partial \mathbf{Q}_i} \left[ \frac{\tilde{D}_{ij} \mathbf{F}_j}{k_B T} \rho \right] + \sum_{ij} \frac{\partial}{\partial \mathbf{Q}_i} \tilde{D}_{ij} \frac{\partial}{\partial \mathbf{Q}_j} \rho. \tag{7.28}$$

When the dilute approximation is applicable, the interesting probability densities depend only on single-particle variables, and the equations above simplify accordingly.

### *7.2.1 How to derive the Smoluchowski equation from the Kramers equation with a multiscale approach*

The transition from the Kramers equation (i.e. Fokker–Planck equation for $\mathbf{Q}$ and $\mathbf{P}$) to the Smoluchowski equation (i.e. Fokker–Planck equation only for $\mathbf{Q}$) can be explained with a simple heuristic argument, for a dilute suspension, in the limit of very large $\zeta$. For the sake of simplicity we consider the one-dimensional case, and assume the possible dependence of $D$ and $\zeta$ on $Q$ to be so weak that it can be neglected.

With these assumptions, the Kramers equation reduces to:

$$\frac{\partial}{\partial t} \rho_c(Q, P, t) = -\left( V \frac{\partial}{\partial Q} + F(Q) \frac{\partial}{\partial P} \right) \rho_c + \zeta k_B T \frac{\partial}{\partial P} \left( \frac{\partial}{\partial P} + \frac{P}{M k_B T} \right) \rho_c, \tag{7.29}$$

[3] The Fokker–Planck equation (sometimes called the Chapman–Kolmogorov equation) determines the time evolution of the probability density function of a Markov process according to the hypothesis that in a small time interval the state of the system (here determined by the variables $\{\mathbf{Q}_i, \mathbf{P}_i\}$) does not change too much; for details see van Kampen (1990). We recall that the Fokker–Planck equation corresponding to a given stochastic dynamics has a form that depends, in general, on how the stochastic integration is defined, either according to Itô or according to Stratonovich, see Gardiner (1990). Here we follow the Itô calculus.

and the underlying dynamics is given by the one-dimensional Langevin equation (7.21):

$$\begin{cases} \mathrm{d}Q/\mathrm{d}t = V \\ \mathrm{d}P/\mathrm{d}t = F(Q) - \zeta V + \sqrt{2k_{\mathrm{B}}T\zeta}\; g. \end{cases} \tag{7.30}$$

The Smoluchowski equation becomes

$$\frac{\partial}{\partial t}\rho = -\frac{D}{k_{\mathrm{B}}T}\frac{\partial}{\partial Q}(F(Q)\rho) + D\frac{\partial^2}{\partial Q^2}\rho; \tag{7.31}$$

and the one-dimensional version of Eq. (7.23) is

$$\frac{\mathrm{d}Q}{\mathrm{d}t} = \frac{D}{k_{\mathrm{B}}T}\, F + \sqrt{2D}\, w, \tag{7.32}$$

where the constancy of $D$ has been taken into account.

A relationship can be readily established between the stochastic equations, (7.30) and (7.32). One considers Eq. (7.30), and notes that the time scale of the velocity variable, which is proportional to $1/\zeta$, is much shorter than that of the spatial variable. If one is interested only in the slower $Q$-evolution, one can treat $V$ as a relaxed variable and one can put formally $\mathrm{d}P/\mathrm{d}t = 0$ in the second equation of (7.30). Therefore one can express $V$ in terms of $Q$ and $g$:

$$V = \frac{1}{\zeta}F(Q) + \sqrt{\frac{2k_{\mathrm{B}}T}{\zeta}}\; g. \tag{7.33}$$

The first equation of (7.30) then yields a Langevin equation for $Q$:

$$\frac{\mathrm{d}Q}{\mathrm{d}t} = \frac{1}{\zeta}F(Q) + \sqrt{\frac{2k_{\mathrm{B}}T}{\zeta}}\, g \tag{7.34}$$

which is the process underlying the Fokker–Planck equation (7.31), if one puts

$$D = \frac{k_{\mathrm{B}}T}{\zeta}. \tag{7.35}$$

This is the celebrated Einstein relation, giving a link between the microscopic and macroscopic levels. We recall that for the validity of this relation, and all the relations involving the fluctuation-dissipation theorem, one assumes that the velocity variables, for $t \gg 1/\zeta$, are described by a statistical equilibrium so that energy equipartition holds.

In spite of the above simple argument, a consistent perturbative derivation of (7.31) from (7.29) was obtained only about 40 years ago. In fact in well-known textbooks one can find an inconsistent derivation, based on a perturbative scheme, in the spirit of the Hilbert approach to the Boltzmann equation (Cercignani 1988,

Gorban *et al.* 2004). In this approach hidden secular terms lead to a secular divergence, i.e. a non-uniform convergence of the expansion for small $1/\zeta$; see van Kampen (1990), section VII.7. The key point is the singular nature of the limit $1/\zeta \to 0$ which does not allow a straightforward application of perturbation theory, see Bocquet (1997) for a detailed discussion.

Let us show how, using the multiscale analysis, one can derive in a consistent way Eq. (7.31) from Eq. (7.29) when $1/\zeta \to 0$. First we introduce new suitable dimensionless variables

$$\tau = t\frac{v_t}{\ell}, \quad v = \frac{V}{v_t}, \quad x = \frac{Q}{\ell}, \quad f = F\frac{\ell}{Mv_t^2}, \quad \zeta_d = \zeta\frac{\ell}{v_t}, \tag{7.36}$$

where $v_t = \sqrt{k_B T/M}$ is the thermal velocity and $\ell$ is a characteristic length scale (e.g. the grain size). The Kramers equation takes the form:

$$\frac{\partial}{\partial v}\left(v + \frac{\partial}{\partial v}\right)p(x, v, \tau) = \frac{1}{\zeta_d}\left(\frac{\partial}{\partial \tau} + v\frac{\partial}{\partial x} + f(x)\frac{\partial}{\partial v}\right)p. \tag{7.37}$$

Following the multiscale procedure we replace $p$ with

$$\begin{aligned} &p_0(x, v, \tau_0, \tau_1, \tau_2, \ldots) + \frac{1}{\zeta_d}p_1(x, v, \tau_0, \tau_1, \tau_2, \ldots) \\ &+ \frac{1}{\zeta_d^2}p_2(x, v, \tau_0, \tau_1, \tau_2, \ldots) + \cdots \end{aligned} \tag{7.38}$$

where $\tau_0 = \tau,\ \tau_1 = \tau_0/\zeta_d,\ \tau_2 = \tau_1/\zeta_d, \ldots$ and

$$\frac{\partial}{\partial \tau} \to \frac{\partial}{\partial \tau_0} + \frac{1}{\zeta_d}\frac{\partial}{\partial \tau_1} + \frac{1}{\zeta_d^2}\frac{\partial}{\partial \tau_2} + \cdots. \tag{7.39}$$

By simple computations one obtains:

$$\mathcal{L}p_0 = 0, \tag{7.40}$$

$$\mathcal{L}p_1 = \left(\frac{\partial}{\partial \tau_0} + v\frac{\partial}{\partial x} + f(x)\frac{\partial}{\partial v}\right)p_0, \tag{7.41}$$

$$\mathcal{L}p_2 = \left(\frac{\partial}{\partial \tau_0} + v\frac{\partial}{\partial x} + f(x)\frac{\partial}{\partial v}\right)p_1 + \frac{\partial}{\partial \tau_1}p_0, \tag{7.42}$$

where

$$\mathcal{L} = \frac{\partial}{\partial v}\left(v + \frac{\partial}{\partial v}\right). \tag{7.43}$$

The zeroth-order equation (7.40) imposes a Maxwellian velocity distribution

$$p_0(x, v, \tau_0, \tau_1, \ldots) = \Phi(x, \tau_0, \tau_1, \ldots)\mathrm{e}^{-v^2/2}, \tag{7.44}$$

where the function $\Phi$ has to be determined. The first-order equation (7.41) gives

$$\mathcal{L}p_1 = \frac{\partial}{\partial \tau_0}\Phi\, e^{-v^2/2} + v\Big(\frac{\partial}{\partial x}\Phi - f(x)\Phi\Big)e^{-v^2/2}. \tag{7.45}$$

The solvability condition imposes

$$\frac{\partial}{\partial \tau_0}\Phi = 0. \tag{7.46}$$

This can be understood in terms of the Fredholm alternative. We recall briefly the basic fact. Consider a linear operator $\mathcal{A}$ with eigenvalues $\lambda_0 = 0$ and $\lambda_n \neq 0$ for $n \neq 0$ and the corresponding eigenvectors $\mathbf{e}_0, \mathbf{e}_1, \mathbf{e}_2, \ldots$. Consider the equation

$$\mathcal{A}\mathbf{z} = \mathbf{w}, \tag{7.47}$$

where $\mathbf{w}$ is known. The following alternatives are given:

(a) $\mathbf{w}$ is not orthogonal to $\mathbf{e}_0$, i.e. $\mathbf{w} = a_0\mathbf{e}_0 + a_1\mathbf{e}_1 + \cdots$, with $a_0 \neq 0$; in such a case Eq. (7.47) does not admit a solution;
(b) $\mathbf{w}$ is orthogonal to $\mathbf{e}_0$, i.e. $\mathbf{w} = a_1\mathbf{e}_1 + a_2\mathbf{e}_2 + \cdots$; in such a case one easily obtains $\mathbf{z} = b_0\mathbf{e}_0 + (a_1/\lambda_1)\mathbf{e}_1 + (a_2/\lambda_2)\mathbf{e}_2 + \cdots$, where $b_0$ is arbitrary.

The eigenfunctions of $\mathcal{L}$ are the functions $H_n(v/\sqrt{2})\exp(-v^2/2)$ where $H_n$ is the $n$th Hermitian polynomial ($H_0 = 1$, $H_1 = v$, $H_2 = 1 - v^2, \ldots$). Since the Maxwellian distribution is associated with the null eigenvalue, the solvability condition of Eq. (7.45) imposes (7.46) and $p_1$ is given by:

$$p_1(x, v, \tau_0, \tau_1, \ldots) = \Psi(x, \tau_0, \tau_1, \ldots)e^{-v^2/2} - v\Big(\frac{\partial}{\partial x}\Phi - f(x)\Phi\Big)e^{-v^2/2}, \tag{7.48}$$

where $\Psi$ must be determined. The equation for $p_2$ becomes

$$\mathcal{L}p_2 = A_0 e^{-v^2/2} + A_1 v\, e^{-v^2/2} + A_2(1 - v^2)e^{-v^2/2}, \tag{7.49}$$

where

$$A_0 = \frac{\partial}{\partial \tau_0}\Psi + \frac{\partial}{\partial \tau_1}\Phi - \frac{\partial}{\partial x}\Big(\frac{\partial}{\partial x}\Phi - f(x)\Phi\Big) \tag{7.50}$$

$$A_1 = \frac{\partial}{\partial x}\Psi - f(x)\Psi \tag{7.51}$$

$$A_2 = -\Big(f - \frac{\partial}{\partial x}\Big)\Big(\frac{\partial}{\partial x}\Phi - f(x)\Phi\Big). \tag{7.52}$$

The solvability condition then requires $A_0 = 0$:

$$\frac{\partial}{\partial \tau_0}\Psi = -\frac{\partial}{\partial \tau_1}\Phi + \frac{\partial}{\partial x}\Big(\frac{\partial}{\partial x}\Phi - f(x)\Phi\Big), \tag{7.53}$$

since the function $\Phi$ does not depend on $\tau_0$, see Eq. (7.46):

$$\frac{\partial}{\partial \tau_0}\Psi = 0. \tag{7.54}$$

In such a way one has a closed equation for $\Phi$:

$$\frac{\partial}{\partial \tau_1}\Phi = -\frac{\partial}{\partial x}\Big(f(x)\Phi\Big) + \frac{\partial^2}{\partial x^2}\Phi. \tag{7.55}$$

The above equation is the Smoluchowski equation (7.31), with a change of variables.

### 7.2.2 Some conceptual and technical remarks

Let us note that the Kramers equation (7.27) describes the colloidal system at a level which is less accurate than the microscopic level ruled by the complete Hamiltonian (7.18). At such a level, which we indicate by the term mesoscopic level I (MeLI), the relevant variables are, in the dilute limit (for the colloidal particles), those of the single colloidal particle $(\mathbf{Q}, \mathbf{P})$.

The Smoluchowski equation (7.28) represents another level of description, mesoscopic level II (MeLII), which describes the system at times larger than the typical time $1/\zeta$ , so that the asymptotic statistical properties of the velocity are well captured by the (stationary) Maxwell–Boltzmann distribution. In the dilute limit, the only relevant variable is the position $\mathbf{Q}$ of the colloidal particle.

Beyond the Kramers and Smoluchowski equations there is another level of description, which rules the evolution of $\rho(\mathbf{Q}, t)$ at very long time and large spatial scale. In order to understand this point, we discuss the case where $\mathbf{Q}$ reduces to a one-dimensional variable and $F(Q)$ is a periodic function of period $L$. We write $F(Q) = -\partial U(Q)/\partial Q$ and we denote by $\{Q_n\}$ the minima of $U$. It is easy to understand that, for small values of the diffusion coefficient $D$, the trajectory $Q(t)$ is a kind of random walk jumping at random times from one minimum $\{Q_k\}$ to one of the two nearest-neighbor minima, i.e. $\{Q_{k+1}\}$ or $\{Q_{k-1}\}$. The difference $t_j$ between two successive jumping times is a stochastic process whose mean value $\langle t_j \rangle$ depends on $D$ and on the shape of $U$. In the limit of small $D$ one has the celebrated Kramers formula (Chandrasekhar 1943, Gardiner 1990):

$$\langle t_j \rangle \simeq \tau_0 e^{\Delta U/D} \tag{7.56}$$

where $\Delta U$ is the difference in potential between the maximum and the minimum and $\tau_0$ is a characteristic time which depends on the second derivatives of $U$ computed at the minimum and the maximum.

Therefore at very large time, $t \gg \langle t_j \rangle$, the field $\widetilde{\rho}(Q, t)$, obtained by a local average over a spatial window of dimension much larger than $L$, evolves according to the Fick equation:

$$\frac{\partial}{\partial t} \widetilde{\rho}(Q, t) = D^{\mathrm{E}} \frac{\partial^2}{\partial Q^2} \widetilde{\rho}(Q, t) \tag{7.57}$$

where $D^{\mathrm{E}}$ depends on $U$ and $D$. In the limit of very small $D$,

$$D^{\mathrm{E}} \sim \frac{L^2}{\tau_0} \mathrm{e}^{-\Delta U/D}. \tag{7.58}$$

In the next section we will discuss the procedure for obtaining (7.57) from the Smoluchowski equation.

Let us conclude this section with a brief summary of the levels of description introduced above:

(1) microscopic level (MiL), describing solvent and colloidal particles (deterministic Hamiltonian equations of the global system (7.19) and Liouville equation (7.24));
(2) mesoscopic level I (MeLI), describing only colloidal particles (stochastic equations (7.20) and Kramers equation (7.27));
(3) mesoscopic level II (MeLII), describing only the positions of colloidal particles (stochastic equations (7.22) and Smoluchowski equation (7.28));
(4) macroscopic level (MaL), large scale and long time description of the spatial distribution of colloidal particles (Fick equation).

In each crossover from a more detailed description to a less precise level one has a loss of information, i.e. a reduction in the number of degrees of freedom and/or a less detailed description in space and/or time resolution.

From MiL $\rightarrow$ MeLI we "forget" the solvent and the description is valid only for times much larger than the typical molecular time $O(10^{-11}\ \mathrm{s})$.

In the transition MeLI $\rightarrow$ MeLII we "delete" the velocity of the colloidal particles. Such a description is valid only for times much larger than $1/\zeta \sim O(10^{-6}\ \mathrm{s})$.

Finally for MeLII $\rightarrow$ MaL we look at the system at very large spatial scale ($\gg L$) and very large time ($t \gg \langle t_j \rangle$).

## 7.3 Diffusion at large scale and eddy diffusivity

In order to show how to use the multiscale technique for the construction of "macroscopic" equations from "microscopic" dynamics, let us briefly discuss the problem of passive scalar transport in incompressible velocity fields (Biferale *et al.* 1995, Majda and Kramer 1999). Taking into account molecular diffusion, the time evolution of a test particle is ruled by the Langevin equation

$$\frac{d\mathbf{x}}{dt} = \mathbf{u}(\mathbf{x}, t) + \sqrt{2D_0}\eta \tag{7.59}$$

where $\mathbf{u}(\mathbf{x}, t)$ is the Eulerian velocity field at the position $\mathbf{x}$ and time $t$, $D_0$ being the molecular diffusion and $\eta$ a white noise vector with independent components, i.e. a Gaussian process with $\langle \eta_i(t) \rangle = 0$ and $\langle \eta_i(t)\eta_j(t') \rangle = \delta_{ij}\delta(t - t')$. The density of tracers $\theta(\mathbf{x}, t)$ evolves according to the transport equation:

$$\frac{\partial}{\partial t}\theta + \nabla \cdot (\mathbf{u}\theta) = D_0 \Delta \theta \tag{7.60}$$

which is the Fokker–Planck equation of the stochastic process (7.59).

To simplify the notation, we assume that $\langle \mathbf{u} \rangle = 0$ and that at initial time $t = 0$ the field $\theta(\mathbf{x}, 0)$ is localized around $\mathbf{x} = 0$. We denote by $U$ the typical speed of the field $\mathbf{u}(\mathbf{x}, t)$ and by $L$ its typical length. At time much larger than the characteristic time $T = U/L$, it is reasonable to expect that the field $\Theta(\mathbf{x}, t)$, obtained by locally averaging $\theta(\mathbf{x}, t)$ over a volume of linear dimension much larger than $L$, evolves according to the Fick equation:

$$\frac{\partial}{\partial t}\Theta(\mathbf{x}, t) = \sum_{i,j} D^{\mathrm{E}}_{ij} \frac{\partial^2}{\partial x_i \partial x_j}\Theta(\mathbf{x}, t). \tag{7.61}$$

It is easy to see that the asymptotic solution of (7.61) is a multivariate Gaussian distribution:

$$\Theta(\mathbf{x}, t) \sim \sqrt{\frac{|\det A|}{(4\pi t)^d}} \exp -\left[\frac{1}{4t}\sum_{i,j} x_i A_{ij} x_j\right], \tag{7.62}$$

where $\mathbf{A} = (\mathbf{D}^{\mathrm{E}})^{-1}$ and $d$ is the spatial dimension. The above solution reveals the physical meaning of $D^{\mathrm{E}}_{ij}$:

$$D^{\mathrm{E}}_{ij} = \lim_{t\to\infty} \frac{1}{2t}\langle x_i(t)x_j(t)\rangle. \tag{7.63}$$

The standard diffusive behavior, i.e. $\langle x(t)^2 \rangle \sim t$ and the Gaussian shape of the probability density function, are simply related to the central limit theorem.

Let us show that for the Fick equation to be valid (and consequently, for the existence of standard diffusion) it is necessary that

(a) the variance of the Lagrangian velocity $v_i = dx_i/dt$ is finite,
(b) the time correlation, between $v_i(t)$ and $v_i(t')$, decays fast enough, i.e. $\langle v_i(t')v_i(t)\rangle \sim |t - t'|^{-\beta}$ with $\beta > 1$.

Writing

$$x_i(t) - x_i(0) = \int_0^t v_i(t')\mathrm{d}t', \tag{7.64}$$

one obtains the following formula (Taylor relation):

$$\langle (x_i(t) - x_i(0))^2 \rangle = \int_0^t \int_0^t \langle v_i(\mathbf{x}(t_1)) v_i(\mathbf{x}(t_2)) \rangle \mathrm{d}t_1 \mathrm{d}t_2 \simeq 2t \int_0^t C_{ii}(\tau)\mathrm{d}\tau, \tag{7.65}$$

where $C_{ii}(\tau) = \langle v_i(\mathbf{x}(\tau)) v_i(\mathbf{x}(0)) \rangle$ is the correlation function of the Lagrangian velocity $\mathbf{v} = \dot{\mathbf{x}}$. If the integral $\int_0^\infty C_{ii}(\tau)\mathrm{d}\tau$ is finite one has

$$D_{ii}^{\mathrm{E}} = \int_0^\infty C_{ii}(\tau)\mathrm{d}\tau, \tag{7.66}$$

which is just a way of writing the Green–Kubo formula. From the above considerations one understands that the Fick equation (7.57) is plausible as an effective equation for long time and large spatial scales of the transport equation (7.60).

The multiscale analysis is a way to rationalize the previous heuristic considerations (Frisch 1995, Majda and Kramer 1999). Let us begin with a presentation of the basic ingredients of the method, that allows us to derive the Fick equation (7.57) from the transport equation (7.60) and to compute the diffusion coefficient $D_{ij}^{\mathrm{E}}$.

Consider the diffusion equation in one spatial dimension:

$$\frac{\partial}{\partial t}\theta = \frac{\partial}{\partial x}\left(D(x)\frac{\partial}{\partial x}\theta\right) \tag{7.67}$$

where $D(x)$ is a periodic function with period $L$. The aim is to write an effective (Fick's) equation valid at long time and large scale (much larger than $L$), i.e. we have to find $D^{\mathrm{E}}$ in terms of $D(x)$. The Langevin equation associated with (7.67) is

$$\frac{\mathrm{d}x}{\mathrm{d}t} = u(x) + \sqrt{2D(x)}\eta, \quad u(x) = \frac{\partial}{\partial x}D(x) \tag{7.68}$$

where the Itô formulation has been used. The stochastic process $x(t)$ ruled by the Langevin equation (7.68) spends a time interval $\Delta t(x) \sim (\Delta x)^2/2D(x)$ in a segment of length $\Delta x$ centered in $x$. Let us now follow $x(t)$ up to the time $t_N$ such that $N$ jumps (among $\Delta x$-segments) occur. Elementary considerations give

$$\langle x(t_N) - x(0) \rangle = 0, \quad \langle (x(t_N) - x(0))^2 \rangle = N\Delta x^2 \tag{7.69}$$

and for $N \gg 1$

$$t_N = \frac{\Delta x^2}{2}\sum_{j=1}^{N}\frac{1}{D(x(t_j))} \simeq \frac{N\Delta x^2}{2}\left\langle \frac{1}{D(x(t_j))} \right\rangle = \frac{N\Delta x^2}{2L}\int_0^L \frac{\mathrm{d}x}{D(x)}, \tag{7.70}$$

where we used the fact that the process $x(t)$ is ergodic and its stationary probability density function is constant. Note that here the phase space coincides with real space so one can exchange the time average with the spatial average. Since $\langle (x(t_N) - x(0))^2 \rangle \simeq 2D^{\mathrm{E}} t_N$, from (7.69) and (7.70) one obtains the eddy diffusion coefficient

$$D^{\mathrm{E}} = \left\langle \frac{1}{D(x(t_j))} \right\rangle^{-1} = \left( \frac{1}{L} \int_0^L \frac{\mathrm{d}x}{D(x)} \right)^{-1}. \tag{7.71}$$

### 7.3.1 Fick equation from the transport equation with a multiscale approach

Here we derive in a more precise way (Frisch 1995), using the multiscale method, the previous results obtained using qualitative arguments. In contrast to the example discussed in Section 7.2.1, now we have to introduce, in addition to the slow time $T$, a large scale spatial variable $X$:

$$T = \epsilon^2 t, \quad X = \epsilon x. \tag{7.72}$$

The procedure is very similar to that described in Section 7.2.1, i.e. we expand $\theta$ in powers of $\epsilon$:

$$\theta(x, X, t, T) = \theta_0(x, X, t, T) + \epsilon \theta_1(x, X, t, T) + \epsilon^2 \theta_2(x, X, t, T) + \cdots. \tag{7.73}$$

The space and time derivatives must be decomposed as follows:

$$\frac{\partial}{\partial x} \to \frac{\partial}{\partial x} + \epsilon \frac{\partial}{\partial X}, \quad \frac{\partial}{\partial t} \to \frac{\partial}{\partial t} + \epsilon^2 \frac{\partial}{\partial T}. \tag{7.74}$$

Using (7.73), (7.74) and (7.67) we obtain the equations:

$$\mathcal{L}\theta_0 = 0 \tag{7.75}$$

$$\mathcal{L}\theta_1 = \frac{\partial}{\partial x}\left( D(x) \frac{\partial}{\partial X} \theta_0 \right) + \frac{\partial}{\partial X}\left( D(x) \frac{\partial}{\partial x} \theta_0 \right) \tag{7.76}$$

$$\mathcal{L}\theta_2 = -\frac{\partial}{\partial T}\theta_0 + \frac{\partial}{\partial x}\left( D(x) \frac{\partial}{\partial X}(\theta_1 + \theta_2) \right) + \frac{\partial}{\partial X}\left( D(x) \left( \frac{\partial}{\partial x}\theta_1 + \frac{\partial}{\partial X}\theta_0 \right) \right), \tag{7.77}$$

where

$$\mathcal{L} = \frac{\partial}{\partial t} - \frac{\partial}{\partial x}\left( D(x) \frac{\partial}{\partial x} \right). \tag{7.78}$$

Now we go on as in Section 7.2.1. Equation (7.75) expresses that $\theta_0$ is the null space of the heat operator $\mathcal{L}$ and, because of the $L$-periodicity, $\theta_0$ will relax to a constant, independent of $x$ and $t$ so that $\partial_x \theta_0 = 0$. Since after a transient $\partial_t \theta_1 = 0$,

Eq. (7.76) can be written

$$D(x)\Big(\frac{\partial}{\partial x}\theta_1 + \frac{\partial}{\partial X}\theta_0\Big) = C, \tag{7.79}$$

dividing by $D(x)$ and using the result $\langle \partial_x \theta_1 \rangle = 0$ (because of the spatial periodicity) we have

$$C\Big\langle \frac{1}{D(x)} \Big\rangle = \frac{\partial}{\partial X}\theta_0, \tag{7.80}$$

where the average is now over the fast variables.

Let us now discuss the equation for $\theta_2$: it can be solved only if a solvability condition, on $\theta_0(X, T) = \langle \theta_0(x, X, t, T) \rangle$, is imposed. Taking the average of Eq. (7.77) one has:

$$\frac{\partial}{\partial T}\theta_0 = \frac{\partial}{\partial X}\Big(\langle D(x)\rangle \frac{\partial}{\partial X}\theta_0\Big) + \frac{\partial}{\partial X}\Big\langle D(x)\frac{\partial}{\partial x}\theta_1 \Big\rangle. \tag{7.81}$$

Using (7.79) and (7.80) one obtains

$$\frac{\partial}{\partial T}\theta_0 = D^{\mathrm{E}}\frac{\partial^2}{\partial X^2}\theta_0, \tag{7.82}$$

with $D^{\mathrm{E}}$ given by (7.71).

The multiscale method can be applied, without particular difficulties, to the more general problem (7.60) in two or three dimensions and with a given generic incompressible velocity field $\mathbf{u}(\mathbf{x}, t)$ (Biferale *et al.* 1995, Majda and Kramer 1999). The computations are rather similar, the unique difference is that now the solvability condition yields an additional equation for an auxiliary vector $\mathbf{w}$,

$$\frac{\partial}{\partial t}\mathbf{w} + (\mathbf{u}\cdot\nabla)\mathbf{w} - D_0\Delta\mathbf{w} = -\mathbf{u}, \tag{7.83}$$

to be solved in such a way that one can express $D^{\mathrm{E}}_{ij}$ in terms of $\mathbf{w}$:

$$D^{\mathrm{E}}_{ij} = D_0\delta_{ij} - \frac{1}{2}\Big[\langle u_i w_j \rangle + \langle u_j w_i \rangle\Big]. \tag{7.84}$$

Now all the, often non-trivial, effects of the advecting field $\mathbf{u}$ on the asymptotic properties of the Lagrangian behavior are included in the eddy diffusivity coefficients $D^{\mathrm{E}}_{ij}$ (Biferale *et al.* 1995, Majda and Kramer 1999).

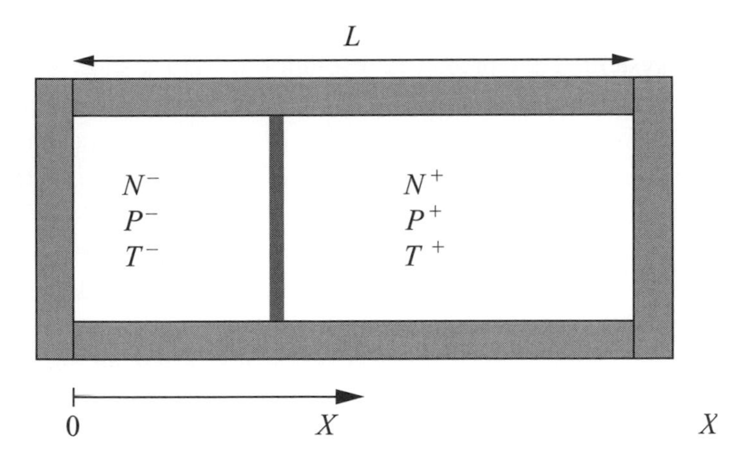


Figure 7.1 Sketch of the adiabatic piston. The superscripts − and + indicate the left and right compartments, respectively.

## 7.4 The adiabatic piston: a system between the microscopic and macroscopic realms

Up to now we have considered issues which are essentially well understood. Now we want to discuss a problem which is still open, and very appropriate, to see how the multiscale procedure works.

The *adiabatic piston problem* refers to a model system composed of $N$ non-interacting point particles of mass $m$ (i.e. an ideal gas) in a container of length $L$ and cross-section $A$, separated in two regions by a movable wall (the piston) of mass $M \gg m$ (see Figure 7.1). The walls of the container are supposed to be perfect insulators preventing any mass or heat exchanges with the exterior. Gas particles undergo purely elastic collisions with the piston and the walls. The piston is constrained to move along the $x$-axis, therefore the problem is essentially one-dimensional. The main problem is to predict the final state to which the system will evolve after releasing a constraint fixing the piston in some prescribed initial state. At this stage, "adiabatic" piston only means "with no internal degree of freedom": when fixed, this piston does not conduct heat. Part of the issue is to determine in what respect the piston still behaves as an adiabatic partition when it is allowed to move. We shall see that the parameter $\alpha \simeq 2m/(M+m)$ and the time scale of the observation dramatically control the answer. Since $\alpha \ll 1$ one has a time scale separation, therefore it is natural to develop a multiscale approach allowing the singular nature of the straightforward perturbation expansion in powers of $\alpha$ to be circumvented.

Although very simple to state, the above problem is far less simple to solve: it is a well-known controversial example *where the two principles of thermostatics (conservation of energy and maximum entropy) are not sufficient* to obtain the final state of the system. They predict an evolution toward a state of mechanical equilibrium but with no way of determining the position of the piston or the final temperatures of the gases (Feynman 1963, Lieb 1999). This point can be foreseen since the final state will depend upon the "friction coefficients" of the piston in the gases, which do not appear in the entropy: *here it is essential to describe the dynamics of the system*. Nevertheless, it took time to reach this understanding since the first formulation of this problem by Rayleigh. During recent decades, various analytical approaches, and direct simulations, gradually led to the now acknowledged but still outstanding conclusion that *the piston, although adiabatic when fixed, becomes diathermal when it is allowed to move.* For a detailed discussion see Curzon (1969), Crosignani *et al.* (1996), Lieb (1999) and Gruber *et al.* (2003). We shall see that the very final state is a state of both mechanical and thermal equilibrium: a slow, fluctuation-driven evolution toward thermal equilibrium follows a fast thermodynamic relaxation toward mechanical equilibrium.

On purpose, this piston model is of the utmost simplicity, allowing us to dissect the different basic mechanisms at work in the relaxation of real systems. The investigations performed on this piston model are exemplary of the subject of this chapter: as sketched below, they show at work all the various approaches and techniques paving the way from a complete deterministic description to various stochastic mesoscopic descriptions to a phenomenological description in the framework of irreversible thermodynamics.

Let us denote by $v$ and $V$ the $x$-components of the velocities of a particle and the piston. Being purely elastic, a collision of a particle with the boundary at $x = 0$ or $x = L$ is associated with the velocity change $v \to v' = -v$. Similarly, from the conservation of kinetic energy and momentum it follows that, after the collision, the velocities are transformed according to:

$$v \to v' = 2V - v + \alpha(v - V), \quad V \to V' = V + \alpha(v - V), \tag{7.85}$$

where

$$\alpha = \frac{2m}{M + m} \approx \frac{2m}{M} \ll 1. \tag{7.86}$$

The relative change in the velocities is directly related to the small parameter according to

$$\frac{|V' - V|}{|v' - v|} \simeq \frac{\alpha}{2} \ll 1. \tag{7.87}$$

The simplicity of the model allows direct simulation of the microscopic dynamics, which is deterministic and reversible, in the whole phase space ($2N + 2$

degrees of freedom). This is done by sampling at random the initial conditions and then implementing the deterministic collision rules (7.85). The role and relevance of numerics are thus to be specially underlined: they guide intuition, help to formulate proper conjectures, allow validation of the assumptions and approximations involved in the analytical resolution and determination of the range of validity of perturbation expansions.

The starting point of the dynamic analysis is the exact deterministic description of the system evolution, namely *Hamilton's equations* for the $2(N+1)$ degrees of freedom. Since we are concerned with the case $N \gg 1$, but we are not interested in the state of the gas molecules, it is natural to adopt a statistical viewpoint using the probability distribution in the $\Gamma$-space, that evolves according to the *Liouville equation*. Focusing on the evolution of the piston, we integrate out particle velocities and positions to reduce the Liouville equation to the relevant degrees of freedom, here the piston velocity $V$ and its distribution $\Phi(V;t)$.[4] This approach is similar to the spirit of the derivation of hydrodynamics from the microscopic level. One can write the evolution equation for $\Phi(V;t)$ in terms of powers of the small parameter $\alpha$, with coefficients $F_k$ related to the moments of order $k$ of the gas particle velocity at the piston surface, in the reference frame of the piston (see the following subsection).

The resulting equation exhibits the flaw encountered in any kinetic hierarchy (e.g. the BBGKY hierarchy), namely it involves higher-order distribution functions, here the joint distributions of the velocity of the piston and of the particles of the gas: the equation has to be supplemented with a *closure relation*. For a detailed discussion of the problem in the framework of kinetic theory see Gruber and Pache (2002), Gruber *et al.* (2002) and Gruber and Morris (2003); recently approaches based on hydrodynamics have been developed by Caglioti *et al.* (2004). See Gruber and Lesne (2006) for a review.

### 7.4.1 About the kinetic hierarchy and Kramers–Moyal expansion

Integrating out the piston position and the $2N$ degrees of freedom of all gas particles in the Liouville equation, then performing a formal expansion in powers of $\alpha \simeq 2m/M$, yields the reduced equation for the velocity distribution $\Phi(V;t)$ of the piston:

$$\partial_t \Phi(V;t) = \alpha A \sum_{k=1}^{\infty} \frac{(-1)^k \alpha^{k-1}}{k!} \frac{\partial^k G_{k+1}(V, \rho^{\pm}_{\text{surf}})}{\partial V^k}, \tag{7.88}$$

[4] We consider only the velocity of the piston as a relevant variable, because its position has a distribution whose evolution can be recovered, once the initial value of the position is given and the evolution of the velocity distribution is known.

where

$$G_k(V, \rho_{\text{surf}}^{\pm}) = \int_V^{\infty} (v - V)^k \rho_{\text{surf}}^{-}(v; V; t)\mathrm{d}v - \int_{-\infty}^{V} (v - V)^k \rho_{\text{surf}}^{+}(v; V; t)\mathrm{d}v \tag{7.89}$$

involves the joint distributions of the piston velocity $V$ and one gas particle velocity $v$ at the piston surface (the signs $-$ and $+$ indicate the left and right surfaces of the piston respectively). The distribution $\Phi(V;t)$ can be formally factorized out of $G_k$ for any $k$. This can be done by introducing a conditional distribution $a^{\pm}(v; V; t)$ such that $\rho_{\text{surf}}^{\pm}(v; V; t) = a^{\pm}(v; V; t)\ \Phi(V; t)$, and then defining the functions $(F_k)_{k\geq 2}$ by the equations

$$G_k(V, \rho_{\text{surf}}^{\pm}) = \Phi(V; t) F_k(V, a^{\pm}). \tag{7.90}$$

At this stage, the unique assumption is that there actually exist *continuous* conditional distributions $a^{\pm}(v; V; t)$. The structure of Eq. (7.88) is quite general and currently encountered in kinetic theory, e.g. in the *Kramers–Moyal expansion*, in the case of a Markov evolution for a variable $z$:

$$\partial_t \Psi(z; t) = \int W(z' \to z)\Psi(z'; t)\mathrm{d}z' \tag{7.91}$$

turning into a perturbation expansion under the assumption that only small jumps $z' - z = O(\alpha)$ are actually observed:

$$\partial_t \Psi = \alpha \sum_{k=1}^{\infty} \frac{\alpha^k}{k!} \frac{\partial^k [\Psi M_k(z, \alpha)]}{\partial z^k}, \qquad M_k(z, \alpha) = \int W(z - \alpha u \to z)\, u^k\, \mathrm{d}u. \tag{7.92}$$

When only the terms for $k = 1$ and $k = 2$ are non-vanishing (or non-negligible), the Kramers–Moyal expansion reduces to a *Fokker–Planck equation* for $\Psi(z, t)$. This corresponds to the case of a continuous Markov process.

### 7.4.2 The time evolution of the piston

Consider first the limiting case $M \to \infty$, taking $L$ fixed, and performing the following limiting procedures:

$$m \quad \text{fixed}, \qquad \frac{A}{M} = \text{constant}, \qquad R^{\pm} = \frac{m}{M} N^{\pm} = \text{constant},$$

where $N^{+}$ and $N^{-}$ are the number of particles in the right and left compartments of the container respectively. In this limit Eq. (7.88) can be exploited at lowest order in $\alpha$, keeping only the term $k = 1$. In this "thermodynamic limit for the piston," the self-consistent assumption $\rho^{\pm}_{\text{surf}}(v; V; t) = a^{\pm}(v; t)\Phi(V; t)$ yields:

$$\Phi(V;t) = \delta(V - \langle V(t)\rangle) \qquad \text{where} \qquad \frac{\mathrm{d}\langle V\rangle}{\mathrm{d}t} = \alpha A F_2(\langle V\rangle; t). \tag{7.93}$$

This expression for the solution validates a posteriori the decoupling assumption, i.e. $a^{\pm}(v; V; t)$ are functions of $v$ and $t$ only. Moreover, the motion of the piston is purely deterministic since $\langle V^n(t)\rangle = \langle V(t)\rangle^n$ for any $n$.

When $L = \infty$, the piston reaches a stationary state with a velocity $\langle V\rangle_\infty$ which is a solution of $F_2(\langle V\rangle_\infty) = 0$, and which is satisfactorily proportional to the pressure difference $P_0^- - P_0^+$.

In the finite size case ($L < \infty$ fixed), the stationary state $\langle V\rangle_\infty$ is observed only transiently, i.e. as long as the boundaries in $x = 0$ and $x = L$ do not yet influence the piston motion, namely at times shorter than the recollision time: $t < t_1 \sim L\sqrt{m/k_B T}$ (the time for a gas particle to hit the piston surface twice). Then recollisions and damping set in: the piston motion, by varying the volume of the compartments, modifies the collision rates of the gas particles on the piston, inducing friction and relaxation toward mechanical equilibrium. The damping is controlled by the dimensionless parameters $R^{\pm} = N^{\pm}m/M$: one observes damped oscillations for $R^{\pm}$ small enough ($R^{\pm} < 4$) whereas the relaxation is overdamped at larger $R^{\pm}$. *This behavior is in agreement with the prediction of thermodynamics and it corresponds to an adiabatic evolution: no heat transfer occurs between the compartments and their thermodynamic entropies both increase monotonously.*

Let us now turn to the less idealized situation where the piston mass $M$ is large but finite. It was observed in numerical simulations that at a first stage, the evolution is described by the above infinite-mass behavior: after a very short transient at constant velocity $\langle V\rangle_\infty$ toward the low pressure side ("phase 0"), the piston relaxes to a state of "mechanical equilibrium" where the pressures are equal but the temperatures different, with or without oscillations according to the value of $R^{\pm}$. In this regime ("phase 1"), the piston behaves as an adiabatic wall, with no heat transfer between the compartments.

Then a second stage ("phase 2") takes place on a time scale $O(M/m)$, during which the piston motion is driven by the asymmetry in the fluctuations felt by the left/right walls, and the temperatures vary very slowly to reach a final equilibrium state where the densities, pressures and temperatures of the two gases are the same. The conclusion, already stated by Feynman (1963) on the basis of qualitative arguments, is that a wall which is adiabatic when fixed becomes heat-conducting when it experiences stochastic motion.

For real *macroscopic* systems, the time (scaling as $M/m$) involved to reach this thermal equilibrium will be several million times the age of universe: the above conclusion is purely theoretical and for all practical purposes at current observation scales, the piston is actually an adiabatic wall relaxing to mechanical equilibrium only. This point should nevertheless be reconsidered at smaller scales, for example for nanoscale devices, feasible today, or biological systems at work inside the cell: the mass ratio $M/m \gg 1$, and hence the thermal equilibration time, decreases by a factor of $O(10^{18})$ when passing from the millimeter scale to the nanometer scale and the second, heat-conducting regime might then become of relevance.

### 7.4.3 Different temporal regimes

From the above qualitative picture of the system evolution, it appears that whereas the first stage is actually described by the leading order $\alpha = 0$ ("thermodynamic behavior" $M = \infty$, in which the evolution is a fast relaxation toward mechanical equilibrium), the fluctuations of order $O(\alpha)$ rule the behavior observed in a second stage, corresponding to a slow relaxation toward thermal equilibrium. A naive perturbation approach cannot give access to *both* regimes; this difficulty is reminiscent of the *boundary-layer problems* encountered in hydrodynamics; the idea is to implement two different perturbation expansions (Gruber and Lesne 2006):

- one at short times, describing a transient regime ruled exclusively by the fast dynamics, during which the order of the observables changes, e.g. initially $V(0) = O(1)$ whereas $V(t) = O(\alpha)$ once $t = O(1/\alpha)$; this transient regime (a "boundary layer" in time) is required to match the "bulk" behavior observed at times $O(1/\alpha)$ with non-compatible initial conditions;
- one for longer times, involving a rescaled time variable $\tau = \alpha t \simeq 2mt/M$.

The perturbation approach is then supplemented with a "*slaving principle*" (see, e.g., Gruber and Lesne (2006)), expressing that at each time of the slow evolution, i.e. at fixed $\tau$, the fast dynamics has reached a local asymptotic state, slaved to the values of the slow observables. Technically, this principle is best implemented using a multiscale method valid once $V$ and $P^- - P^+$ have decreased to values $O(\alpha)$. Basically, this slaving principle ensures that the fast dynamics do not interfere with the slow dynamics and thus prevents the emergence of secular divergences in the perturbation expansion for the slowly varying observables. The initial conditions are set on the first-stage solution, at $t = 0$. The initial conditions of the second regime, at $\tau = 0$, match the asymptotic behavior $t \to \infty$ of the first-stage solution ("*matching condition*") (Gruber and Lesne 2006).

The perturbation approach for the piston problem is singular in two ways and hence requires two different methods: a *matched expansion* to bridge the first and second regimes (namely to bridge values of different orders in $\alpha$ for $V$ and $P^+ - P^-$), and a *multiscale method* in the second regime, to express the superimposition of a slow evolution and a fast relaxation slaved to the slowly evolving temperatures.

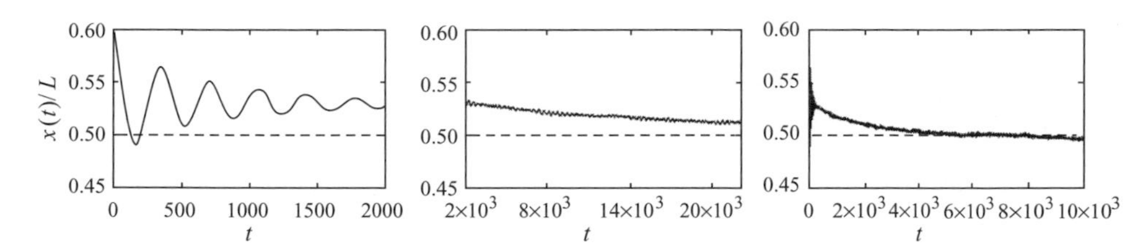


Figure 7.2 Evolution of the piston position $x(t)$ in the simulation of an ideal gas with $N^- = N^+ = 1000$, $M = 100$, $m = 1$ and $L = 2000$, corresponding to $R = 10$, the initial state is set as $T_l = 40$, $T_r = 60$ and $x = 0.6L$. In order to obtain a clear curve we performed an average over 100 independent realizations.

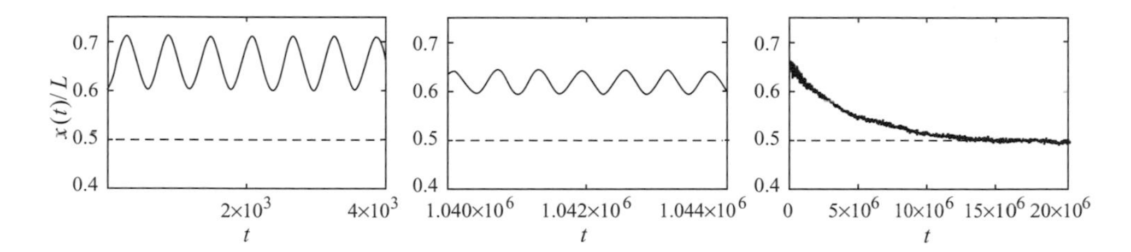


Figure 7.3 As in Figure 7.2 with $M = 5000$, corresponding to $R = 0.2$, the initial state is set as $T_l = 150$, $T_r = 50$ and $x = 0.6L$.

The evolution equation (7.88) of the velocity distribution function of the piston $\Phi(V;t)$ is not exploited directly but is replaced with the hierarchy of equations describing the evolution of its moments $\langle V^n \rangle$. Whereas it is exact in the thermodynamic limit, the Boltzmann-like decorrelation property $\rho^{\pm}_{\text{surf}}(v; V; t) = a^{\pm}(v;t)\,\Phi(V;t)$ now has to be introduced as an assumption.[5] As an additional assumption one identifies the densities and temperatures of the fluids at the piston surface with their bulk values.

Results following from the multiscale analysis of these moment equations are shown in Figures 7.2 and 7.3, and they can be summarized as follows.

**Phase 0** The evolution begins with a short transient, of duration shorter than the recollision time $t_1 \sim L\sqrt{m/k_B T}$, during which the piston motion is a fast relaxation to a stationary velocity $\langle V \rangle$ (what is observed steadily if $M = \infty$ and $L = \infty$).

**Phase 1** At times $t \sim O(t_1)$, the motion of the piston is no longer at constant velocity and the system evolves to a state of mechanical equilibrium where the pressures on both sides are approximately equal, but the temperatures are different.

[5] This is similar to the Boltzmann *stossahlansatz* $f_2(x, v_1; x, v_2; t) = f_1(x, v_1; t) f_1(x, v_2; t)$ introduced in the kinetic theory of a single dilute species, leading to the celebrated Boltzmann kinetic equation and supported by molecular chaos.

The evolution is *adiabatic*, similar to the case $M = \infty$. In particular, it is independent of $M$ for $M$ sufficiently large. The motion is essentially *deterministic*: $\Phi(V;t) = \delta(V - \langle V(t) \rangle)$.

**Phase 2** Then on a much larger time scale, $t_2 \sim M t_1/m$, the piston evolves *stochastically* and with *heat transfer* to a state of thermal equilibrium, where the temperatures (and still the pressures) on both sides of the piston are equal. Thermal fluctuations of the piston velocity have developed and now the temperature $T_P$ of the piston can no longer be ignored (it coincides with the common temperature of the gases at the end of this phase).

**Phase 3** Finally, one observes an indirect relaxation of the gases toward thermal equilibrium (Maxwellian distributions), induced by a common interaction with the moving piston.

## 7.5 Remarks and perspectives

We conclude the chapter with some general remarks on the multiscale method, its connections with other important issues and a short discussion of some open problems.

In the mathematical literature the multiscale method is also known by the name of homogenization technique. The origin of this term is the fact that one has a replacement of a heterogeneous context (associated with the presence of $\mathbf{u}(\mathbf{x}, t)$ in (7.60) or the spatial dependence of $D(x)$ in (7.67)) with a homogeneous one, associated with Eq. (7.57) which describes the system at large scale and long time.

Let us underline some technical and conceptual resemblances between the multiscale treatment of ordinary differential equations (Section 7.1) and the homogenization method for partial differential equations (Sections 7.2.1 and 7.3.1). In both cases the "non-trivial part" of the problem (i.e. the non-linear contribution for ordinary differential equations and the presence of $\mathbf{u}(\mathbf{x}, t)$ in (7.60)) induces asymptotically a "renormalization" of relevant parameters.

For ordinary differential equations we saw that with the multiscale analysis one replaces the original non-linear equation (7.3) with an effective linear one

$$\frac{d^2 x}{dt^2} + \omega^2(\epsilon) x = 0 \tag{7.94}$$

which is also valid at very large time, where $\omega(\epsilon)$ depends both on $\epsilon$ and on the initial conditions; in the example discussed in Section 7.1 one has $\omega(\epsilon) = 1 + 3\epsilon/8 + O(\epsilon^2)$. In a similar way the asymptotic (in both space and time) behavior of the transport problem (including an advection velocity term) is described by a Fick's equation. Now the eddy diffusion coefficients take into account the original inhomogeneity due to the presence of the advection field (and/or non-constant $D$) in the starting problem.

### 7.5.1 Chapman–Enskog and multiple time scale method

We want to stress that the multiple time scale method is rather close in spirit to the Chapman–Enskog procedure for the Boltzmann equation (Cercignani 1988, Gorban *et al.* 2004). One can interpret the Chapman–Enskog method as an ad hoc formal expansion of the time derivative in powers of a small parameter $\epsilon$ (the Knudsen number, i.e. the ratio between the collision length and the typical scale of the hydrodynamic field). Let us recall briefly the key points of the method: the Boltzmann equation can by written in the form

$$\left(\frac{\partial}{\partial t} + \mathcal{D}\right) f = \frac{1}{\epsilon} J(f|f) \tag{7.95}$$

where

$$\mathcal{D} = \mathbf{v} \cdot \frac{\partial}{\partial \mathbf{x}} + \frac{\mathbf{F}}{m} \cdot \frac{\partial}{\partial \mathbf{v}}, \tag{7.96}$$

$\mathbf{F}$ is the external force and $J(f|f)$ is the collision integral. One expands $f$ in a power series

$$f = f_0 + \epsilon f_1 + \epsilon^2 f_2 + \cdots \tag{7.97}$$

and the time derivative is replaced by

$$\frac{\partial^{(0)}}{\partial t} + \epsilon \frac{\partial^{(1)}}{\partial t} + \epsilon^2 \frac{\partial^{(2)}}{\partial t} + \cdots \tag{7.98}$$

where the $\partial^{(n)}/\partial t$ are defined by the $n$th approximation of the conservation theorem, e.g. $\partial^{(0)}/\partial t$ is given by the equation

$$\frac{\partial^{(0)}}{\partial t}\rho + \frac{\partial}{\partial \mathbf{x}}\left(\rho \mathbf{u}\right) = 0, \tag{7.99}$$

where $\rho$ is the mass density and $\mathbf{u}$ is the hydrodynamic velocity field. The physical motivation of (7.98) is the assumption that $f$ depends on $t$ only through the hydrodynamical field $\rho$, $\mathbf{u}$ and the energy density.

With (7.95), (7.96) and (7.97) one obtains a chain of equations:

$$J(f_0|f_0) = 0 \tag{7.100}$$

$$\left(\frac{\partial^{(0)}}{\partial t} + \mathcal{D}\right) f_0 = J(f_0|f_1) + J(f_1|f_0) \tag{7.101}$$

$$\left(\frac{\partial^{(0)}}{\partial t} + \mathcal{D}\right) f_1 = -\frac{\partial^{(1)}}{\partial t} f_0 + J(f_0|f_2) + J(f_2|f_0) + J(f_1|f_1), \tag{7.102}$$

and so on. The resemblance to the multiscale procedure discussed in Section 7.2.1 is evident. The zeroth-order equation (7.100) gives for $f_0$ the usual Maxwellian, $f_1$ is determined by the solvability condition (via the Fredholm alternative). With the first correction $f_1$ of the Chapman–Enskog method it is possible to write the dissipative hydrodynamics (e.g. Navier–Stokes and Fourier equations) where the kinetic coefficients are expressed in terms of interaction of the microscopic particles. This result was the first "true" success of the Boltzmann equation since it made it possible to derive the macroscopic equations without a priori guessing.

### 7.5.2 On standard and anomalous diffusion

One can wonder what is the reason for the specific choice (7.72) for the slow variables, that is a special case of $T = \epsilon^s t$ , $X = \epsilon x$. The value $s = 2$ is due to the fact that at long time one expects standard diffusion, i.e. $\langle x^2(t) \rangle \sim t$. This is somehow the usual behavior as shown by Avellaneda and Majda (1989): if the molecular diffusion coefficient $D_0$ is positive, in an incompressible velocity field the standard diffusion holds when the infrared contribution to **u** is not too strong, i.e.

$$\int \mathrm{d}\mathbf{k} \frac{\langle |\hat{\mathbf{u}}(\mathbf{k})|^2 \rangle}{k^2} < \infty, \tag{7.103}$$

where $\langle \cdot \rangle$ indicates the time average and $\hat{\mathbf{u}}(\mathbf{k})$ is the Fourier transform of $\mathbf{u}(\mathbf{x})$.

However, one can have anomalous diffusion, in particular super-diffusion, i.e. $\langle x^2(t) \rangle \sim t^{2\nu}$ with $\nu \geq 1/2$ (this corresponds to $D^{\mathrm{E}} = \infty$), and a non-Gaussian shape of the probability density function. The result of Avellaneda and Majda (1989) implies that only two possible origins exist for this super-diffusion:

I $D_0 > 0$ and strong infrared contributions to the velocity field, i.e. a violation of relation (7.103);

II $D_0 = 0$ and strong temporal correlations in the Lagrangian velocity, i.e. $\langle v_i(t')v_i(t) \rangle \sim |t - t'|^{-\beta}$ with $\beta \leq 1$, in such a way that $D^{\mathrm{E}}$ is formally infinite, see Eq. (7.66).

We do not enter here into a detailed discussion of this difficult issue. As an example of anomalous diffusion according to mechanism I we mention the remarkable result of Matheron and de Marsily (1980) who showed, in a rigorous way, the existence of super-diffusion in certain random shear flows. For instance, if in Eq. (7.60) $\mathbf{u} = (u(y), 0)$, where $u(y)$ is a random function obtained with a spatial Brownian motion process, then one has $\nu = 3/4$ and a non-Gaussian probability density function. The mechanism II is much more delicate and can be obtained only in systems with rather peculiar dynamics, for example deterministic chaotic systems with fine tuning of the control parameters (Zaslavsky *et al.* 1993, Leboeuf 1998).

Now we discuss briefly two aspects of the diffusion problem wich are of general interest. First we note that the Fick equation (7.61), which describes the asymptotic evolution of the transport equation (7.60), is an approximation in the sense that it is not able to catch the small spatial scales and short time behavior. Nevertheless Eq. (7.61), where the $D^E_{ij}$ are computed according to (7.83) and (7.84), is extremely precise in its asymptotic context; i.e. it is able to describe the possible delicate (and often counterintuitive) dependence of the eddy diffusion on $D_0$ and on the control parameters of the field $\mathbf{u}$ (Castiglione *et al.* 1999). As an example, we consider the diffusion in a two-dimensional field $\mathbf{u}(x, y, t) = (-\partial\psi/\partial y, \partial\psi/\partial y)$ where the stream function $\psi(x, y, t)$ is

$$\psi(x, y, t) = \psi_0 \sin\Big(x + B\sin(\omega t)\Big)\sin(y). \tag{7.104}$$

This simple model is able to capture the essential features of Rayleigh–Bénard convection, since the term $B\sin(\omega t)$, giving the lateral oscillation of the rolls, takes into account the oscillatory instability. For a generic value of $\omega$ (and $D_0 > 0$) one has standard diffusion (Castiglione *et al.* 1999). On the other hand one observes anomalous diffusion (with $\nu > 1/2$) for $D_0 = 0$ and special values of $\omega$ (resonance with the characteristic frequency of the test particle in the steady case limit). For these values of $\omega$ the eddy diffusivity $D^E$ diverges in the limit $D_0 \to 0$, i.e. $D^E \sim D_0^{-a}$ with $a > 0$. The multiscale approach is able to reproduce in a perfect way all these non-trivial behaviors of $D^E$ as a function of $\omega$ and $D_0$.

It is rather natural to wonder about the effective equation which rules the asymptotic evolution, i.e. at large spatial scales and long time, in the case of anomalous diffusion. In other words, one would like to know what is the equation replacing the Fick one. As far as we know this is still an open (interesting) problem.

The simplest (not so interesting from a physical point of view) example of anomalous diffusion is the Lévy flight, for which the probability density function of $x$ at time $t$ is the $\alpha$-stable Lévy distribution function (Bouchaud and Georges 1990). The asymptotic behavior at large $|x|$ is

$$P(x, t) \sim \left(\frac{x}{t^{1/\alpha}}\right)^{-(1+\alpha)} \tag{7.105}$$

with $1 \leq \alpha < 2$. Note that $\langle |x(t)|^q \rangle \sim t^{q/\alpha}$ for $q < \alpha$ and $\langle |x(t)|^q \rangle = \infty$ for $q \geq \alpha$, so that the "typical" value of $|x(t)|$ is $O(t^{1/\alpha})$ and therefore one can consider the Lévy flight as an example of anomalous diffusion. In such a case the time evolution of $P(x, t)$ is obtained simply by replacing the operator $\partial^2_{xx}$ in (7.61) with $-(-\partial^2_{xx})^{\alpha/2}$, i.e. the probability density function evolves according to:

$$\partial_t P(x,t) = -D(-\partial^2_{xx})^{\alpha/2} P(x,t). \tag{7.106}$$

The fractional derivative $(-\partial^2_{xx})^{\alpha/2}$ indicates an integral operator whose representation in the Fourier space is just a multiplication by $|k|^\alpha$.

In more interesting cases it is not at all simple to find the evolution rule which replaces the Fick equation, and knowing the scaling exponent $\nu$ (given by $\langle x(t)^2\rangle \sim t^{2\nu}$) is not enough to find the shape of $P(x,t)$. Still assuming a scaling relation

$$P(x,t) \sim \frac{1}{t^\nu} F(x/t^\nu), \tag{7.107}$$

the function $F(\cdot)$ is not determined by $\nu$ (Bouchaud and Georges 1990). Other forms have been proposed in specific cases, in terms of fractional time and/or spatial derivative, for which the interested reader can see Zaslavsky (2002).

Additional difficulties arise in the presence of the so-called *strong anomalous diffusion* (Castiglione *et al.* 1999), i.e. when one has

$$\langle |x(t)|^q\rangle \sim t^{q\nu(q)} \tag{7.108}$$

where $\nu(q)$ is not constant. In such a case it is evident that Eq. (7.107) cannot hold.

### 7.5.3 Remarks and perspectives on the piston problem

Despite, or rather because of, its highly idealized character, the piston problem has many pedagogical virtues. Using this problem one can understand the following points.

(a) The relative nature of an *equilibrium state*, depending on the time scales and on the observed variables. Here three different equilibrium states have been revealed:
  - for the piston at short times (mechanical equilibrium, $P^+ = P^-$);
  - for the piston at long times (thermal equilibrium, $T^+ = T^- = T_\mathrm{P}$);
  - for the whole system, at still longer times (complete thermodynamic equilibrium).

(b) As also discussed in Chapter 5, *irreversibility* arises from the fact that the system has been prepared in some special, *non-typical state*, i.e. the initial (macroscopic) state $(P_0^-, T_0^-, P_0^+, T_0^+, X_0)$ is in a region of the whole phase space (position–velocity of the $N$ particles and the piston) which is very small compared to the region associated with mechanical (and thermal) equilibrium.

(c) The piston is a typical instance where a high-dimensional deterministic system ($N \gg 1$) produces an effective stochastic motion for a reduced number of variables, here the piston position $X$ and velocity $V$.

(d) In a variant of the model used in Section 7.4, the ideal gas is replaced by a *hard-sphere gas* (Mansour *et al.* 2005). In such a case one also has particle–particle collisions which are generally (as soon as the density is not vanishingly small) more efficient than the indirect coupling of the particles (via collision with the piston). Therefore the relaxation of the gas particle distribution toward the Maxwell distribution (phase 3 above) occurs sooner, at pace with the evolution of the temperatures. In addition to the technical aspects, an interesting fact is that, at least if the initial probability distribution of the particle velocity is Maxwellian, the time evolution of the piston (apart from phase 3) does not depend too much on the model of the gas. So we have two models (perfect gas and hard spheres) which are very different from a dynamical point of view (i.e. integrable and chaotic) and produce the same macroscopic behavior. This is rather similar to the results discussed in Chapter 3 for diffusive behavior and it is further evidence of the small role of chaos in systems with many degrees of freedom.

(e) A more general conclusion, actually relevant for the whole chapter, is that a multiscale approach is required as soon as the phenomenon involves different scales that cannot be decoupled; in particular, plain averaging or naive perturbation procedures are invalid. The signature of such a situation is a singular dependence with respect to the small parameter quantifying the scale separation, here $\alpha \simeq 2m/M$. This means that the behavior for $\alpha \to 0$ is qualitatively different from the behavior observed for $\alpha = 0$, with no possible way to match continuously (far less in a perturbative way) these behaviors.

## References

Arnold, V. I. (1976). *Méthodes Mathématiques de la Mécanique Classique*, Moscow: Mir.

Avellaneda, M. and Majda, A. J. (1989). Stieltjes integral representation and effective diffusivity bounds for turbulent transport, *Phys. Rev. Lett.* **62**, 753.

Bender, C. and Orszag, S. A. (1978). *Advanced Mathematical Methods for Scientists and Engineers*, New York: McGraw-Hill.

Biferale, L., Crisanti, A., Vergassola, M. and Vulpiani, A. (1995). Eddy diffusivities in scalar transport, *Phys. Fluids* **7**, 2725.

Bocquet, L. (1997). High friction limit of the Kramers equation: the multiple time-scale approach, *Am. J. Phys.* **65**, 140.

Bouchaud, J. P. and Georges, A. (1990). Anomalous diffusion in disordered media – statistical mechanisms, models and physical applications, *Phys. Rep.* **195**, 127.

Caglioti, E., Chernov, N. and Lebowitz, J. L. (2004). Stability of solutions of hydrodynamic equations describing the scaling limit of a massive piston in an ideal gas, *Nonlinearity* **17**, 897.

Callen, H. B. (1963). *Thermodynamics*, New York: Wiley.

Castiglione, P., Mazzino, A., Muratore-Ginanneschi, P. and Vulpiani, A. (1999). On strong anomalous diffusion, *Physica D* **134**, 75.

Cercignani, C. (1988). *The Boltzmann Equation and its Applications*, Berlin: Springer.

Cercignani, C. (1998). *Ludwig Boltzmann: the Man who Trusted Atoms*, Oxford: Oxford University Press.

Cercignani, C., Illner, R. and Pulvirenti, M. (1994). *The Mathematical Theory of Dilute Gases*, Berlin: Springer.

Chandrasekhar, S. (1943). Stochastic problems in physics and astronomy, *Rev. Mod. Phys.* **15**, 1.

Crosignani, P., Di Porto, P. and Segev, M. (1996). Approach to thermal equilibrium in a system with adiabatic constraints, *Am. J. Phys.* **64**, 610.

Curzon, A. E. (1969). A thermodynamic consideration of mechanical equilibrium in the presence of thermally insulating barriers, *Am. J. Phys.* **37**, 404.

E, W. and Engquist, B. (2003). Multiscale modeling and computation, *Notices Am. Math. Soc.* **50**, 1062.

Español, P. (2003). Statistical mechanics of coarse-graining, in *Novel Methods in Soft Matter Simulations*, ed. Karttunen, M., Vattulainen, I. and Lukkarinen, A., vol. 140, *Lecture Notes in Physics*, p. 69, Berlin: Springer.

Feynman, R. P. (1963). *The Feynman Lectures on Physics*, New York: Addison-Wesley.

Frisch, U. (1995). *Turbulence*, Cambridge: Cambridge University Press.

Frisch, U., Hasslacher, B. and Pomeau, Y. (1986). Lattice gas automata for the Navier–Stokes equations, *Phys. Rev. Lett.* **56**, 1505.

Gardiner, C. W. (1990). *Handbook of Stochastic Methods for Physics, Chemistry and the Natural Sciences*, Berlin: Springer.

Gruber, C. and Morriss, G. P. (2003). A Boltzmann equation approach to the dynamics of the simple piston, *J. Stat. Phys.* **113**, 297.

Gorban, A. N., Karlin, I. V. and Zinovyev, A. Y. (2004). Constructive methods of invariant manifolds for kinetic problems. *Phys. Rep.* **396**, 197.

Gruber, C. and Lesne, A. (2006). Adiabatic piston in *Encyclopedia of Mathematical Physics*, ed. Francoise, J. P., Naber, G. and Tsun, T. S., Amsterdam: Elsevier.

Gruber, C. and Pache, S. (2002). The controversial piston in the thermodynamic limit, *Physica A* **314**, 345.

Gruber, C., Pache, S. and Lesne, A. (2002). Deterministic motion of the controversial piston in the thermodynamic limit, *J. Stat. Phys.* **108**, 669.

Gruber, C., Pache, S. and Lesne, A. (2003). Two-time-scale relaxation towards thermal equilibrium of the enigmatic piston, *J. Stat. Phys.* **112**, 1177.

Leboeuf, P. (1998). Normal and anomalous diffusion in a deterministic area-preserving map, *Physica D* **116**, 8.

Lieb, E. H. (1999). Some problems in statistical mechanics that I would like to see solved, *Physica A* **263**, 491.

Majda, A. J. and Kramer, P. R. (1999). Simplified models for turbulent diffusion: theory, numerical modelling, and physical phenomena, *Phys. Rep.* **314**, 238.

Mansour, M. M., Van den Broeck, C. and Kestemont, E. (2005). Hydrodynamic relaxation of the adiabatic piston, *Europhys. Lett.* **69**, 510.

Matheron, G. and de Marsily, G. (1980). Is transport in porous media always diffusive? A counter-example, *Water Resour. Res.* **16**, 901.

Mehra, J. (2001). *The Golden Age of Theoretical Physics*, Singapore: World Scientific.

van Kampen, N. G. (1990). *Stochastic Processes in Physics and Chemistry*, Amsterdam: North-Holland.

Zaslavsky, G. M. (2002). Chaos, fractional kinetics, and anomalous transport, *Phys. Rep.* **371**, 461.

Zaslavsky, G. M., Stevens, D. and Weitzner, H. (1993). Self-similar transport in incomplete chaos, *Phys. Rev. E* **48**, 1683.

# 8

# Renormalization-group approaches

> If we study the history of science we see produced two phenomena which are, so to speak, each the inverse of the other. Sometimes it is simplicity which is hidden under what is apparently complex; sometimes, on the contrary, it is simplicity which is apparent, and which conceals extremely complex realities.
>
> *Henri Poincaré*

Devoting only a short chapter to the renormalization group (RG) is a challenge owing to the wealth of both fundamental concepts and computational tools associated with this term. The RG is indeed encountered in many different domains of theoretical physics, ranging from quantum electrodynamics to second-order phase transitions to fractal growth and diffusion processes – one should rather speak of renormalization group*S*! We refer to Brown (1993) and Fisher (1998) for an historical account, to Goldenfeld (1992) and Lesne (1998) and references therein for an overview of the domains of application and variants of RGs.

We here emphasize that the RG is a way, maybe the most successful and with no doubt the most systematic and constructive, to *derive effective low-dimensional descriptions that capture large-scale and/or long-time behavior.* The RG can be extended far beyond the specific scope of critical phenomena, to an *iterated multiscale approach* allowing the construction of robust and minimal macroscopic models describing the universal large-scale features and asymptotics of a complex system. This generalized viewpoint brings out the close logical and even technical connections that bridge, within a unified framework, perturbative RG for singular series expansions, spin-block RG and momentum-shell RG for critical phenomena, RG for the asymptotic analysis of differential and partial differential equations, and probabilistic RG for the derivation of statistical laws and limit theorems.

## 8.1 Renormalization group(s): a brief overview

### *8.1.1 Renormalization-group philosophy*

The RG resembles a chameleon, changing appearance when passing from one domain of application to another: the connection between the RG in quantum electrodynamics, the RG describing the onset of chaos, and the RG for polymer conformations, for instance, is not obvious, to say the least. Let us summarize the main features, both conceptual and technical, common to all RGs.

- RGs are implemented in situations involving (infinitely) *many different scales* when moreover there is *no decoupling* of these scales: when arbitrarily small-scale features or events might have major consequences at arbitrarily large scales, that reflects in *singular behaviors and divergences*, e.g. of susceptibilities and response functions. RGs then offer an operational and systematic way of taming or circumventing these singularities.
- RGs involve the determination of *effective terms*, for instance *effective parameters*, accounting in an integrated and physically meaningful way, e.g. directly measurable in some experiment, for many indirect or smaller-scale contributions. An important benefit of this procedure is to take advantage of cancellations that happen between large or even diverging contributions, when their singularity is a feature of the (idealized) mathematical model but not of physical reality, rather associated with the resulting effective (and finite) terms. An historical example is the renormalization of the electron mass in quantum electrodynamics: the bare mass $m_0$ arising in the original theory is involved in an unbounded number of successive $(e^+, e^-)$ paired annihilations and creations, of arbitrarily short durations. These events, occurring at arbitrarily high frequencies (equivalently high energies) introduce divergences in the theory: the so-called *ultraviolet divergences*. But in fact, most of these events cancel out when taken all together and the remaining contribution is physically well defined: it yields the measurable mass $m$ of the electron. By contrast, the bare mass $m_0$ appears only as an auxiliary parameter in a mathematical picture; actually, the *individual* physical reality of the elementary events parametrized by $m_0$ might be questioned. Other examples will be detailed in the following sections, for instance the renormalization treatment of singular perturbation series encountered in celestial mechanics and non-linear physics.
- RG methods describe *macroscopic behaviors*, at long times and large spatial scales (equivalently at small frequencies and wave vectors, hence also named *infrared behaviors*). RGs are required when these behaviors are singular insofar as they cannot be obtained by a mere averaging over microscopic scales or projection techniques. Typical instances are critical phenomena where long-range correlations build up a singular macroscopic behavior, but we shall see that RGs are relevant in many other contexts.
- A common idea of RG methods is to investigate the *links between behaviors at different scales* in time and space. Along this line, one way to handle divergences and extract meaningful information from diverging behaviors is to determine finite relations (e.g. finite ratios or finite differences) between infinite quantities relative to different levels; another way is to determine the rate of divergence of an observable quantity $X(a, N)$ as

the system size or evolution duration $N$ goes to infinity, under the form of a *scaling law* with respect to the control parameter $a$ and size $N$: $X(a, N) \sim N^\alpha f[(a - a_c)N^\gamma]$, termed a *finite-size scaling* relation. Compared to other approaches that are merely phenomenological, RGs prove to be a constructive method insofar as they allow the values of the exponents $\alpha$ and $\gamma$, and even in some instances the universal function $f$, to be computed explicitly, thus *demonstrating* the existence of a scaling behavior and its expression, with no need for preliminary guesses or scaling assumptions.

- RGs currently involve *joint rescalings*, acting as zooms or magnifying glasses and allowing a straightforward comparison of features at different scales: one here recovers the standard and intuitive way to reveal quantitatively *self-similarity* and *scaling laws*.

### 8.1.2 Renormalization transformations

Any RG centrally involves a *renormalization transformation* $\mathcal{R}$ (an operator) expressing the effect of coarse graining or cutoffs and rescalings on the quantities ruling the evolution or the equilibrium statistics: the evolution law or map, the Hamiltonian, the set of transition probabilities, a partial differential equation, to quote the most current cases. The main ingredients of $\mathcal{R}$ are the following.

- *A change in our viewpoint on the system*, following from a cutoff, a coarse graining, a modification in the definition of elementary units or in the decomposition between zeroth-order and perturbation terms. Indeed, defining a model always implies splitting the system into a core and a remaining part (fine structure, noise, perturbation), and the first step of renormalization is to modify this subjective partition: that amounts to considering another modeling of the same system.
- *A corresponding change in the set of variables* describing the system state.
- *A modification of the structure and evolution* rules themselves, now expressed as functions of the new variables to fit with our modified viewpoint. For instance, changing the scale of our description amounts to coarse graining the state variables, and the renormalized model should express the evolution or equilibrium equations at this coarser level. In particular, this step embeds the replacement of parameters by effective ones, accompanied in most cases by an extension of the parameter space since the original parametrized form of the evolution or equilibrium equations ( e.g. the parametrized form of the Hamiltonian) is in general not preserved upon the action of $\mathcal{R}$, and additional terms involving extra parameters arise.
- *Additional rescalings* to preserve normalizations or physical invariants (the physical system is indeed unchanged upon renormalization) and to make self-similarity appear as a fixed-point property.

The renormalization operator $\mathcal{R}$ acts at the level of models; it expresses *how our modeling of a given system has to be modified when we change our viewpoint on the system*. We emphasize that the underlying physical reality (or mathematical reality in the case of series expansions or probabilistic theorems) is *not* modified:

renormalization acts *upon its representations* at different scales. Consistency relations, called *renormalization equations*, arise that express this fact precisely: that we are dealing with various models corresponding to different views of the *same* system.

### *8.1.3 Renormalization groups*

The renormalization transformation $\mathcal{R}$ is then to be *iterated*, thus generating a discrete group structure: $\mathcal{R}^j \circ \mathcal{R}^m = \mathcal{R}^{j+m}$. Here the so-called *renormalization group* appears, actually a *semi-group* since $\mathcal{R}$ is not invertible in most if not all cases, owing to the loss of information accompanying coarse graining. A more general and less trivial group structure arises in the dependence with respect to the rescaling factors, say $(k_0, k_1, \ldots, k_n)$, involved in the renormalization:

$$\mathcal{R}_{k_0,k_1,\ldots,k_n} \circ \mathcal{R}_{k'_0,k'_1,\ldots,k'_n} = \mathcal{R}_{k_0k'_0,k_1k'_1,\ldots,k_nk'_n}. \tag{8.1}$$

This group relation expresses the consistency of the model transformations upon changes of viewpoints or description scales. When the rescaling factors can take any positive real values (and under some regularity conditions) the RG is endowed with a continuous group structure: a *Lie semi-group* in mathematical terms. It is then fruitful to consider the infinitesimal transformations, called the *infinitesimal generators* of the group, providing a basis of the Lie algebra of the group:

$$\mathcal{T}_i \equiv \left(\frac{\partial R}{\partial k_i}\right)_{k_0=\cdots=k_n=1}. \tag{8.2}$$

Most often, the $(n+1)$ rescalings are not independent but on the contrary are performed jointly: $k_0 \equiv k$, $k_1 = k^{\alpha_1}, \ldots, k_n = k^{\alpha_n}$, what is called a *scaling limit* when $k \to \infty$. The basic rescaling factor $k$ is usually that associated with linear spatial extension: each iteration of $\mathcal{R}$ amounts to increasing the scale of the description by a factor of $k$. In consequence, *the asymptotics of the RG flow, generated by the action of $\mathcal{R}$ in the space of models, corresponds to effective large-scale models.* Accordingly, the exponents $(\alpha_1, \ldots, \alpha_n)$ leading to a non-trivial fixed point under the iterated action of $\mathcal{R}_{k,k^{\alpha_1},\ldots,k^{\alpha_n}}$ are the scaling exponents describing the critical or anomalous asymptotic behavior of the system (or ratios of such exponents). This fact will be worked out explicitly in Sections 8.3.2 and 8.3.3.

Since RG techniques and achievements are too numerous even to be listed, we shall present only a small sample of applications, chosen to illustrate how the RG achieves a multiscale analysis, as argued in the introduction.

## 8.2 Renormalization groups to cure singular perturbation expansions

Several RG techniques have been developed since the seminal works of Lindstedt and Rayleigh to obtain *uniformly valid* perturbation expansions allowing the investigation of asymptotic behaviors (Lindstedt 1882, Rayleigh 1917). In its simplest formulation, the RG demanding problem is encountered when the evolution of the system is described by a differential equation which depends on a small parameter $\epsilon$ *in a singular way*. Typically, the singularity is associated with *turning points* (i.e. when some leading term involves a vanishingly small multiplicative parameter $\epsilon$) or with *resonances* between forcing terms of amplitude $\epsilon$ and the linearized evolution (see the discussion in Chapter 7 on the ensuing *secular divergences*).

It is always possible to look for a solution in the form $u(t) = \sum_{n\geq 0} \epsilon^n u_n(t)$ and identify term-wise the successive powers of $\epsilon$ in the evolution equation. It is another matter to prove that the components $\{u_n(t)\}$ obtained in this way yield a well-behaved solution: this requires assessing good convergence properties for the perturbation expansion $\sum_{n\geq 0} \epsilon^n u_n(t)$. One gets a regular behavior on a time interval $[t_0, t_1]$ as soon as each component is bounded: $\sup_{[t_0,t_1]} |u_n(t)| \leq M_n$ and the series $\sum \epsilon^n M_n$ converges for $\epsilon < \epsilon^*$ (with a finite radius of convergence $\epsilon^* > 0$). Mathematical theorems then ensure that the solution $u(t)$ inherits the regularity properties of the components $\{u_n(t)\}$. But difficulties often appear while investigating the asymptotic behavior of the system since it requires control of the expansion of $u(t)$ for arbitrarily long times ($t_1 = +\infty$). Singular behaviors are expected when the condition of uniform boundedness fails to be true, e.g. $\lim_{t\to\infty} u_n(t) = \infty$ for some $n$ (at the very least, this failure already spoils the consistency of the method since the perturbation order no longer reflects the ordering of the term strengths). In this section we show how the RG allows such singularities to be cured.

### 8.2.1 Lindstedt method of strained parameters

The basic idea goes back to Lindstedt (1882). Consider a weakly non-linear oscillator:

$$\frac{\mathrm{d}^2 u}{\mathrm{d}t^2} + \omega_0^2 u = \epsilon f\left(u, \frac{\mathrm{d}u}{\mathrm{d}t}\right) \quad \text{with} \quad \epsilon \ll 1, \tag{8.3}$$

the effect of the anharmonicity $\epsilon f(u, \mathrm{d}u/\mathrm{d}t)$ can be accounted for by a redefinition of the original frequency $\omega_0$ into an effective one:

$$\omega(\epsilon) = \omega_0 + \epsilon\omega_1 + \epsilon^2\omega_2 + O(\epsilon^3). \tag{8.4}$$

The time variable is accordingly rescaled into a dimensionless variable:

$$\tau = t\omega(\epsilon). \tag{8.5}$$

The benefit of this procedure is to make valid the determination of the solution as a perturbation series $u(\tau) = u_0(\tau) + \epsilon u_1(\tau) + \epsilon^2 u_2(\tau) + O(\epsilon^3)$. The key step is the self-consistent determination of the coefficients $\{\omega_i\}$ so as to prevent the appearance of secular terms in the expansion for $u(\tau)$ and get a uniformly convergent series (Nayfeh 1973). Today one would see this technique as a renormalization method, calling $\omega_0$ the *bare* frequency and $\omega(\epsilon)$ the *renormalized* frequency. As underlined in Section 8.1.1, the experimentally measurable parameter is the renormalized and not the bare parameter, which appears only as an auxiliary quantity in the original mathematical model.

Compared to the multiscale approach presented in Chapter 7, the redefinition of the time variable $t$ as $\tau = t\omega(\epsilon)$ is here performed not at the level of the evolution equation (as in Chapter 7), but within the straightforward perturbation expansion of its solution. In consequence, the consistency conditions determining $\omega(\epsilon)$ involve algebraic equations instead of differential ones, which is obviously a technical benefit. Moreover, *no prior guess of what would be the proper rescaled times is involved*, thus circumventing the failure of the multiscale approach when an intuitively unexpected rescaled variable enters the scene (e.g. $t_1 = \sqrt{\epsilon}t_0$ and $t_2 = \epsilon t_0$ while one would have naively introduced $t_1 = \epsilon t_0$ and $t_2 = \epsilon^2 t_0$).

### *8.2.2 Renormalization group alternative to the multiscale approach: an example*

To illustrate further the application of the RG to singular perturbation analysis, and to single out its principles compared with multiscale analysis, let us detail its implementation for an example encountered in Chapter 7: the Duffing oscillator with given initial conditions $(a_0, b_0)$,

$$\frac{\mathrm{d}^2 u}{\mathrm{d}t^2} + u + \epsilon u^3 = 0, \quad u(t_0) = a_0, \quad \frac{\mathrm{d}u}{\mathrm{d}t}(t_0) = b_0. \tag{8.6}$$

The general idea is first to implement the straightforward perturbation scheme, simply plugging the expansion $u(t) = \sum_{n \geq 0} \epsilon^n u_n(t)$ into the evolution equation; *then* the secular divergences arising in the resulting series are cured by replacing the bare parameters $a_0$ and $b_0$ with slowly varying renormalized values $a_1(t, \epsilon)$ and $b_1(t, \epsilon)$, in such a way that good convergence properties are recovered at fixed $a_1$ and $b_1$. Following this program, one first gets at the lowest order:

$$u_0(t) = a_0 \cos t + b_0 \sin t = \mathrm{Re}[A_0\, \mathrm{e}^{\mathrm{i}t}] \quad \text{with} \quad A_0 = a_0 - \mathrm{i}b_0. \tag{8.7}$$

The equation at next order is written:

$$\mathrm{d}^2 u_1/\mathrm{d}t^2 + u_1 = -\, u_0^3. \tag{8.8}$$

The coefficients appearing in its solution are fully determined given the bare complex parameter $A_0$:

$$u_1(t) = \mathrm{Re}[B\mathrm{e}^{\mathrm{i}t} + C\mathrm{e}^{3\mathrm{i}t} + Dt\mathrm{e}^{\mathrm{i}t}] \quad \text{with} \quad C = \frac{A_0^3}{32}, \qquad D = \frac{3\mathrm{i}A_0|A_0|^2}{8}, \tag{8.9}$$

and

$$B = -\mathrm{Re}(C) + \mathrm{i}\ \mathrm{Re}(D) - 3\mathrm{i}\ \mathrm{Im}(C).$$

This expression for $u_1(t)$ clearly reveals the singular nature of the perturbation analysis, due to the presence of the secular term $Dt\mathrm{e}^{\mathrm{i}t}$. At this stage, the RG principle is to replace $A_0$ by a *slowly varying renormalized parameter* $A_1(t, \epsilon)$ in the expression (8.7):

$$\widetilde{u}_0(t) = \mathrm{Re}[A_1(\epsilon, t)\, \mathrm{e}^{\mathrm{i}t}] \tag{8.10}$$

so as to regularize the equation for the next component $\widetilde{u}_1(t)$ by eliminating the source of the secular divergence. By imposing the condition that the coefficient of $\mathrm{e}^{\mathrm{i}t}$ on the right-hand side of this equation vanishes, we get

$$-\frac{\mathrm{i}}{2}\frac{\mathrm{d}^2 A_1}{\mathrm{d}t^2} + \frac{\mathrm{d}A_1}{\mathrm{d}t} = \epsilon D(A_1) + O(\epsilon^2) \quad \text{where} \quad D(A) = \frac{3\mathrm{i}A|A|^2}{8}. \tag{8.11}$$

In other words, the consistency of the regularization, i.e. the fact that the secular term is fully absorbed in the time dependence of the effective initial condition $A_1(t, \epsilon)$, is expressed by a *renormalization equation* for $A_1(\epsilon, t)$, that appears as the analog of the solvability condition in the multiscale approach.

The very presence of $\epsilon$ on the right-hand side of (8.11) ensures that the function $A_1(t, \epsilon)$ is a slowly varying function of time, hence better expressed as a function of $\tau = \epsilon t$: $A_1(t) = \widetilde{A}_1(\tau)$ so that $\mathrm{d}A_1/\mathrm{d}t = \epsilon \mathrm{d}\widetilde{A}_1/\mathrm{d}\tau$, $\mathrm{d}^2 A_1/\mathrm{d}t^2 = \epsilon \mathrm{d}^2\widetilde{A}_1/\mathrm{d}\tau^2$. Since only the lowest order in $\epsilon$ is relevant (and consistent), this scaling of the derivatives allows us to ignore $\mathrm{d}^2 A_1/\mathrm{d}t^2$ (it would contribute to equations for higher-order components $u_n(t)$ with $n > 1$), leading to:

$$\mathrm{d}\widetilde{A}_1/\mathrm{d}\tau = D[\widetilde{A}_1]. \tag{8.12}$$

Turning back to real-valued quantities by writing $A_1 = a_1 - \mathrm{i}b_1$ yields:

$$\begin{cases} \dfrac{\mathrm{d}a_1}{\mathrm{d}\tau} = \dfrac{3b_1}{8}(a_1^2 + b_1^2) \\[2ex] \dfrac{\mathrm{d}b_1}{\mathrm{d}\tau} = -\dfrac{3a_1}{8}(a_1^2 + b_1^2). \end{cases} \tag{8.13}$$

It is straightforward to conclude from (8.13) that $|A_1|$ does not vary with time: $|A_1|^2 = (a_1^2 + b_1^2) = r^2 =$ constant. The evolution for the phase $\theta$ that follows from (8.13) finally yields the following first-order solution, with $A_1(\epsilon, t) = |A_1|\, e^{i\theta(\epsilon,t)}$:

$$\theta = \theta_0 + \epsilon\omega(t - t_0) = \theta_0 + \omega(\tau - \tau_0) \quad \text{where} \quad \omega = \frac{3r^2}{8} \quad \text{and} \quad \tau = \epsilon t, \tag{8.14}$$

recovering the first-order solution given in Chapter 7 when plugged into (8.10).

### 8.2.3 Perturbative renormalization groups

The previous example illustrates the general idea of *perturbative RGs*: first one performs formally a straightforward perturbation expansion, as if the problem were regular, with the advantage of using a standard and fully determined procedure. Then the coefficients are renormalized in such a way that convergence properties (at fixed values for the renormalized coefficients) are improved.

This second step in fact applies to any singular series $\sum_{n\geq 0} \epsilon^n u_n(t, a_0)$ not necessarily associated with some ordinary differential equation. Such a series is said to be *renormalizable* when it is possible to turn it into a uniformly valid expansion $\sum_{n\geq 0} \epsilon^n \widetilde{u}_n(t, a)$, for $\epsilon < \epsilon^*$ small enough, by transforming the bare parameters $a_0$ of the original expansion into renormalized parameters $a(a_0, \epsilon, t)$. The formulation possibly embeds a redefinition of the time variable $t \to b(\epsilon)t$, associated with a bare parameter $b_0 = 1$. Both the suitable time rescaling $b(\epsilon)$ and the other renormalized parameters $a(a_0, \epsilon, t)$ are to be *determined in a self-consistent way*, at successive orders in the expansion parameter $\epsilon$, typically by imposing a uniform bound $\sup_t |\, \widetilde{u}_n[t, a(a_0, \epsilon, t)]\, | < \infty$ and a slow time dependence $da/dt = O(\epsilon a)$. Let us underline that such a *self-consistent* determination is ubiquitous in renormalization procedures: the proper redefinitions and rescalings are the only ones leading to a non-trivial and well-behaved solution, endowing the RG with a constructive character, requiring *no prior guess or scaling assumptions*.

Still more generally, let us consider a function $F(a_0, t, \epsilon)$ exhibiting some singular part $\epsilon F_{\text{sing}}$, typically a perturbative correction whose amplitude remains small and controlled by a parameter $\epsilon \ll 1$ in finite time, but asymptotically becomes an overwhelming influence. The general perturbative RG scheme amounts to replacing the bare parameter $a_0$ with an unknown function $a(a_0, t, \epsilon)$ of time and perturbation parameter $\epsilon$, to be determined by the following regularization condition:

$$F(a_0, t, \epsilon) \equiv F_{\text{reg}}(a_0, t, \epsilon) + \epsilon F_{\text{sing}}(a_0, t, \epsilon) = F_{\text{reg}}[a(a_0, t, \epsilon), t, \epsilon] \; + \; \text{h.o.} \tag{8.15}$$

As already mentioned above in the more restricted instance of a series, what is termed *renormalizability* of $F$ is the possibility of absorbing all its singularities in such a redefinition of some parameters or variables. The very definition of $a(a_0, t, \epsilon)$ ensures that $F_{\text{reg}}(a, t, \epsilon)$ remains regular as $t \to \infty$ and $\epsilon \to 0$ at fixed $a$, and reduces to a limiting function $F_0(a)$. Accordingly, it now becomes possible to describe the leading behavior of $F$ in these limits:

$$F(a_0, t, \epsilon) \sim F_0[a(a_0, t, \epsilon)] + \text{h.o.} \quad (t \to \infty, \; \epsilon \to 0). \tag{8.16}$$

The singularity of the asymptotic behavior is now embedded in the correspondence $t \to a(a_0, t, \epsilon)$. The renormalized coefficient $a(a_0, t, \epsilon)$ appears as a "macroscopic" parameter integrating all the underlying, "microscopic" influences involved in $F_{\text{sing}}(a_0, t, \epsilon)$ and building up its singularities. The consistency of this effective description, absorbing the singular part $F_{\text{sing}}$ into a regular renormalized component $F_0[a(a_0, t, \epsilon)]$, is expressed through the following *renormalization equations*:

$$\frac{\partial F_0}{\partial a} \cdot \frac{\partial a}{\partial t} = \frac{\partial F_{\text{sing}}}{\partial t}, \quad \frac{\partial F_0}{\partial a} \cdot \frac{\partial a}{\partial \epsilon} = \frac{\partial F_{\text{sing}}}{\partial \epsilon} \quad \text{and} \quad a(a_0, t, \epsilon = 0) = a_0. \tag{8.17}$$

These equations re-express the solvability condition arising in the multiscale approach. The RG thus appears as an extension of multiscale methods with the main advantage of providing a constructive method: there is no need to guess the proper rescaled variables since they arise naturally in solving RG equations.

The RG ideas presented in this section, first implemented for singular perturbation series arising in mechanics (three-body problem, driven or anharmonic oscillators), non-linear physics and kinetic theory, have been applied to critical phenomena and phase transitions, introducing our next section. In this context, the time $t$ and the perturbation parameter $\epsilon$ are replaced by the size $N$ (number of particles) and temperature difference $T - T_c$ between $T$ and the critical temperature $T_c$: indeed, the issue encountered in the perturbation approach of critical phase transitions lies in the non-uniformity with respect to $N$ of the temperature expansion of the free energy $F(T, N)$ around $T = T_c$, preventing interchange of the expansion in powers of $T - T_c$ with the thermodynamic limit $N \to \infty$.

## 8.3 A multiscale and constructive approach to capture critical behavior

The RG developed in the context of statistical mechanics (in the 1970s) and dynamical systems (in the 1980s) initially appeared as a framework specially designed to investigate *universal large-scale features of critical phenomena* (Stanley 1999). This is at the same time a restriction, insofar as such an RG can only capture asymptotic and self-similar features, and a strength: by focusing on the critical features and their origin, namely strong correlations at all scales, the RG allows us to *prove* the existence of scaling laws describing the behavior around the critical points and to *compute* the numerical values of their exponents, the so-called *critical exponents*. The RG is here to be seen as an efficient alternative approach when scale separation and mean-field arguments fail to apply. We shall see in Section 8.4 how this RG and its central principle of iterated coarse graining in fact extend into a more general technique for asymptotic analysis, far beyond the scope of critical behaviors.

### *8.3.1 Scaling theories and critical exponents*

Criticality is a multiform notion acknowledged in many domains of the natural sciences, and the best way to grasp its meaning and content is to describe the signatures of a critical state (also termed a *critical point*, in the control parameter space).

- The range of the correlations that take place between subsystems is unbounded in space and time, which is expressed quantitatively in the *divergence of the correlation length* $\xi \to \infty$ *and time* $\tau_c \to \infty$.
- Accordingly, a system in a critical state exhibits *slow relaxation* properties, instead of an exponential decrease as $e^{-t/\tau_c}$ toward its steady state when $\tau_c < \infty$ (the relaxation times are indeed related to the correlation time and diverge together). In the same spirit, a system in a critical state exhibits a *singular response* to perturbations: response functions diverge at the critical point of the system.
- The strongly correlated statistics near a critical point, say $T = T_c$ if the control parameter is the temperature, is also reflected in *scaling laws* $A(T) \sim |T - T_c|^\alpha$ for the correlation functions and order parameters (actually any macroscopic observable $A$ in the thermodynamic limit) as a function of the temperature difference $T - T_c$.
- Fluctuations at all scales (in amplitude, spatial extension and duration) are observed at a critical point.

These features have been known for a long time in the context of *second-order phase transitions*, observed at the so-called *critical point* $(P_c, T_c)$, a single isolated point in the phase diagram, ending a line of phase coexistence and first-order transition points. For instance, in the case of the liquid–gas transition in a simple fluid, the presence at the critical point of fluctuations at all spatial scales can be visualized directly as a sudden milky appearance, called *critical opalescence*; its origin is the diffraction of visible light on large density inhomogeneities, indicating that their size is of the order of visible wavelengths (roughly half of a micrometer), much larger than the typical molecular size of the density fluctuations observed far from the critical point. Another experimental evidence is the direct observation under the microscope of large micrometer-size domains of differing concentrations. Nevertheless, when crossing the critical point by varying $T$ at fixed $P = P_c$ or varying $P$ at fixed $T = T_c$, the average density (called the *order parameter* of the system) exhibits a continuous variation and no phase coexistence occurs: we, macroscopic observers, see one and the same fluid experiencing only slight morphological changes, by contrast to the liquid–vapor transition just below the critical temperature. Whereas no discontinuity occurs in the order parameter at the critical point, singularities appear in its derivatives, and then markedly, associated with a divergence: such a transition is termed a *second-order phase transition*. In the same way, all the susceptibilities, specific heat, and response functions also diverge at the critical point, endowing it with enough striking experimental signatures to be located very accurately, as an isolated point, in the control parameter space $\{(P, T)\}$. The same features arise in various second-order phase transitions, for instance at the Curie point $T = T_c$ (transition between ferromagnetic and paramagnetic behavior in a magnet) with mesoscopic domains of differing spin orientations, a continuous average magnetization $M(T)$ and divergence of the magnetic susceptibility $\chi(T)$ at $T_c$.

The first major step in understanding critical phenomena was the recognition of their *scale invariance*, originating[1] in the divergence of the correlation ranges $\xi$ and $\tau_c$ and the associated absence of any finite characteristic scale. A *scaling theory* was developed to express and exploit this remarkable property. This theory *assumes* (on experimental grounds) the existence of scaling laws, describing for example the divergence of the specific heat, $C_v(T) \sim |T - T_c|^{-\alpha}$, the divergence of the correlation length $\xi(T) \sim |T - T_c|^{-\nu}$ measured as the size of inhomogeneities or domains, or the behavior of the order parameter $M(T) \sim |T - T_c|^{\beta}$ at the critical point $T = T_c$. Beyond providing a simple analytical fit for the relevant observables, this theory predicts some relations between the exponents of the scaling laws; it was thus shown in the early 1960s that *only two critical exponents are independent.*

What took more time, and was achieved only in the 1970s or even later, was to compute the anomalous[2] values for the critical exponents. The main challenge was to understand their remarkable universality: for instance, the same values are observed within experimental accuracy for the liquid–gas transition and transitions occurring in binary mixtures or metallic alloys. The main problem was the

[1] On mathematical grounds, the connection between scale invariance and the divergence $\xi \to \infty$ of the correlation range lies in the fact that the generic alternative to an exponential decay $e^{-x/\xi}$ is provided by power laws $x^{-\gamma}$.

[2] One speaks of *anomalous scaling laws* or *anomalous exponents* (from the Greek *anomalos*, meaning "contrary to custom") when they differ from those obtained by a mere dimensional analysis. For instance, a fractal object (e.g. a porous medium) is anomalous since its mass $M(r)$ does not plainly scale with its volume $r^3$ (where $r$ is the linear size of the sample), but scales as $r^{d_f}$ with an exponent $d_f < 3$ called its *fractal dimension*.

indisputable evidence that exponent values computed within a mean-field approach[3] were wrong. This was shown both by experimental values established with an exquisite accuracy and by an analytically exact calculation achieved by Onsager for the Ising model in dimension $d = 2$ yielding an exponent $\beta = 1/8$ whereas the mean-field prediction, independent of the system and its dimension $d$, is $\beta = 1/2$.

The quantitative understanding of critical phenomena remained an outstanding puzzle for many years; whose difficulty was clearly related to the singularities arising in thermodynamic functions at critical points. From the mathematical viewpoint, critical points appeared as situations where a plain expansion of the free energy, order parameter and other statistical properties in powers of $T - T_c$ does not work, reflecting the *non-commuting character* of the thermodynamic limit $N \to \infty$ (where $N$ is the number of particles or more generally the number of degrees of freedom of the system) and the limit $T \to T_c$. It is at least clear that scales do not decouple in a system at or near a critical point as they do far away, and that, in consequence, all scales should be jointly handled using some non-trivial multiscale procedure in order to account for critical behavior.

### *8.3.2 Iterated coarse graining and renormalization-group transformation*

The major breakthrough in solving the puzzle was the idea of *iterated coarse graining* proposed by Kadanoff in the context of spin lattices and the critical ferromagnetic transition at the Curie point (Kadanoff 1966). The starting point was the investigation of a regular spin lattice of dimension $d$ (square lattice for $d = 2$, cubic lattice for $d = 3$) where neighboring spins $\vec{s}_i$ and $\vec{s}_j$ on the lattice are coupled through a ferromagnetic interaction $-J\,\vec{s}_i.\vec{s}_j$, with a coupling constant $J > 0$. The equilibrium properties of such a spin lattice are fully prescribed by its Hamiltonian $H$, where the pair interactions are supplemented with the influence of the local magnetic field $\vec{h}_i$ at site $i$, namely

$$H = -\sum_{\langle i,j \rangle} J\,\vec{s}_i.\vec{s}_j - \sum_i \vec{h}_i.\vec{s}_i.$$

The first idea was to describe the system *with a coarser resolution*, considering blocks of $2^d$ spins as basic units instead of the initial description at the level of single spins, and to compute *the effective interactions between these spin-blocks* as

[3] In brief, a mean-field approach amounts to identifying the local environment, which is in general fluctuating and spatially inhomogeneous (e.g. the local magnetic field generated by neighboring spins in a spin lattice model) with the average environment (a spatial average or equivalently a statistical average in the limit as the system size tends to infinity). This average local field is identified with a homogeneous mean field depending only on the average order parameter of the whole sample, hence the established name of the method. We know now that mean-field approaches are valid only in high enough spatial dimension $d > d_c$, when they yield the exact critical exponents (becoming dimension independent in this regime); they fail dramatically for $d < d_c$. The threshold $d_c$, called the *critical dimension*, depends on the class of the critical phenomenon: $d_c = 4$ for the liquid–vapor transition or the ferromagnetic transition, whereas $d_c = 6$ for percolation.

the net result of spin–spin interactions between adjacent blocks. The interactions between spins of the same block (internal to a block) contribute to the renormalized local magnetic field (at the scale of a block).

The second idea underlying the design of the RG transformation $\mathcal{R}$ was *to express self-similar properties in terms of a functional invariance upon renormalization*, provided $\mathcal{R}$ involves adequate rescalings, as explained in Section 8.1.3. The coarse graining and associated transformation of the model are thus supplemented with a spatial rescaling $x \to x/2$ in each space direction, in order to *restore the apparent value of the lattice cell size*. Accordingly, the apparent correlation length decreases by a factor of 2, indicating that the critical character of the rescaled system of spin-blocks is weakened compared with the initial system of spins. Another requirement supporting the introduction of such rescalings and prescribing their actual values is the *conservation of physical invariants upon renormalization*: indeed, renormalization amounts to changing the model as the standing point of the observer changes but *without changing the underlying real system*. In a spin lattice, the physical invariant to be preserved is the value of the partition function since it is directly related to the free energy of the system. With this aim, an additional spin rescaling, depending upon the dimension $d$ and the lattice geometry, has to be included in the renormalization procedure, leading to a certain expression $\vec{s}'_b = 2^{-\alpha} \sum_{j \in \text{block}(b)_i} \vec{s}_j$ for the renormalized spin associated with block $b$.

The third idea was to *iterate* the whole renormalization procedure. This provides a constructive way, now known as a *renormalization group*, to integrate out recursively all the small-scale features (occurring at the spin level, inside blocks) into their increasingly large-scale consequences (occurring at the block level, for blocks increasing in size at each iteration of the renormalization), at scales increasing up to infinity (in the thermodynamic limit).

To implement these ideas operationally, a first technical point is to work with generalized Hamiltonians $\mathcal{H} = H/kT$ encapsulating the temperature dependence. Another more decisive step has to be taken: to extend the initial space of Hamiltonians, or equivalently the initial space of coupling parameters, into a more general infinite-dimensional space. This is essential because the RG transformation generates new terms in the effective Hamiltonians, that cannot be cast in a mere modification of the initial coupling parameters: it is a "trajectory" in a whole set of possible Hamiltonians that should be constructed upon repeated action of $\mathcal{R}$ and analyzed. This step was achieved by Wilson (1971, 1975), with the following key results (see Figure 8.1).

(1) Starting from a model $\mathcal{H}_c$ at its critical point, the limiting behavior of this trajectory $(\mathcal{R}^n \mathcal{H}_c)_{n \geq 0}$ under the RG action is a hyperbolic fixed point $\mathcal{H}^*$ of $\mathcal{R}$ describing an "ideally critical" system in the sense that it is exactly scale invariant, by the very construction of $\mathcal{R}$. The fixed-point condition $\mathcal{R}\mathcal{H}^* = \mathcal{H}^*$ requires that $\xi(\mathcal{R}\mathcal{H}^*) \equiv \xi(\mathcal{H}^*)/2 = \xi(\mathcal{H}^*)$,

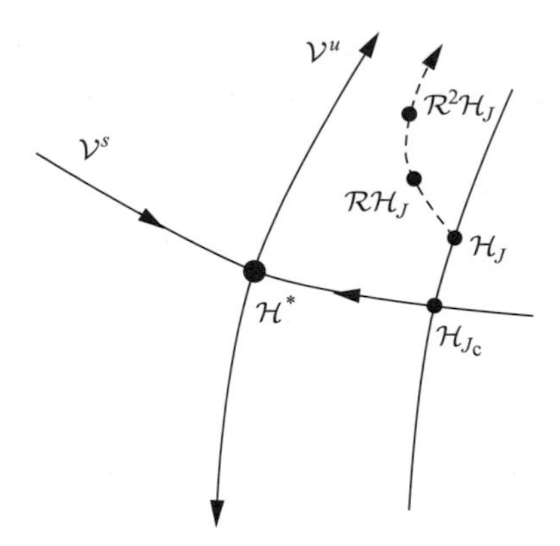


Figure 8.1 RG flow around a hyperbolic fixed point $\mathcal{H}^*$, in a space of models $\{\mathcal{H}\}$ (for instance a space of Hamiltonians). The stable manifold $\mathcal{V}^s$ is the locus of models at their critical point; the unstable eigenvalues are directly related to the critical exponents (for the sake of simplicity, only one stable direction and one unstable direction are represented here).

with two possible solutions $\xi^* = 0$ and $\xi^* = \infty$. Critical fixed points correspond to $\xi^* = \infty$. Fixed points with $\xi^* = 0$ correspond to trivially self-similar systems, typically the statistically homogeneous and uncorrelated systems obtained at infinite temperature.

(2) The linear stability analysis of the RG transformation $\mathcal{R}$ around the critical fixed point $\mathcal{H}^*$, i.e. determination of the eigenvalues of $D\mathcal{R}(\mathcal{H}^*)$, gives access to the critical exponents, thus providing a constructive and systematic method to compute their values. To understand this point, let us consider the simplified case of a parametrized Hamiltonian $\mathcal{H}_J$ such that the renormalization is expressed at the leading order as a transformation $r_k$ of the coupling constant $J$ (i.e. a transformation acting in the control parameter space) according to $\mathcal{R}_k\mathcal{H}_J \approx \mathcal{H}_{r_k(J)}$. The critical Hamiltonian in this one-parameter family $\{\mathcal{H}_J\}$ will be $\mathcal{H}_{J_c}$ where $J_c$ is an approximate fixed point of $r_k$ in the parameter space. Taking together the approximate fixed-point equation, the renormalization $\xi[r_k(J)] = \xi(J)/k$ of the correlation length, the scaling law $\xi(J) \sim |J - J_c|^{-\nu}$ and the linear expansion $r_k(J) \approx J_c + r_k'(J_c)(J - J_c)$ yields the value of the exponent:

$$\nu = \frac{\log k}{\log |r_k'(J_c)|} \tag{8.18}$$

where $k$ is the length rescaling factor (often $k = 2$). The RG group structure ensures that the resulting exponent $\nu$ does not depend on $k$. In the typical instance, $J$ is directly related to the temperature $T$ in a monotonous way, thus yielding scaling laws with variable $T - T_c$. As illustrated in Figure 8.1, the relation $\mathcal{R}_k\mathcal{H}_J \approx \mathcal{H}_{r_k(J)}$ defining the map $r_k$, hence the fixed-point equation $r_k(J_c) \approx J_c$, is only valid at the leading order, provided $\mathcal{H}_{J_c}$ is close enough to the actual fixed point $\mathcal{H}^*$ of $\mathcal{R}$. They would be exact only for a one-parameter family embedded in the unstable manifold $\mathcal{V}^u$. The idea here is that the RG action upon a transverse one-parameter family $\{\mathcal{H}_J\}$ is essentially the same as its action within the unstable manifold. This argument and the above sequence of relations have to be made more rigorous by working, as underlined above, in a space of Hamiltonians instead of restricting to a parametrized form $J \to \mathcal{H}_J$, and by carefully handling and controlling the higher-order terms, but the reasoning and the formula remain basically the same. Detailed examples can be found in Goldenfeld (1992), Plischke and Bergersen (1994), Lesne (1998) and in Section 8.3.3 below.

(3) Each critical fixed point $\mathcal{H}^*$ of $\mathcal{R}$ defines a *universality class* gathering systems whose critical behaviors are described by the same set of critical exponents. This universality class is related to the basin of attraction of $\mathcal{H}^*$ that contains all the critical Hamiltonians sharing the same critical features. More precisely, any one-parameter family $\mathcal{H}_J$ crossing the stable manifold of $\mathcal{H}^*$ for some $J_c$ will exhibit the same scaling laws with the same critical exponents, as derived in (2).

(4) The unstable and stable eigenvectors of $D\mathcal{R}(\mathcal{H}^*)$ correspond respectively to *essential* and *inessential perturbations* of the Hamiltonian, modifying or not its degree of criticality (it decreases when departing from $\mathcal{H}^*$ along unstable directions).

(5) Finally, perturbations of the initial Hamiltonian such that the ensuing trajectory upon the action of $\mathcal{R}$ converges to a different critical fixed point, hence modifying the universality class of the original systems, are called *crossover terms*.

When the definition of the renormalization transformation involves a rescaling of the system size (number $N$ of time steps, of particles, of spatial cells), the RG also yields a quantitative account of the size dependence of the collective behavior of the system, typically in the form already mentionned in Section 8.1.1 of *finite-size scaling relations*, i.e. scaling laws involving $N$ among the variables. Specifically, a scaling law $X \sim (a - a_c)^{-\nu}$ becomes $X_N \sim N^\alpha f[(a - a_c)N^\gamma]$ in finite size, typically with $f(z) \sim z^{-\nu}$ when $z \to \infty$. Consistency yields $\alpha = \nu\gamma$; hence finite-size scaling analysis, by plotting $N^\alpha X_N$ as a function of $N^\gamma(a - a_c)$, gives access to the scaling exponent $\nu$. The RG can be seen as a form of finite-size scaling analysis, an interpretation that will be met again in Section 8.5.1.

### 8.3.3 The renormalization group time analog: the example of scenarios to chaos

Kadanoff's idea of iterated coarse graining, initially introduced and described above in a spatial setting at thermal equilibrium, was later developed in a purely *temporal setting* (among others by Kadanoff himself) for discrete-time dynamical systems. The position $\vec{r}$ in real space and the time $t$ might indeed be treated on the same footing: the examples presented in this subsection will reveal the parallel between the spatial RGs developed in equilibrium statistical mechanics, aimed at determining thermodynamic behavior, and the dynamic RGs aimed at determining asymptotic behavior.

One of the most striking achievements of the temporal RG methods is the computation of the universal number $\delta \approx 4.66920$ describing the accumulation of bifurcation values $\{\mu_j\}$ toward the value $\mu_c$ corresponding to the onset of chaos in the *period-doubling scenario:* $\mu_c - \mu_j \sim \delta^{-j}$, where $\mu$ is a control parameter of the evolution law that typically quantifies the strength of its non-linearities, as exemplified by the normal form[4] $x_{n+1} = f_\mu(x_n) \equiv 1 - \mu x_n^2$. In this scenario, a period-doubling bifurcation occurs in the dynamics at each value $\mu = \mu_j$, during which a $2^j$-cycle loses its stability and continuously gives birth to a stable $2^{j+1}$-cycle. The onset of chaos at $\mu = \mu_c$ appears as the *time analog of a critical point*; in particular, the attractor exhibits remarkable self-similar properties at this point. Several scaling laws around $\mu = \mu_c$ (for instance for the envelope of the Lyapunov exponent or the typical time for reaching the attractor) reinforce the analogy (Lesne 1998).

In this context, the relevant class of models is the set $\mathcal{F}$ of unimodal (i.e. single-humped) maps of the interval $[-1, 1]$ having a quadratic maximum, e.g. in $x = 0$:

$$\mathcal{F} = \left\{ \begin{array}{c} f\ :\ I \mapsto I,\ \text{even, analytical,}\ \ f(0) = 1 \\ f'(0) = 0 \text{ and } f \text{ is strictly increasing in } [-1, 0[ \\ Sf < 0 \text{ in } [-1, 0[ \cup ]0, 1] \ \text{ where } \ Sf = f'''/f' \ - \ 3/2\ \left(f''/f'\right)^2 \end{array} \right\}. \tag{8.19}$$

Translating the block-spin RG principles in this temporal setting, we consider time steps twice as long and an appropriate normalization so that the renormalized maps still take their values in $[-1, 1]$ with a maximum value 1 reached at $x = 0$. The appropriate renormalization transformation is thus written

$$\mathcal{R}f(x) = \frac{1}{\lambda_f} f \circ f(x\lambda_f) \quad \text{with} \quad \lambda_f = f[f(0)] = f(1). \tag{8.20}$$

In other words, the RG exploits the similarity betweeen $f$ and its iterate $f^2 = f \circ f$ around the point $x = 0$ after a truncation to the interval $[1/\lambda_f, -1/\lambda_f]$ and a suitable rescaling by a factor of $\lambda_f^{-1} < 0$. One then shows that this renormalization operator $\mathcal{R}$ admits a unique hyperbolic[5] fixed point $\varphi$, and that $\delta$ is the unique unstable eigenvalue ($\delta > 1$) of $D\mathcal{R}(\varphi)$ (Feigenbaum 1978, Collet and Eckmann 1981). Let us detail this proof since it provides an exemplary RG implementation and shows that its strength lies in the relation between critical exponents and unstable eigenvalues of the linearized renormalization operator around fixed points.

[4] A simple computation shows that this even map in $[-1, 1]$ is equivalent to the so-called *logistic map* $g_a(z) = az(1 - z)$ in $[0, 1]$ with a one-to-one increasing correspondence between the control parameters $\mu \in [0, 2]$ and $a \in [0, 4]$.

[5] A fixed point of a discrete flow is said to be *hyperbolic* if it has stable and unstable directions with well-separated stable and unstable eigenvalues: $\sup |\lambda^s| < 1 < \inf |\lambda^u|$.

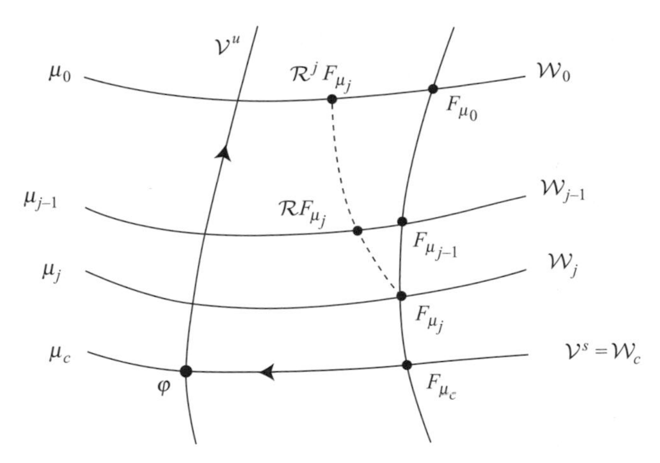


Figure 8.2 Given any one-parameter family $(F_\mu)_\mu$ in $\mathcal{F}$, transverse to the stable manifold $\mathcal{V}^s$, the action of the renormalization operator $\mathcal{R}$ is expressed in the relations $\lim_{n\to\infty} \mathcal{R}^n F_{\mu_c} = \varphi$ (with $F_{\mu_c} \in \mathcal{V}^s$) and $\mathcal{R}F_{\mu_j} \in \mathcal{W}_{j-1}$ (with $F_{\mu_j} \in \mathcal{W}_j$). Linearization of the approximate relation $\mathcal{R}F_{\mu_j} \approx F_{\mu_{j-1}}$ yields the scaling law $\mu_c - \mu_j \sim \delta^j$ where $\delta > 1$ is the unique unstable eigenvalue of $D\mathcal{R}(\varphi)$, roughly quantifying the action of $\mathcal{R}$ along the unstable manifold $\mathcal{V}^u$ (see the text).

The renormalization operator $\mathcal{R}$ given in (8.20) is well defined only in a subset $\mathcal{D}_0$ of $\mathcal{F}$ whose elements remain in $\mathcal{F}$ upon the action of $\mathcal{R}$, yielding

$$\mathcal{D}_0 = \{f \in \mathcal{F}, \lambda_f < 0,\ f(\lambda_f) > 0,\ f^2(\lambda_f) < -\lambda_f\} \subset \mathcal{F}. \tag{8.21}$$

The boundaries of $\mathcal{D}_0$ are the manifolds $\mathcal{S}_0 = \{f \in \mathcal{F},\ \lambda_f = 0\}$ and $\mathcal{U}_0 = \{f \in \mathcal{F},\ f^2(\lambda_f) + \lambda_f = 0\}$. One then introduces the subset $\mathcal{W}_0$ of maps at a pitchfork bifurcation point (in particular $f_{\mu_0} \in \mathcal{W}_0$). Similar definitions are introduced for the iterates $\mathcal{R}^j$, with straightforward relations $\mathcal{D}_j = \mathcal{R}^{1-j}[\mathcal{D}_0]$, $\mathcal{S}_j = \mathcal{R}^{1-j}[\mathcal{S}_0]$, $\mathcal{U}_j = \mathcal{R}^{1-j}[\mathcal{U}_0]$ and $\mathcal{W}_j = \mathcal{R}^{1-j}[\mathcal{W}_0]$ (see Figure 8.2). The elements of $\mathcal{S}_j$ possess a stable cycle of period $2^{j+1}$, whereas the $2^j$th iterate of an element of $\mathcal{W}_j$ experiences a pitchfork bifurcation. Since

$$\mathcal{R}^n f(x) = \frac{1}{\Lambda_n} f^{2^n}(\Lambda_n x) \quad \text{with} \quad \Lambda = f^{2^n}(0), \tag{8.22}$$

the asymptotic behavior $n \to \infty$ of the trajectories $\{x_n = f^n(x_0)\}$ is directly related to the asymptotic behavior $j \to \infty$ of the RG trajectory $\{\mathcal{R}^j f\}$. In particular, given two maps $f$ and $g$, comparison of the RG trajectories $\{\mathcal{R}^j f\}$ and $\{\mathcal{R}^j g\}$ provides a direct comparison of the asymptotic properties of the flows generated by $f$ and $g$, considered as a whole.

A central feature of the RG flow asymptotic behavior is the existence and stability of fixed points $\varphi$ of $\mathcal{R}$. By plugging the analytical expansion of a putative fixed point $\varphi$ in equations $\lambda = \varphi(1)$ and $\mathcal{R}\varphi = \varphi$, solving term-wise and checking the good convergence properties of the resulting expansion, it is shown that $\mathcal{R}$ admits a unique non-trivial fixed point. The linearized renormalization operator is written

$$[D\mathcal{R}(\varphi).g](x) = \frac{1}{\lambda}\,[g(1)[x\varphi' - \varphi(x)] + g[\varphi(\lambda x)] + \varphi'[\varphi(\lambda x)]g(\lambda x)]. \tag{8.23}$$

One shows that it possesses a unique unstable eigenvalue $\delta > 1$ and strictly stable ones (that is, $\varphi$ is a hyperbolic fixed point). The stable manifold can be characterized alternatively as the subset where $\mathcal{R}$ can be iterated indefinitely

$$\mathcal{V}^{\mathrm{s}} = \cap_{j=0}^{\infty}\mathcal{D}_j = \lim_{j\to\infty} \mathcal{D}_j \tag{8.24}$$

(the second equality follows from the inclusion $\mathcal{D}_j \subset \mathcal{D}_{j-1}$). It is also straightforward to check that $\mathcal{V}^{\mathrm{s}} = \lim_{j\to\infty}\mathcal{W}_j = \lim_{j\to\infty}\mathcal{S}_j = \lim_{j\to\infty}\mathcal{U}_j$. Manifolds $\{\mathcal{W}_j\}$ comprise a bundle accumulating on $\mathcal{V}^{\mathrm{s}}$ (see Figure 8.2). The manifolds $\{\mathcal{S}_j\}$ intermingle in between whereas the manifolds $\{\mathcal{U}_j\}$ are located on the other side of $\mathcal{V}^{\mathrm{s}}$.

Let $F_\mu$ be a one-parameter family of maps in $\mathcal{F}$. The critical value $\mu_{\mathrm{c}}$ of the control parameter follows from the condition $F_{\mu_{\mathrm{c}}} \in \mathcal{V}^{\mathrm{s}}$, from which it follows that

$$\lim_{j\to\infty} \mathcal{R}^j[F_{\mu_{\mathrm{c}}}] = \varphi. \tag{8.25}$$

This means that $F_{\mu_{\mathrm{c}}}$ is the evolution map of a dynamical system at the onset of chaos. By definition of $\mathcal{W}_j$, the sequence $\{\mu_j\}$ of bifurcation values is obtained according to the condition $F_{\mu_j} \in \mathcal{W}_j$. By construction, one has $\mathcal{R}[F_{\mu_j}] \in \mathcal{W}_{j-1}$, close to $F_{\mu_{j-1}}$. When $F_{\mu_{\mathrm{c}}}$ is close to $\varphi$ and for large enough $j$, it is possible to perform linear analysis of the approximate relation $\mathcal{R}[F_{\mu_j}] \approx F_{\mu_{j-1}}$. With this aim, one introduces the projection $P^{\mathrm{u}}$ on the unstable direction $e^{\mathrm{u}}$ of $D\mathcal{R}(\varphi)$ associated with the unique unstable eigenvalue $\delta > 1$. It satisfies $P^{\mathrm{u}} \circ D\mathcal{R}(\varphi) = \delta.P^{\mathrm{u}}$, from which it follows that $P^{\mathrm{u}}[\mathcal{R}F_{\mu_j} - \varphi] \approx \delta.P^{\mathrm{u}}[F_{\mu_j} - \varphi]$ when quadratic and higher-order terms are neglected. The action of $P^{\mathrm{u}}$ on the relation $\mathcal{R}[F_{\mu_j}] \approx F_{\mu_{j-1}}$ yields $\delta.P^{\mathrm{u}}[F_{\mu_j} - \varphi] \approx P^{\mathrm{u}}[F_{\mu_{j-1}} - \varphi]$. Plugging in this relation the linear-order estimate

$$P^{\mathrm{u}}[F_\mu - \varphi] \approx P^{\mathrm{u}}[F_\mu - F_{\mu_{\mathrm{c}}}] \approx (\mu - \mu_{\mathrm{c}})\, P^{\mathrm{u}}\left(\frac{\partial F_\mu}{\partial \mu}\right)_{\mu=\mu_{\mathrm{c}}} + O(\mu - \mu_{\mathrm{c}})^2, \tag{8.26}$$

allows us to conclude that $\mu_{j-1} - \mu_{\mathrm{c}} \approx \delta(\mu_j - \mu_{\mathrm{c}})$. To be fully rigorous, this reasoning requires the transversality condition $P^u(\partial F_\mu.\partial\mu)_{\mu=\mu_c} \neq 0$ to be satisfied and the non-linear terms to be controlled so as to ensure that they do not spoil the conclusion.

Other universality classes are obtained when the space of models is extended and one considers the sets $\mathcal{F}_\epsilon$ of single-humped maps that behave as $|x|^{1+\epsilon}$ near

$x = 0$ for some $\epsilon > 0$. The universal scaling factor $\delta(\epsilon)$ computed in a similar way to $\delta = \delta(1)$ is the unstable eigenvalue of $D\mathcal{R}(\varphi_\epsilon)$ (Lesne 1998).

Another nice application of this temporal RG concerns the *intermittency scenario* of chaos, describing the destabilization of a low-dimensional and fully predictable dynamics through bursts of irregular evolution, unpredictable in both amplitude and duration. This scenario occurs via a saddle-node bifurcation (that is, in which a stable fixed point disappears by merging with an unstable point) and the issue is to determine the scaling law of the persistence time of the laminar regime.

For simplicity, we shall consider the simplest discrete-time family exhibiting saddle-node intermittency: $f_a(x) = [-a + x - x^2]_a$. Two fixed points $x_a^{\pm} = \pm\sqrt{-a}$ are present for $a < 0$ where $x_a^+$ is stable and $x_a^-$ is unstable as soon as $a > -1$. They coalesce in $a_c = 0$ with $x_{a_c}^{\pm} = 0$ and $f'_{a_c}(0) = +1$. For $a > 0$, there is no longer a fixed point, either stable or unstable, but the dynamics is nevertheless slowed down near $x = 0$. Indeed, the trajectory is trapped in the channel delineated by the diagonal $y = x$ and the graph $y = f_a(x)$ of the evolution map, during a time depending on the entrance point in the channel.

The RG idea for computing the average time $\tau(a)$ spent in the channel, corresponding to the duration of regular phases in the observed dynamics, is to perform a temporal coarse graining by a factor, say, of 2. This amounts to comparing the original discrete dynamics with $\Delta t = 1$ and the dynamics observed at the larger time scale $\Delta t = 2$, namely comparing the evolutions generated respectively by $f$ and $f \circ f$. As in the case of the period-doubling scenario, the corresponding RG transformation is completed with a rescaling $\mathcal{R}f = \lambda^{-1} f \circ f(\lambda x)$ so as to maximize the similarity between $f$ and $\mathcal{R}f$: here the criterion is identification of the coefficient of the monomial $x$ in these maps yielding $\lambda = 1/2$. Taking together the renormalization action $\tau(\mathcal{R}f) = \tau(f)/2$ on the persistence time and the approximate action $\mathcal{R}f_a \approx f_{4a}$ of $\mathcal{R}$ on the parametrized family (that is, $\mathcal{R}f_a \approx f_{r(a)}$ with $r(a) = 4a$) leads to the scaling law:

$$\tau(a) \sim \frac{1}{\sqrt{a}}, \quad \text{i.e.} \quad \tau(a) \sim a^{-\mu} \quad \text{with} \quad \mu = \frac{\log 2}{\log |r'(0)|} = 1/2. \tag{8.27}$$

The action of $\mathcal{R}$ on $f_a$ in fact generates extra cubic and quartic terms beyond $f_{r(a)}$, and a more rigorous RG analysis has to be performed in a whole functional space instead of being restricted to a given one-parameter family and its one-dimensional parameter space. The main step is the determination of the fixed points of $\mathcal{R}$ and associated eigenvalues, while controlling the higher-order terms in order to assess the robustness of the result beyond linear RG analysis. This shows that the scaling law for $\tau(a)$ does not depend on the details of the maps but only on their belonging to the universality class of intermittency, whose normal form is the family $f_a(x) = [-a + x - x^2]_a$.

As in the case of the period-doubling scenario, it is worth noticing that other universality classes can be determined by extending the action of $\mathcal{R}$ to an enlarged functional space, obtained by considering monomials $|x|^{1+\epsilon}$ instead of $x^2$; a different exponent $\mu(\epsilon)$ is obtained for each value of $\epsilon$ (Lesne 1998).

## 8.4 Renormalization groups: a multiscale approach for asymptotic analysis

Moving beyond the study of critical phenomena, the RG has been extended more recently to an *iterated multiscale approach* allowing the determination in a systematic and constructive way of *effective equations* describing the large-scale behavior, in both real space and time, of multiscale systems and the investigation of their structural stability (see e.g. Chen *et al.* (1994), Oono (2000), Mazzino *et al.* (2004), Cencini *et al.* (2006)). RG approaches are specially relevant and fruitful for *partial differential equations*, encountered for instance in hydrodynamics, growth phenomena, or diffusion in complex media (Goldenfeld 1992), that are singular insofar as the asymptotic and large-scale behavior of their solutions cannot be determined straightforwardly by a simple averaging procedure, due to resonances (generating secular divergences) or essential non-linearities (inducing mode coupling).

### *8.4.1 Asymptotic analysis and structural stability*

In the previous subsections, we have seen that the RG devised to cure non-uniformities and secular divergences in perturbation series unifies various singular perturbation techniques, including the multiscale approach presented in Chapter 7. A main advantage of the RG is that *it does not require any a priori guess of the proper rescalings*: the slow-time, large-scale components of the solution arise naturally through a slow dependence in time and space of the renormalized parameters, without invoking any arbitrary choice in their description. The discussion of RG principles for critical phenomena and chaos, in Section 8.3, has allowed us to understand better the meaning and significance of secular divergences: they reflect persistent large-scale consequences of microscopic perturbations; in this regard they are closely related to criticality.

Our point is now to emphasize that *only the secular terms are relevant when dealing with asymptotics*, exactly like the mechanisms generating long-range correlations are the only relevant ones to give a full account of the large-scale behavior at a critical point. Accordingly, in asymptotic analysis, one can ignore other contributions (*inessential terms*) and reduce the dynamics to its secular terms. This reveals the bridge between the perturbative RG devised to handle and regularize secular terms in perturbation series and the RG related notion of universality of large-scale features at a critical point. Whatever the context, *the essential terms are all those, and only those, with non-trivial behavior under the renormalization action.* This signature provides constructive access to the robustness properties of asymptotic behavior, at the cost of devising the proper renormalization operator and analyzing the associated RG flow.

We have seen in Section 8.3.2 that the RG philosophy to capture universal critical features relies on their invariance upon coarse graining, rescaling and associated transformation of parameters into effective parameters, translated into an *invariance upon renormalization* and a *fixed-point property* for the Hamiltonian or evolution law. We shall see now that this philosophy can be extended far beyond the scope of self-similar critical phenomena to achieve a systematic extraction of asymptotic features and to investigate their structural stability, that is, to determine the class of models sharing these features. The central RG step is the *self-consistent determination of an adapted transformation* (the renormalization transformation) of the model that leaves invariant its asymptotics. Determining that the original model and its renormalized version share the same large-scale behavior puts constraints on the renormalization transformation, for instance relations between the different rescaling factors or the very expression of renormalized parameters. Accordingly, *devising an RG amounts to capturing quantitatively the underlying inter-level relationships and the multiscale organization inducing this behavior.* This explains why the renormalization operator encapsulates quantitative features of the asymptotic behavior and how they can be unraveled, "with no magic," by the analysis of the RG flow.

### 8.4.2 Green's function approach and general RG equations

Here we show how general RG principles apply to determine the asymptotic behavior of evolution equations (ordinary or partial differential equations) in the case *when their solutions depend in a singular way on initial conditions.* One here recovers the early developed Green function RG approach (Stueckelberg and Petermann 1953, Jona-Lasinio 2001). For simplicity we restrict the discussion to ordinary differential equations, with only one space or time variable $z$, and consider a given solution,

$$f(z|z_0, f_0) \quad \text{with} \quad f(z_0|z_0, f_0) = f_0 \tag{8.28}$$

with special emphasis on boundary/initial conditions in $z = z_0$. As an illustration think for instance of an expression $f(z|z_0, f_0) = f_0(z/z_0)^\alpha$ with $\alpha > 0$. The singularity is expressed as a divergence when $z \to +\infty$ at fixed $z_0$, that can be handled upon renormalization and tamed by a redefinition of initial conditions and associated integration constant, as explained in Section 8.2.2. But the singularity is also expressed as a divergence when $z_0 \to 0$ at fixed $z$, hinting at another method of regularization, by introducing the multiplicative analog of a cutoff. Namely, regularization might be achieved by changing the initial point $z_0$ into $\lambda z_0$ with $\lambda \geq 1$ and by an adapted change of the initial condition $f_0$ into $f(\lambda z_0|z_0, f_0)$:

$$f(z|z_0, f_0) = f[z|\lambda z_0, f(\lambda z_0|z_0, f_0)]. \tag{8.29}$$

These joint changes are nothing but a renormalization transformation,

$$f(\lambda z_0|z_0, f_0) = R(\lambda z_0, z_0, f_0), \tag{8.30}$$

endowed with a bona fide RG structure through the semi-group relation

$$R(\lambda_2 z_0, \lambda_1 z_0, R(\lambda_1 z_0, z_0, f_0)) = R(\lambda_2 z_0, z_0, f_0) \quad \text{with} \quad \lambda_2 \geq \lambda_1 \geq 1. \tag{8.31}$$

The RG equation follows, that expresses the validity of (8.31) for any intermediary scaling factor; in particular, for $\lambda_2 z_0 = z$ and $\lambda_1 = 1$, it reduces to

$$\partial_2 R(z, z_0, f_0) + \partial_3 R(z, z_0, f_0).\partial_1 R(z_0, z_0, f_0) = 0. \tag{8.32}$$

The scaling behavior, if any, is expressed in a relation which is no longer dependent on the initial point $z_0$ (scale invariance here appears as a homogeneity property):

$$f(z|z_0, f_0) = F(z/z_0, f_0) \quad (\text{for any } z \geq z_0) \quad \text{and} \quad F(1, f_0) = f_0. \tag{8.33}$$

In this case, there is a direct link between the singularities as $z \to +\infty$ and $z_0 \to 0$ and associated renormalizations, both expressed in the following relation:

$$F(x, f_0) = F(x/\lambda, F(\lambda, f_0)) \quad (\text{for any } \lambda \geq 1 \quad \text{and any} \quad x \geq \lambda \geq 1). \tag{8.34}$$

The explicit singularity as $x \to +\infty$ is reduced upon dividing $x$ by $\lambda > 1$ while its integrated effect on the overall asymptotics is conserved, by means of the redefinition of the initial condition $f_0$, replaced with $F(\lambda, f_0)$. RG consistency requires that the right-hand side should not depend on $\lambda$, yielding a so-called *RG equation* describing the regularized dependence of asymptotics on the initial conditions,

$$x \frac{\partial F}{\partial x}(x, f_0) = a(f_0) \frac{\partial F}{\partial f_0}(x, f_0) \quad \text{with} \quad a(f_0) = \frac{\partial F}{\partial x}(1, f_0), \tag{8.35}$$

recovering (8.32) in the special case where $R(z, z_0, f_0) = F(z/z_0, f_0)$. A change of variable $t = \log z$ casts the above multiplicative framework into an additive one. In particular, a scale invariance for $z$ turns into a translation invariance for $t$; a singularity when $z_0 \to 0$ then corresponds to a singularity when $t_0 \to -\infty$. In the additive framework, the singularity is circumvented by introducing an intermediary time $t_1$ and decomposing $(t - t_0) = (t - t_1) + (t_1 - t_0)$. This auxiliary time $t_1 > t_0$ plays the same regularizing role as a cutoff, considering the evolution over $[t_1, +\infty[$

instead of $[t_0, +\infty[$. The RG procedure accounts for the asymptotic consequences of the singular transient $[t_0, t_1[$ into a modification of the initial conditions:

$$f_1 = \mathcal{R}(t_1, t_0, f_0) = R(\mathrm{e}^{t_1}, \mathrm{e}^{t_0}, f_0). \tag{8.36}$$

In other words, renormalization encapsulates the only meaningful way to perform the limit $t_0 \to -\infty$: it requires *jointly* an adapted variation of $f_0$, following the variation of $t_0$ according to relation (8.36), at fixed $t_1$ and $f_1$. The relations $x = \mathrm{e}^{t_1 - t_0}$ and $\mathcal{R}(t_1, t_0, f_0) = F(x, f_0)$ provide the link with the previous multiplicative formulation (8.34). As above, the consistency condition that the solution should not depend on the arbitrarily chosen intermediary time $t_1$ is equivalent to the requirement of a *semi-group structure* ensuring that replacing $(t_1, f_1)$ with the intermediary time $t_2$ and associated initial condition $f_2$ still yields one and the same solution: $\mathcal{R}(t_2, t_0, .) = \mathcal{R}(t_2, t_1, .) \circ \mathcal{R}(t_1, t_0, .)$.

We illustrate in the next section how the RG equations (8.35) provide effective large-scale hydrodynamic equations, starting from the Boltzmann kinetic equation, and refer to Goldenfeld (1992), Chen *et al.* (1994), and Oono (2000) for other applications.

### 8.4.3 Renormalization-group derivation of hydrodynamics from the Boltzmann equation

A basic idea in the RG derivation of effective large-scale models (if any) is that the essential dynamics at macroscopic scales corresponds to the slow evolution of a low-dimensional manifold, invariant upon the short-term dynamics and termed the *slow manifold*. At each fixed macroscopic time, a quasi-stationary approximation quenches the slow manifold into an invariant manifold. The complete dynamics is thus decomposed into a fast component, prescribing the invariant manifold, and a slow evolution of this manifold, described in the space spanned by quantities parametrizing the manifold. This viewpoint gives a precise meaning to the *macroscopic* dynamics of a complex system: a slow evolution taking place in a reduced space compared to the original microscopic phase space. As an illustration, we now show how this RG scheme can be exploited to derive Navier–Stokes equations from the Boltzmann kinetic equation (Hatta and Kunihiro 2002, Kunihiro and Tsumura 2006), and how it is closely related to the initial condition renormalization presented in the previous section. Taking as a basis the observed existence of a regular macroscopic behavior, the initial value is determined self-consistently so as to encapsulate the appropriate counter-terms, ensuring cancellation of the secular terms (arising in the naive perturbation treatment of the influence of this initial condition) and giving access to the macroscopic evolution.

The starting point is the Liouville equation describing the evolution of the overall probability distribution function $f_N(\vec{r}_1, \vec{v}_1, \ldots, \vec{r}_N, \vec{v}_N, t)$ of an $N$-particle system upon its Hamiltonian dynamics. The Boltzmann equation is derived by integrating over $N-1$ particles and using the Boltzmann decorrelation ansatz (see Chapter 5) $f_2(\vec{r}, \vec{v}_1, \vec{r}, \vec{v}_2, t) \approx f_1(\vec{r}, \vec{v}_1, t) f_1(\vec{r}, \vec{v}_2, t)$ to obtain a close evolution equation for the one-particle distribution function $f \equiv f_1$:

$$\frac{\partial}{\partial t} f(\vec{r}, \vec{v}, t) + \vec{v}.\vec{\nabla} f(\vec{r}, \vec{v}, t) = J[f](\vec{r}, \vec{v}, t) \tag{8.37}$$

where the right-hand side of the above equation is the collision integral

$$\begin{aligned} J[f](\vec{r}, \vec{v}, t) = \int & \mathrm{d}^3\vec{v}_1 \mathrm{d}^3\vec{v}_2 \mathrm{d}^3\vec{v}_3 \; B(\vec{v}, \vec{v}_1 \mid \vec{v}_2, \vec{v}_3) \\ & \times [f(\vec{r}, \vec{v}_2, t) f(\vec{r}, \vec{v}_3, t) - f(\vec{r}, \vec{v}, t) f(\vec{r}, \vec{v}_1, t)]. \end{aligned} \tag{8.38}$$

This collision term (where $B$ contains the details of the binary collisions) describes the particle interactions at the level of the marginal distribution $f$.

In order to adapt the RG procedure acting at the level of initial conditions, Section 8.4.2, to the Boltzmann equation, the first trick is to give a special role to $\vec{v}$ turning $f(\vec{r}, \vec{v}, t)$ into a family $[f_{\vec{v}}(\vec{r}, t)]_{\vec{v}}$ of functions with variable $z = (\vec{r}, t)$. Given the initial condition $f_{\vec{v}}(\vec{r}, t_0)$, we denote by $F_{\vec{v}}(\vec{r}, t, t_0)$ the solution of the Boltzmann equation around $t = t_0$ and consider situations where the fluid motion is slow with a macroscopic characteristic scale in space, that is

$$\vec{v}.\vec{\nabla}_{\vec{r}} F_{\vec{v}} = O(\epsilon) \quad \text{with} \quad \epsilon \ll 1. \tag{8.39}$$

Defining the rescaled variable $\vec{R} = \epsilon\vec{r}$ explicitly introduces a small parameter $\epsilon \ll 1$ in the Boltzmann equation, since the term $\vec{v}.\vec{\nabla}_{\vec{r}} F_{\vec{v}}$ is now written $\epsilon\vec{v}.\vec{\nabla}_{\vec{R}} F_{\vec{v}}$. This hints at a perturbation analysis, implemented by plugging the straightforward expansion $F_{\vec{v}} = \sum_{n\geq 0} \epsilon^n F_{\vec{v}}^{(n)}$ in the Boltzmann equation and solving term-wise. Investigating the slow motion means at order 0 restricting to the stationary regime, $\partial F_{\vec{v}}^{(0)}/\partial t = 0$, yielding (back in the original variable $\vec{r}$):

$$F_{\vec{v}}^{(0)}(\vec{r}, t, t_0) = n(\vec{r}, t_0) \left( \frac{m}{2\pi k T(\vec{r}, t_0)} \right)^{3/2} \exp\left[ -\frac{m[\vec{v} - \vec{u}(\vec{r}, t_0)]^2}{2\pi k T(\vec{r}, t_0)} \right] \tag{8.40}$$

which is independent of $t$ by construction. The physically relevant solution should be robust in the sense that its large-scale behavior should no longer depend on the time $t_0$ at which the initial condition has been prescribed; in particular, at $t = t_0$,

$$\left( \frac{\partial F}{\partial t_0} \right)_{t_0 = t} = 0. \tag{8.41}$$

Higher orders essentially contribute in this equation, as detailed below. Varying $t_0$ requires an adapted change of the initial value, that will be embedded in the expression of $F$, with a semi-group structure since changing $t_0$ into $t_2$ should coincide with changing $t_0$ into $t_1$, then $t_1$ into $t_2$. We here recover the Green function RG given in an abstract formulation in Section 8.4.2, and (8.41) is nothing but a standard RG equation. It also amounts to requiring that the initial condition belongs to the slow manifold, so as to avoid secular divergences generated at any finite perturbation order by the transverse components of the initial condition.

The expression of the zero-order solution involves five parameters $n(\vec{r}, t_0)$, $\vec{u}(\vec{r}, t_0)$ and $T(\vec{r}, t_0)$ whose hydrodynamic meaning becomes apparent when computing the moments of the initial distribution:

$$n(\vec{r}, t_0) = \int F_{\vec{v}}^{(0)}(\vec{r}, t, t_0) d\vec{v} \qquad \text{(density)}, \tag{8.42}$$

$$n\vec{u}(\vec{r}, t_0) = \int \vec{v} F_{\vec{v}}^{(0)}(\vec{r}, t, t_0) d\vec{v} \qquad \text{(stream velocity)}, \tag{8.43}$$

$$3nkT(\vec{r}, t_0) = \int mv^2 F_{\vec{v}}^{(0)}(\vec{r}, t, t_0) d\vec{v} \quad \text{(temperature)}. \tag{8.44}$$

They span the invariant manifold arising in the quasi-stationary approximation (equivalent to restricting the solution to the zero-order stationary regime). Reduced macroscopic dynamics, corresponding to the slow drift of this manifold, describes the time evolution of these auxiliary variables generated by the higher-order convection term $\vec{v}.\vec{\nabla}_{\vec{r}} F$ in the Boltzmann equation. The first order is written

$$(\partial_t - \mathcal{A}) F^{(1)} = -\vec{v}.\vec{\nabla}_{\vec{r}} F^{(0)}, \tag{8.45}$$

where $\mathcal{A}$ is the linearized collision integral:

$$J(F^{(0)} + \epsilon F^{(1)}) = J(F^{(0)}) + \epsilon \mathcal{A} F^{(1)}. \tag{8.46}$$

$\mathcal{A}$ basically acts on the $\vec{v}$-dependence, coupling the components $F_{\vec{v}}$ at fixed position $\vec{r}$ and time $t$. The kernel of $\mathcal{A}$ is spanned by 1, $\vec{v}$ and $v^2$, and all other eigenvectors are stable (associated with eigenvalues with a negative real part) hence do not contribute to the slow manifold (note that such spectral features support the existence of a slow manifold). Denoting by $\mathcal{P}$ the projection onto the kernel of $\mathcal{A}$, and making use of the explicit expression for the first-order contribution to the solution,

$$F_{\vec{v}}^{(1)} = -(t - t_0)\mathcal{P}\vec{v}.\vec{\nabla}_{\vec{r}} F_{\vec{v}}^{(0)} + \mathcal{A}^{-1}(1 - \mathcal{P})\vec{v}.\vec{\nabla}_{\vec{r}} F_{\vec{v}}^{(0)}, \tag{8.47}$$

turns the RG equation (8.41) into the closed equation

$$\frac{\partial F_{\vec{v}}^{(0)}}{\partial t_0} + \epsilon \mathcal{P}\vec{v}.\vec{\nabla}_{\vec{r}} F_{\vec{v}}^{(0)} = 0, \tag{8.48}$$

from which follow the equations governing the evolution of $n(\vec{r}, t_0)$, $\vec{u}(\vec{r}, t_0)$ and $T(\vec{r}, t_0)$. The resolution in fact amounts to considering that the deformation $\epsilon F^{(1)}$ of the equilibrium unperturbed solution $F^{(0)}$ (that is, the deformation of the invariant manifold of the unperturbed motion into a slow manifold) belongs to the kernel of $\mathcal{A}$. This is fully relevant when investigating the asymptotic behavior, since it is only influenced by the slow modes associated with eigenvectors spanning the kernel of $\mathcal{A}$. The other modes, although ultimately decreasing to 0, generate singular behavior when plugged into a finite perturbation series; only a proper resummation of the whole perturbation series can recover regular behavior.

At this stage, the variable $t_0$ might be treated like the current time variable and simply written $t$, yielding the usual hydrodynamic equations. Using the ideal gas state equation $nkT = P$ to define the pressure leads to the non-dissipative Euler equations (with $\rho = mn$):

$$\partial_t \rho + \vec{\nabla}(\rho \vec{u}) = 0 \tag{8.49}$$

$$\partial_t(\rho u_i) + \partial_j(\rho u_j u_i) + \partial_i P = 0. \tag{8.50}$$

Actually the first-order solution exhibits a distortion from local equilibrium $F_{\vec{v}}^{(0)}$ and gives rise to dissipation. Solving the RG equation at the second order allows us to account for this dissipation in the resulting macroscopic equation, now yielding the Navier–Stokes equations, written:[6]

$$\vec{\nabla}.\vec{u} = 0 \tag{8.51}$$

$$\partial_t \vec{u} + \vec{u}.\vec{\nabla}\vec{u} = -\frac{\vec{\nabla}P}{\rho} + \nu \Delta \vec{u} \tag{8.52}$$

in the incompressible case where $\rho =$ constant. The transport coefficient $\nu$ (kinematic viscosity) arises as an effective parameter following from the RG procedure, that can be expressed explicitly as a function of the collision kernel $B$, Eq. (8.38).

The microscopic phase space for the Boltzmann equation is the functional space of the one-particle distribution functions $f(\vec{r}, \vec{v}, t)$. The slow manifold is spanned by five hydrodynamic quantities, $\rho(\vec{r}, t)$, $\vec{u}(\vec{r}, t)$ and $P(\vec{r}, t)$, defined as moments of $f(\vec{r}, \vec{v}, t)$ over $\vec{v}$. These quantities are also involved in the parametrization of the initial condition, bridging the present derivation and the invariant-manifold viewpoint with the Green function RG presented in Section 8.4.2. It is also closely reminiscent of the RG approach implemented in Section 8.2.2 for the example of the Duffing oscillator, with two differences: the time variable is extended to a space-time dependence and the renormalization is not applied to the integration

[6] Note that such Navier–Stokes equations are obtained while neglecting higher spatial derivatives of the velocity field; these contributions would lead to the so-called "Burnett terms."

constant but directly at the level of the initial condition. In the example of the Duffing oscillator, the invariant manifold was represented by $u_0(t) = \mathrm{Re}(A(\epsilon, t)\mathrm{e}^{\mathrm{i}t})$ and the idea was to encapsulate within the dependence $t_0 \to A(t_0)$ of the integral constant upon the initial time the consistency conditions ensuring that we get proper macroscopic behavior; the resulting slow evolution of $A$, Eq. (8.12), provided the desired effective macroscopic equation.

## 8.5 Probabilistic viewpoint on renormalization groups

Since the very beginning of RG theory, a parallel probabilistic viewpoint on RG techniques and their deep theoretical rationale has been developed, mainly by Jona-Lasinio and collaborators (a short review is given in Jona-Lasinio (2001)). This reveals, from a more formal and mathematical point of view, how RGs are specially designed to manage with long-range correlations. So doing, it *unravels the statistical nature of criticality:* critical phenomena appear as "statistical catastrophes," differing dramatically from plain collective behaviors associated with the celebrated law of large numbers and central limit theorem, for which standard mean-field arguments apply for passing from microscopic descriptions to macroscopic predictions. This probabilistic viewpoint also unravels other statistical anomalies, for instance individual singularities, disorder or geometric frustration, *at the origin of anomalous large-scale behaviors*. It explains how the quantitative features of RGs are derived, in particular their anomalous exponents, and delineates their universality.

Conversely, we shall see how probability theory benefits from RG methods: iterated coarse graining and joint rescalings provide a *systematic procedure to establish statistical laws*, bridging knowledge of elementary ingredients and their stochastic features with prediction of their typical collective behavior.

### 8.5.1 Statistical laws

*Statistical laws* describe the appearance of macroscopic features from assemblies of stochastic elements. Their study gives deep insights into the ingredients and parameters that control the emergence of reproducible behaviors from highly fluctuating individual behaviors. Statistical laws provide typical examples of *emergent properties*, not directly nor obviously foreseeable from the observation of individual behaviors.

Two basic statistical laws, playing a central role in statistical mechanics, are the *law of large numbers* and the *central limit theorem*, both relative to the asymptotic behavior $N \to \infty$ of a sum $\sum_{i=1}^{N} X_i$ of *independent and identically distributed* (i.i.d.) random variables $\{X_i\}$ of finite variance $\sigma^2$. The law of large numbers states the convergence with probability 1 of $(\sum_{i=1}^{N} X_i)/N$ toward the value $\langle X \rangle$, whereas the central limit theorem states the convergence of the probability distribution function of $(\sum_{i=1}^{N} X_i - \langle X \rangle)/\sqrt{\sigma^2 N}$ toward a centered Gaussian distribution with unit variance.[7]

The central limit theorem accounts for the *normal diffusion law*,

$$R^2(t) \equiv \left\langle \left( \sum_{i=1}^{t/\tau} X_i \right)^2 \right\rangle \sim 2Dt \qquad \text{(assuming here } \langle X_i \rangle \equiv 0\text{)}, \tag{8.53}$$

describing the time dependence of the mean square displacement $R^2(t)$ of an *ideal random walk*, that is, with independent, centered and identically distributed increments $\{X_i\}$ of duration $\tau$. The diffusion coefficient is given by $D = a^2/2\tau$ where $a^2$ is the square average length of a step $X_i$. The theorem stated with $N = t/\tau$ also gives the asymptotic distribution $p(x, t)$ of the walker position $x$ along the line at large time, namely $p(x, t) = 1/\sqrt{4\pi Dt}\, \mathrm{e}^{-x^2/4Dt}$.

Statistical laws appear as a *special instance of asymptotic analysis* since the issue is to extract robust *global* features of an assembly of $N$ elements, $N \to \infty$. They provide finite-size approximations for $N$ large enough, but then their validity should be checked by estimating the strength of finite-size effects, typically by computing higher-order terms (in particular, at lowest order, the central limit theorem allows us to estimate the fluctuations around the average behavior described by the law of large numbers).

The link with RG motivations is obvious. We shall see that the RG approach to statistical laws is possible because of the existence of relations between the properties at different scales, typically some kind of scale invariance or hierarchical structure that rules the collective behavior and controls its features. The examples proposed in this section will illustrate the achievements of the RG when implemented in this probabilistic context:

(1) to integrate short-range correlations, moderate individual singularities or weak disorder into *effective parameters* in the asymptotic statements;
(2) to derive *anomalous scaling* with respect to $N$ of the emerging properties originating in long-range correlations, if any, or other statistical "pathology" (like strong disorder or extreme events with non-vanishing probability) strong enough to prevent the law of large numbers and the central limit theorem holding;
(3) to discriminate between the *essential ingredients* and irrelevant details having no consequence on the statistical behavior, that is, to investigate the *robustness* of this behavior (also termed its *structural stability*);
(4) to determine the *universality class* associated with a given statistical law. The expected universal aspect of probabilistic limit theorems deserves to be underlined: these

[7] In mathematical terms, the law of large numbers states an *almost sure convergence* toward $\langle X \rangle$ and the central limit theorem a *convergence in law* toward the normal distribution.

theorems assess some limiting statistical behavior in large classes of random variable sequences, i.e. *statistical ensembles* defined by a few constraints on the nature (dimension, distribution shape) of their elements and their correlations, but none on the physical entities or quantities they represent.

### 8.5.2 Coarse graining random variable sequences

The basic probabilistic interpretation of the RG was introduced more than three decades ago by Jona-Lasinio. It now appears as the probabilistic formulation of the Kadanoff–Wilson block-spin RG (Section 8.3.2) and more generally as real-space RGs. For pedagogical purposes, the technical benefits and unifying power offered by this more abstract and mathematical formulation are best captured in the simplest framework of real random variable sequences $\vec{X} = \{X_i\}$. They could be one-dimensional arrays of spins ($i$ then being the space-cell label) or sequences of increments of a random walk ($i$ then being a time label) where each elementary unit (spin or increment) is described by a real-valued random variable $X_i$. The issue, connected with those discussed in Section 8.5.1, is to determine the collective behavior $N \to \infty$ of the sum $\sum_{i=1}^{N} X_i$ of $N$ such variables, and the conditions for its occurrence.

Extension to vector-valued variables $X_i \in \mathbb{R}^d$ corresponding to $d$-component spins or random walk in a $d$-dimensional space is straightforward. Extensions to continuous sequences (i.e. *stochastic processes*, with $i \in \mathbb{R}$) and multivariate sequences (i.e. *random fields*, with $i \in \mathbb{R}^d$) can also be considered; the bridge between discrete and continuous-time statistical laws will be implemented explicitly, as a side benefit of their RG derivation, in Section 8.5.7.

RG principles teach us that a natural way to investigate the asymptotic behavior $N \to \infty$ of the sum $S_N = \sum_{i=1}^{N} X_i$ is to proceed recursively by performing *nested coarse grainings*. This amounts to investigating the relation $S_{N_n} \to S_{N_{n+1}}$ between the successive terms of a subsequence where $N_{n+1}/N_n = k$ is a given rescaling factor (most often $k = 2$). More explicitly, following Jona-Lasinio (1975, 2001), one defines a coarse-grained sequence $\{B_j\}$ according to:

$$B_j(\vec{X}, \nu, k) \equiv \frac{1}{k^\nu} \left( \sum_{i=1}^{k} X_{k(j-1)+i} \right) \quad (\nu > 0,\ k \text{ integer}). \tag{8.54}$$

This coarse graining amounts to lumping together $k$ random variables and rescaling the resulting behavior by $k^\nu$ and the apparent time by $k$. It defines a bona fide RG insofar as it exhibits a (semi)group structure upon composition:

$$\mathcal{R}_{\nu,k}(\vec{X}) \equiv \vec{B}(\vec{X}, \nu, k) \quad \text{with} \quad \mathcal{R}_{\nu,k_1} \circ \mathcal{R}_{\nu,k_2} = \mathcal{R}_{\nu,k_1 k_2}. \tag{8.55}$$

A key point is the choice of the exponent $\nu$: it is to be *determined self-consistently in the course of the RG procedure* so as to give non-trivial asymptotic behavior for the RG flow generated by (8.55). This requirement ensures a non-trivial limit as $k \to \infty$ for (8.54) since $\mathcal{R}^n_{\nu,k} = \mathcal{R}_{\nu,k^n}$: the group structure implies the equivalence of iterating $\mathcal{R}_{\nu,k}$ at fixed $k$ and increasing $k$. The recursive formulation is more efficient in practice since it offers a systematic procedure, namely the investigation of the RG flow, its fixed points, their basin of attraction and unstable directions. It thus provides a constructive method to determine the limit $k \to \infty$, while the convergence property is nothing but the expression of the statistical law we are looking for. The adequate rescalings $(k, k^\nu)$ reflect the unique way to rescale state variables and time jointly so as to compensate for the integrated effect of small-scale details and obtain a well-defined large-scale behavior. This is the core of the RG principle since it gives access to the exponent $\nu$ and its possibly anomalous value.

In the case of i.i.d. variables with finite variance, the limiting behavior as $k \to \infty$ is well known: the law of large numbers is recovered with $\nu = 1$, whereas considering the centered sequence $\vec{X} - \langle X \rangle$ and $\nu = 1/2$ yields the central limit theorem. Anomalous behaviors with different values for $\nu$ are expected if the conditions of validity of the central limit theorem fail, namely if

(1) the variance of the variables $\{X_i\}$ is infinite,
(2) the variables $\{X_i\}$ are correlated,
(3) the duration $\tau_X(i)$ associated with $X_i$ depends on $i$ and $\{\tau_X(i)\}$ exhibits an anomalous distribution, typically according to a power law, with $\langle \tau_X \rangle = \infty$, instead of the normal exponential decay associated with a Poisson distribution,
(4) the variables $\{X_i\}$ are not independently distributed but some variability is present in their individual statistical features; such a "disorder" is generally accounted for by considering these features themselves as random variables, introducing a second level of stochasticity ("probability of probabilities").

These situations can be substantiated for random walks, yielding various anomalous diffusions laws $R^2(t) \sim t^{2\nu}$ (in other words, novel limit theorems), respectively:

(1) super-diffusion with $\nu > 1/2$, as observed in the Lévy flight model,
(2) persistent (with $\nu > 1/2$) or anti-persistent (with $\nu < 1/2$) motions,
(3) sub-diffusion, with $\nu < 1/2$, due to trapping phenomena,
(4) anomalous diffusion in disordered media, with possibly a time dependence more complicated than a mere power law $R^2(t) \sim t^{2\nu}$.

We shall detail case (1) in Section 8.5.4 and case (2) in Section 8.5.5; all have been already discussed in Chapter 7, hence it is meaningful to reconsider the discussion with a probabilistic RG viewpoint. For pedagogical purposes, it is fruitful first to present the RG at work in the simplest situation, namely the derivation of the central limit theorem. Anomalous statistical laws embed the central limit theorem as a special instance, and their RG derivation will follow the same line.

### 8.5.3 Renormalization-group derivation of the central limit theorem

We here consider sequences $\vec{X} = \{X_i\}$ of real *independent and identically distributed* (i.i.d.) random variables of *finite variance* $\sigma_X^2 < \infty$. In practice, the "block-RG" described in Section 8.5.2 is implemented on the probability distribution functions: choosing $k = 2$, the basic renormalization transformation should describe the passage from the elementary probability distribution function $p$ of any $X_i$ to the probability distribution function $\mathcal{R}p$ of the coarse-grained variables $B_j(\vec{X}, \nu, k = 2)$ obtained by grouping the original variables in blocks of $k = 2$ units. We thus consider the operator acting upon probability distribution functions in the following way:

$$\mathcal{R}_{\nu,k=2}\, p(x) = 2^\nu \int_{-\infty}^{\infty} p(y)p(2^\nu x - y)\mathrm{d}y \equiv 2^\nu \; p \otimes p\,(2^\nu x). \qquad (8.56)$$

This obviously preserves the positivity $p(x) \geq 0$ and the normalization to 1 of the probability distribution functions. The rationale underlying the choice of the exponent $\nu$ is to render $p$ and $\mathcal{R}_{\nu,2}\, p$ as similar as possible, so as to optimize the occurrence of a well-defined asymptotic behavior for the RG flow, namely the convergence to a fixed point. Taking into account the i.i.d. character of the elementary variables, and the ensuing i.i.d. character of the blocks as regards the subsequent iterations of the renormalization procedure, the condition that $\mathcal{R}$ preserves the variance (a necessary condition for the existence of a fixed point of $\mathcal{R}$ with finite variance) prescribes $\nu = 1/2$. The renormalization operator $\mathcal{R} \equiv \mathcal{R}_{\nu=1/2,k=2}$ acts in the space of univariate probability distribution functions and the associated discrete flow admits a *family of Gaussian fixed points*:

$$p_\sigma^*(x) = \frac{1}{\sqrt{2\pi}\,\sigma}\, \mathrm{e}^{-x^2/2\sigma^2} \quad \text{satisfying} \quad \mathcal{R}p_\sigma^* = p_\sigma^*. \qquad (8.57)$$

According to the general RG scheme (see Figure 8.1 in Section 8.3.2) one has to *investigate the stability and the basin of attraction* of each of these fixed points. Showing the convergence toward $p_\sigma^*$ would at the same time prove the central limit theorem, put forward its conditions of validity and its robustness, and determine the selection rule for the value of the parameter $\sigma$. Indeed, direct examination shows that $\mathcal{R}^n p$ is the probability distribution function of $\sum_{i=1}^N X_i/\sqrt{N}$ with $N = 2^n$ (it is a mere consequence of the group relation $\mathcal{R}^n_{\nu,k=2} = \mathcal{R}_{\nu,k=2^n}$ and independence of variables). Setting aside the technicalities required to interpolate between integers of the form $2^n$ and fully assess the convergence as $N \to \infty$, the central limit theorem is here formulated as the weak convergence of $\mathcal{R}^n p$ to $p^*_{\sigma(p)}$ for any probability distribution function $p$ of finite variance $\sigma^2(p) < \infty$.

To summarize the procedure from the perspective of its generalization, we are thus led to *investigate the RG flow seen as a dynamical system in the functional space* of probability distribution functions on **R** with finite variance (i.e. such that $x^2 p(x)$ is integrable). An important property underlying the proof of the required convergence is the existence of *three conservation laws* (normalization to 1, centered character and variance):

$$\int_{-\infty}^{\infty} p(x)\mathrm{d}x = 1, \quad \int_{-\infty}^{\infty} xp(x)\mathrm{d}x = 0, \quad \int_{-\infty}^{\infty} x^2 p(x)\mathrm{d}x = \sigma^2. \tag{8.58}$$

By construction of $\mathcal{R}$, these three integral relations are preserved upon renormalization: we here recover the requirement that the RG should preserve the symmetries and invariances of the initial system or setting. In particular, the third relation in (8.58) shows that if convergence occurs to a fixed point $p^*_\sigma$, its variance $\sigma^2$ is determined by the variance $\sigma^2(p)$ for the identically distributed elementary units with distribution $p$. Necessary convergence conditions follow from the local stability analysis of the fixed point, considering the action of $\mathcal{R}$ on $p = p^*_\sigma(1 + \epsilon q)$ to first order in $\epsilon$. It reduces to study of the linearized operator $D\mathcal{R}(p^*_\sigma)$ or equivalently the linear operator $\mathcal{L}_\sigma$ such that $D\mathcal{R}(p^*_\sigma) = p^*_\sigma \, \mathcal{L}_\sigma$,

$$\mathcal{R}[p^*_\sigma(1 + \epsilon q)] = p^*_\sigma(1 + \epsilon \mathcal{L}_\sigma q + O(\epsilon^2)), \tag{8.59}$$

namely, using the symmetry $p^*_\sigma(x) = p^*_\sigma(-x)$ of the Gaussian fixed point $p^*_\sigma$:

$$\mathcal{L}_\sigma q(x) = \frac{2}{\sqrt{\pi}\sigma} \int_{-\infty}^{\infty} \mathrm{e}^{-y^2/\sigma^2} h\left(y + \frac{x}{\sqrt{2}}\right) \mathrm{d}y. \tag{8.60}$$

A direct computation shows that the eigenvalues of $\mathcal{L}_\sigma$ are:

$$\lambda_n = 2^{1-n/2} \quad \text{for all integers } n \geq 1. \tag{8.61}$$

hence $\lambda_1 = \sqrt{2}$, $\lambda_2 = 1$, $\lambda_3 = 1/\sqrt{2}$ and so on, in a decreasing sequence. The relevant stability analysis is in fact to be restricted to $\mathcal{R}$-invariant subsets of probability distribution functions satisfying the above three conservation laws (8.58), defining one subset for each value of $\sigma$. Accordingly, one can verify that the components of the relevant probability distribution function perturbations $q$ onto the first eigenvectors ($n = 1$ and $n = 2$) vanish owing to these conservation laws. Only the eigenvalues with $n \geq 3$ are actually considered, which proves the linear stability of the fixed point $p^*_\sigma$ since $\lambda_n < 1$ for any $n \geq 3$. A non-linear analysis should then be performed to ensure that non-linear contributions do not modify this conclusion and to determine the basin of attraction of $p^*_\sigma$. Again, the situation is strongly constrained by the conservation laws: one shows that the basin of attraction of $p^*_\sigma$ is the whole set of probability distribution functions satisfying these three laws with the same value $\sigma$. Finally, estimating the distance between $\mathcal{R}^n p$ and its limit as a function of $n$ would allow control of the error, that is, higher-order terms in finite-size approximations $\mathcal{R}^n\, p \approx p^*_{\sigma(p)} +$ h.o.

It is possible, at least as a training exercise, to implement this RG approach for vector variables $\vec{X}_{\vec{i}}$ (with $D$ components), in higher underlying dimension $d > 1$ (meaning that labels $\vec{i}$ now span $\mathbf{Z}^d$), and involving larger blocks of $k^d$ elementary units (rescaling factor $k$ instead of 2). The RG map is then written:

$$\mathcal{R}_k\, p(\mathbf{x}) = k^{d/2}\, p^{\otimes k^d}\, (k^{d/2}\, \mathbf{x}) \quad (\vec{x} \in \mathbb{R}^D) \tag{8.62}$$

(where $\otimes$ denotes the convolution, see (8.56)) similarly possessing a family of now $D$-dimensional Gaussian fixed points.

This RG proof of the central limit theorem illustrates a general procedure, allowing us to derive limit theorems in more complex instances when the classical probabilistic approach does not provide a simpler way to get the result. The RG *shifts the analysis from the level of random variables to that of probability distribution functions* by devising and analyzing transformations acting on these probability distribution functions. We are now in a position to elaborate on this point, and show moreover that it leads to the notion of continuous self-similar processes, in three typical instances: when elementary events are independent but exhibit an infinite variance (Section 8.5.4), when they are long-range coupled (Section 8.5.5), and when they are long-range correlated (Section 8.5.6).

### 8.5.4 Stable laws and associated limit theorems

The starting point in this section will be the relevance of the RG in the context of *convolution equations*. The idea is to reformulate the convolution equation as a fixed-point equation $\mathcal{R}p^* = p^*$ for an appropriate RG operator $\mathcal{R}$. First, this yields recursive access to the solution $p^*$ (namely, $p^* = \lim_{n\to\infty} p_n$ with $p_n = \mathcal{R}p_{n-1} = \mathcal{R}^n\, p_0$) which could be more efficient than other methods of resolution. In addition, unfolding the RG fixed-point property into a convergence property allows us to establish *novel limit theorems for i.i.d. variables* whose individual distribution $p$ belongs to the basin of the fixed-point $p^*$ and to assess their validity conditions.

In this perspective, the RG fixed-point equation $\mathcal{R}p^* = p^*$ with $\mathcal{R}$ defined in (8.56) expresses that $p \otimes p$ has the *same functional form* as $p$, up to a suitable transformation of its variables and numerical parameters. This question about stability upon convolution was answered a long time ago: the solutions are a family of probability distribution functions known as *stable laws* (or *Levy distributions*), all of infinite variances except the special subfamily of Gaussian laws. Their explicit expression is easier to write in the conjugate space, considering characteristic functions $\varphi(u) = \langle e^{iuX} \rangle$, and restricting to the centered laws for simplicity:

$$\widehat{L}_{a,\nu}(u) = e^{-a|u|^{1/\nu}} \tag{8.63}$$

with $1/2 \leq \nu < 1$; the Gaussian subfamily is recovered for $\nu = 1/2$. The bounds on $\nu$ are required to get a bona fide probability distribution function $L_{a,\nu}$ by inverse Fourier transform of $\widehat{L}_{a,\nu}(u)$. Indeed, it is necessary that $\widehat{L}_{a,\nu}(u = 0) = 1$ and $\widehat{L}'_{a,\nu}(u = 0) = i \langle X \rangle = 0$, requiring $\nu < 1$ (actually $\nu \geq 1$ is still meaningful if the existence of a well-defined first moment is also relaxed). To be a characteristic function, $\widehat{L}_{a,\nu}$ also has to be of positive type,[8] requiring $\nu \geq 1/2$. In particular, this shows that the Gaussian distribution is the only solution with finite variance. The positive real $a$ is related to the width of the distribution $L_{a,\nu}$; indeed, $a^{-\nu}$ gives a characteristic size in the conjugate space, hence in the direct space the distribution will be more spread out as $a$ is large.

These stable laws, first defined through their remarkable behavior upon convolution, also appear to play a central role in generalized limit theorems extending the central limit theorem to infinite-variance but still independent variables. Let us present the RG derivation of this assertion. Denoting by $p(x, t)$ the probability distribution function of the sum $S(t) = \sum_{i=1}^{t} X_i$ of identical random variables $\{X_i\}$, the RG transformation (8.55) is written

$$\mathcal{R}_{\nu,k} p(x, t) = k^{\nu} p(k^{\nu} x, kt). \tag{8.64}$$

This transformation can be expressed equivalently in terms of characteristic functions $\varphi(u, t) \equiv \langle e^{iuS_t} \rangle$, the characteristic function $\mathcal{T}_{\nu,k}\varphi$ that corresponds to $\mathcal{R}_{\nu,k} p$ being:

$$\mathcal{T}_{\nu,k}\varphi(u, t) = \varphi(k^{-\nu} u, kt). \tag{8.65}$$

Extension to continuous time values is straightforward, allowing random walks and stochastic processes to be encapsulated in a unified framework, as detailed in Section 8.5.7. The fixed-point equation $\mathcal{R}_{\nu,k}(p^*) = p^*$ defines *self-similar processes* $p^*(x, t)$. Since it should be satisfied for all $k > 0$, it can be written with $k = 1/t$, yielding the probability distribution function fixed-point functional form and the corresponding expression for the characteristic function:

$$p^*(x, t) = t^{-\nu} \; P(xt^{-\nu}) \quad \text{or} \quad \varphi^*(u, t) = \phi(ut^{\nu}). \tag{8.66}$$

[8] A function $\varphi(u)$ is said to be of positive type if for any integer $n$, for any $n$-tuplet of variable values $u_1, \ldots, u_n$ and any $n$-tuplet of complex numbers $c_1, \ldots, c_n$, it satisfies $\sum_{i,j=1,\ldots,n} c_i \bar{c_j} \varphi(u_i - u_j) \geq 0$.

Self-similarity of this fixed point is revealed by the involvement of a single scaling variable $xt^{-\nu}$ (respectively $ut^{\nu}$).

In the case of *independent* variables of characteristic function $\varphi(u)$, it is straightforward to show that the characteristic function of the aggregated variable $S(t) = \sum_{i=1}^{t} X_i$ is $\varphi(u)^t$. Plugging this in (8.66) yields $\Phi(z) \sim e^{-a|z|^{1/\nu}}$, that is, we recover the characteristic function $\widehat{L}_{a,\nu}$ of a stable law $L_{a,\nu}$. We here meet again a general RG principle: the adequate rescalings involved in the RG transformation precisely reflect the unique way of jointly rescaling space and time so that the RG flow exhibits a non-trivial limiting behavior. In this way, by construction, for any $a$, the law $L_{a,\nu}$ is a fixed point of $\mathcal{R}_{\nu,k}$ (for any $k$), and the RG fixed-point equation expresses the self-similarity of the collective variables $B_\nu(t)$ constructed from an i.i.d. sequence $\{X_i\}$ of random variables following the stable law $L_{a,\nu}$. A straightforward scaling argument yields the associated *diffusion law*, anomalous as soon as the stable law has an infinite variance (for $\nu > 1/2$):

$$R^2(t) \equiv \langle S(t)^2 \rangle \equiv\sim t^{2\nu}. \tag{8.67}$$

More generally, one shows the stability of the fixed point $L_{a,\nu}$ in the space of probability distribution functions that behaves as $p(x) \sim |x|^{-(1+1/\nu)}$ at large $x$. Accordingly, this stability ensures the convergence of the RG flow to $L_{a,\nu}$. Translated back in probabilistic terms, it yields an *anomalous central limit theorem*, stating that if the original probability distribution function behaves as $p_0(x) \sim |x|^{-(1+1/\nu)}$ for $|x| \to \infty$ with $\nu < 1$, then $B_{a,\nu}(t) \equiv t^{-\nu} \sum_{i=1}^{t} X_i$ converges in law toward a stable law $L_{a,\nu}$. It means that these stable laws describe the asymptotic collective behavior of sequences of i.i.d. variables of infinite variance. We here recover, using RG concepts and arguments, the limit theorems of probability theory for the sum of independent variables, as investigated by Lévy (1954) and Gnedenko and Kolmogorov (1954).

### 8.5.5 Renormalization-group analysis of the hierarchical model

A different illustration of departure from the normal behavior associated with the central limit theorem is given by the so-called *hierarchical model*, namely a class of abstract models specially devised to exhibit direct (physical) couplings at all scales, in an explicitly workable way. The Hamiltonian $H_{n,c}$ of the model with parameter $c$, with $1 < c < 2$, and $2^n$ real-valued degrees of freedom is defined recursively (see Figure 8.3):

$$\begin{aligned} H_n(x_1, \ldots, x_{2^n}) = {} & H_{n-1}(x_1, \ldots, x_{2^{n-1}}) \\ & + H_{n-1}(x_{2^{n-1}+1}, \ldots, x_{2^n}) - \left(\frac{c}{2}\right)^n \left(\sum_{i=1}^{2^n} x_i\right)^2 . \end{aligned} \tag{8.68}$$

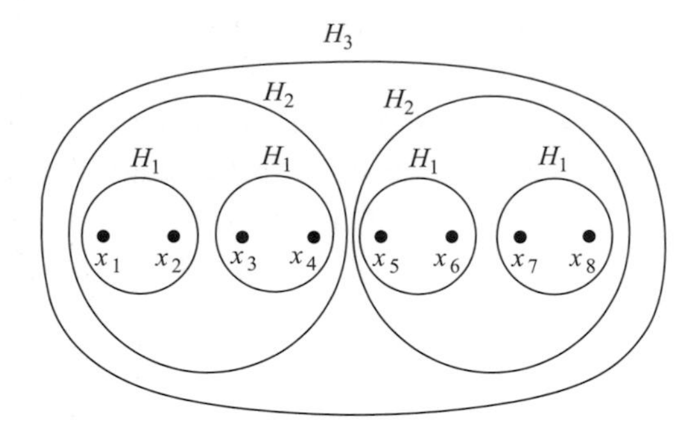


Figure 8.3 Hierarchical model up to $n = 3$. Each circle represents a term $-(c/2)^n(\sum_{i=1}^{2^n} x_i)^2$ where the sum runs over all the points contained within the circle: $H_3(x_1, \dots, x_8) = H_2(x_1, \dots, x_4) + H_2(x_5, \dots, x_8) - (c/2)^3(x_1 + \dots + x_8)^2$, $H_2(x_1, \dots, x_4) = H_1(x_1, x_2) + H_1(x_3, \dots, x_4) - (c/2)^2(x_1 + \dots + x_4)^2$ and $H_1(x_1, x_2) = H_0(x_1) + H_0(x_2) - (c/2)(x_1 + x_2)^2$.

It thus encapsulates a quadratic coupling $-J_{ij}(c, n)x_i x_j$ between any pair $(i, j)$ of degrees of freedom (appearing twice for $i \neq j$) whose strength $J_{ij}$ is given by:

$$J_{ij}(c, n) = \sum_{q=n_0(i,j)}^{n} \left(\frac{c}{2}\right)^q \quad \text{with} \quad 2^{n_0-1} \leq |j - i| < 2^{n_0}. \tag{8.69}$$

Provided[9] $c < 2$ and $n$ is large enough, (8.68) yields:

$$H_n(x_1, \dots, x_{2^n}) \approx - \sum_{i,j} J_{i,j}(c) x_i x_j \quad (n \to \infty) \tag{8.70}$$

where[10]

$$J_{i,j}(c) = J_{i,j}(c, n \to \infty) \sim \frac{J_0(c)}{|j - i|^{\alpha(c)}} \tag{8.71}$$

with

$$\alpha(c) = 1 - \log_2(c) \quad \text{and} \quad J_0(c) = \frac{1}{1 - c/2}. \tag{8.72}$$

These expressions provide a qualitative understanding of the thermodynamic behavior. For $1 < c < 2$, $0 < \alpha(c) < 1$ hence the model exhibits a divergence of the *coupling* range. It is accordingly expected to exhibit a divergence of the correlation range and ensuing critical features in a domain $[0, T^*(c)[$ of low temperatures (or equivalently a domain $]\beta^*(c), \infty[$ in terms of inverse temperature $\beta$) where

[9] If $c < 1$, the coupling is sufficiently short range that $\sum_j J_{ij} < \infty$ for any fixed $i$, and the behavior is the same as for a plain ferromagnetic Hamiltonian (same universality class). If $c > 2$, $J_{ij}(c, n)$ diverges for $n \to \infty$ and the model is thermodynamically unstable. Consistently, the bounds $c = 1$ and $c = 2$ are related to special values $\alpha(1) = 1$ and $\alpha(2) = 0$ of the exponent $\alpha(c) = 1 - \log_2 c$, see (8.71) and (8.72).

[10] More precisely, $2^{-\alpha(c)}|j - i|^{-\alpha(c)} J_0(c) < J_{i,j}(c, n \to \infty) \leq |j - i|^{-\alpha(c)} J_0(c)$.

thermal fluctuations are dominated by the couplings. By comparison, the ferromagnetic Ising model (Section 8.3.2) exhibits a divergence of the *correlation* range at an isolated critical temperature $T^*(c)$ while the couplings remain restricted to nearest neighbors. In both cases, criticality arises at the "balance temperature" $T^*(c)$ where competing influences of bare couplings (toward order) and bare thermal fluctuations (toward disorder) match, inducing an anomalous distribution for the resulting statistical fluctuations.

An RG approach can be developed to assess quantitatively this qualitative picture, mainly to determine the threshold $\beta^*(c)$, to understand the complex subthreshold interplay between thermal fluctuations and long-range couplings (inducing a strong trend toward full order but at the same time enhancing thermal fluctuations by propagating them to the whole system) and finally to describe the criticality at $\beta = \beta^*(c)$ and determine its universality class. The macroscopic consequences of the couplings present at all scales will be determined by a recursive integration of their influence into block distributions. Let us now work out this program explicitly, best implemented on the probability distribution function of the order parameter $\sum_{i=1}^{2^n} x_i$.

Starting with $H_0 = 0$ and $p_0(x) = 1/\sqrt{2\pi}\, \mathrm{e}^{-x^2/2}$, we consider the Boltzmann distribution for the model with $2^n$ degrees of freedom and Hamiltonian $H_n$ and denote by $p_n$ the ensuing probability distribution function for the aggregated and rescaled variable $s_n \equiv 2^{-n/2}(\sum_{i=1}^{2^n} x_i)$. It is straightforward to determine the relation between $p_n$ and $p_{n+1}$:

$$p_{n+1}(s) = g_n(s^2)\, \mathcal{R}p_n(s) \tag{8.73}$$

where $\mathcal{R}$ has already been encountered in Section 8.5.3 and $g_n$ is a rescaling function:

$$\mathcal{R}p(s) = \sqrt{2}\,(p \otimes p)(s\sqrt{2}) \quad \text{and} \quad g_n(s^2) \sim \mathrm{e}^{\beta(c/2)^n s^2}. \tag{8.74}$$

Such a recursion is somehow peculiar and departs from the usual RG since it defines a *non-autonomous RG operator*, depending explicitly on the number $n$ of coarse grainings, that plays the role of time in the dynamical system analogy for the RG flow. It yields an increasing variance, that integrates part of the couplings:

$$\sigma_n^2 = \frac{1}{1 - 2\beta \sum_{q=1}^{n} (c/2)^q}. \tag{8.75}$$

As long as $\beta < \beta^*(c) = 1/c - 1/2$, this effective variance $\sigma_n^2$ tends toward a finite value:

$$\sigma_{\mathrm{eff}}^2(\beta, c) = \frac{1}{1 - \frac{2\beta c}{2-c}} = \frac{1}{1 - \beta/\beta^*(c)} \geq 1. \tag{8.76}$$

This effective variance diverges if $\beta = \beta^*(c)$, reflecting the critical behavior of the model and the failure of the central limit theorem at that point. The central limit theorem holds only at large temperature, namely $\beta < \beta^*(c)$. Nevertheless, collective amplification of fluctuations implies that $\sigma_{\text{eff}}(\beta) > 1$ as soon as $\beta > 0$: this result is not intuitive since it means that long-range coupling actually enhances the consequences of thermal fluctuations, by propagating the fluctuation of some $x_i$ to any other variable $x_j$. Note finally that the enhancing factor $g_n(s^2)$ tends to 1 as $T \to \infty$ ($\beta \to 0$) at fixed $s$ and $n$ and $\sigma_{\text{eff}}(\beta = 0, c) = 1$, so that the plain recursion valid for i.i.d. variables is consistently recovered at infinite temperature.

At $\beta = \beta^*_{(c)}$, a stronger rescaling of the block variables is required to compensate for the couplings, that overwhelm thermal fluctuations above this threshold, and to get a well-defined renormalized behavior as $n \to \infty$. The proper coarse-grained variable happens to be $\widetilde{s}_n \equiv (\sqrt{c}/2)^n(\sum_{i=1}^{2^n} x_i)$ instead of $s_n \equiv 2^{-n/2}(\sum_{i=1}^{2^n} x_i)$. Considering a different coarse graining leads to the definition of a different recursion:

$$\widetilde{\mathcal{R}}_{c,\beta}\, p(s) \equiv \Lambda(p, c)\, \mathrm{e}^{\beta s^2}\, (p \otimes p)(2s/\sqrt{c}) \tag{8.77}$$

where $\Lambda(p, c)$ is a normalization constant ensuring that $\widetilde{\mathcal{R}}_{c,\beta}\, p$ is a probability distribution function. The determination of the rescaling factor $2/\sqrt{c} > \sqrt{2}$, instead of $\sqrt{2}$, amounts to the determination of the proper rescaling exponent $\nu$ discussed above in Sections 8.5.2 and 8.5.3: it is chosen self-consistently so that the RG flow converges toward a non-trivial distribution. This limiting distribution obviously has to be a fixed point of $\widetilde{\mathcal{R}}_{c,\beta}$, and the convergence of the RG flow is nothing but the expression of a statistical law. We here shift from $\nu = 1/2$ to an anomalous value $1 - \log_2 \sqrt{c} = 1/2 + \alpha(c)/2 > 1/2$, reflecting the strong collective behavior arising at (and above) the critical threshold $\beta^*(c)$. By comparison, $\nu = 1$ in the case of fully coherent behavior of the elementary units (this case is also observed when reaching the lower bound $c = 1$). The recursion (8.77) now defines a bona fide renormalization operator, and we might carry on the RG program. $\widetilde{\mathcal{R}}_{c,\beta}$ admits a Gaussian fixed point:

$$\widetilde{p}_{\mathrm{G},c,\beta}(s) = \sqrt{\frac{1}{2\pi a^2(c,\beta)}}\, \mathrm{e}^{-s^2/2a^2(c,\beta)} \quad \text{where} \quad a^2(c,\beta) = \frac{2-c}{2\beta c} = \frac{\beta^*(c)}{\beta} \leq 1 \tag{8.78}$$

$a(c, \beta^*_{(c)}) = 1$ hence one recovers a normal distribution at $\beta = \beta^*(c)$.

For $2 > c > \sqrt{2}$, the linear stability analysis of this fixed point shows that it has a non-empty domain of attraction (a *stable manifold* in the language of dynamical systems, termed a *critical manifold* in the RG context, see Figure 8.1). Translated back in statistical terms, it means that variables $\{x_i\}$ coupled according to the hierarchical model with $\beta = \beta^*_{(c)}$ and $2 > c > \sqrt{2}$ follow a generalized central limit theorem with a rescaling exponent $\nu = 1/2 + \alpha(c)/2$, that is, involving the variables $\widetilde{s}_n$. We might speak of an *"anomalous central limit theorem"*.

At lower temperatures, for $\beta > \beta^*(c)$, this theorem holds with moreover a renormalized variance $a^2 < 1$ (smaller than the individual variance) reflecting the overwhelming ordering influence of the couplings. As expected, complete order is recovered at zero temperature since the variance then vanishes $a(c, \beta = \infty) = 0$.

For $1 < c < \sqrt{2}$, the Gaussian fixed point $\widetilde{p}_{G,c,\beta}$ is now unstable and an exchange of stability occurs with a hyperbolic *non-Gaussian* fixed point. Thus a novel limit theorem arises for $c < \sqrt{2}$, associated with this new fixed point and accordingly describing non-Gaussian asymptotic behavior. We are now faced with an *"anomalous and non-Gaussian central limit theorem."*

Let us summarize the results of this study. The hierarchical model exhibits long-range correlations in the whole low-temperature domain $\beta \geq \beta^*_{(c)}$ following from long-range couplings built into the Hamiltonian, reflected in an anomalous exponent $\nu > 1/2$: the variance of the order parameter $\sum_{i=1}^{N} x_i$ increases as $N^{2\nu}$, faster than $N$. By contrast, at high temperatures ($\beta < \beta^*(c)$), the influence of the couplings does not spoil the normal central limit theorem, since they can be fully integrated in a renormalized variance $\sigma^2_{\text{eff}}(\beta, c)$, but it enhances the consequences of thermal fluctuations: $\sigma^2_{\text{eff}}(\beta, c) > 1$.

Once the proper renormalization operator and its fixed points have been determined, the RG analysis can be carried on as described in Section 8.3.2 and Figure 8.1. In particular, the unstable eigenvalues of the linearized RG operator (associated with eigenvectors transverse to the critical manifold) will be directly related to critical exponents, as detailed in Sections 8.3.2 and 8.3.3.

### 8.5.6 Limit theorems for correlated variables

The hierarchical model analyzed in the previous subsection leads us to a caveat: the difference between (physical) *couplings* and (statistical) *correlations*. In the hierarchical model, the anomalous behavior was somehow rooted directly in the Hamiltonian owing to the presence of direct couplings at all scales. But critical phenomena arise as soon as long-range correlations at all scales are present, and can occur in a system of short-range coupled or even non-interacting elements, think for instance of a percolation network. Moreover, whereas couplings are fully prescribed by the Hamiltonian, correlations are prescribed by the probability distribution function of the model, and hence depend also on the temperature in a system at thermal equilibrium.

The same distinction prevails in the dynamic context. Anomalous asymptotics might originate from *long-term memory* prescribed at the basic level of evolution equations. For instance, in the context of random walks, this situation occurs if the walker remembers at all times the sites already visited and avoids them; this is called a *self-avoiding walk* and is used to model polymer conformations. But anomalous asymptotics might already arise, without any memory, from strong enough *time correlations*; such correlations are observed, for instance, in a Markov chain whose transition matrix is quasi-reducible, with eigenvalues very close to 1 hence associated with slowly relaxing modes. In this case, time correlations are emergent features of the whole stationary phase-space distribution of the dynamics.

Let us investigate in the present subsection this general situation by considering a sequence $\{X_i\}$ of correlated variables, notwithstanding the origin of their correlations. In the case of correlated variables, one of the validity conditions of the standard central limit theorem is not satisfied hence we cannot a priori expect normal behavior: the asymptotic behavior of $\sum_{i=1}^{N} X_i$ has to be reconsidered and it will possibly exhibit anomalous features. We show here how the RG allows us to derive rigorously novel limit theorems in this context while revealing the criterion separating normal and critical behavior.

To quantify the correlations, a meaningful and efficient notion is that of *correlation length* $\xi$ or correlation time in dynamic contexts. It is defined as the characteristic length of the correlation function, typically given by the exponential decay rate of correlations in the case of a statistically homogeneous sequence:

$$C(j-i) \equiv \langle X_i X_j \rangle - \langle X_i \rangle \langle X_j \rangle \sim C_0 \, \mathrm{e}^{-|i-j|/\xi}. \tag{8.79}$$

$\xi$ can be understood qualitatively as a *coherence length*, that is, the characteristic size of regions that behave as a whole, each like a single block, as regards their relaxation to equilibrium, their response to external forcings or perturbations, and their interactions with adjacent regions.

If $\xi$ is finite, the correlation function actually exhibits an exponential decay over a distance of order $\xi$ and correlations are accordingly termed *short range*. $\xi$ gives the relevant scale for defining the elementary units in the most meaningful and efficient way: the best choice is to identify the elementary units with subsystems of linear size $\xi$. Smaller subsystems are strongly correlated and never seen in isolation; larger ones would involve unnecessary averaging and associated loss of resolution and information. A typical example is encountered in conformational studies of a linear polymer chain: the monomers are identified with segments of the size of the *Kuhn length* (accounting on purely geometrical grounds for the combinatorial variability of the angles between successive monomers along the chain) or at a larger scale, aimed at a more integrated effective description, with segments of the size of the *persistence length* (accounting for the thermal fluctuations of the monomer relative orientations). Below we recover this interpretation of $\xi$ more rigorously by computing the renormalized step-length of a short-range correlated random walk.

If $\xi$ is infinite, the system contains correlated regions of any size. In consequence, there is no way to reduce the system to an assembly of independent or short-range correlated units by choosing a proper scale at which to define the units. At each scale, strong correlations remain between the blocks and the RG should enter the scene, integrating these correlations in a recursive way into the definition of larger blocks and their effective features, up to the whole system.

Although intuitive, the distinction between finite versus infinite correlation lengths does not exactly match the boundary between normal and anomalous statistical behaviors, rather it is related to the integrability of the correlations. Indeed, considering a sequence $\{X_i\}$, stationary in the sense that the covariance of $X_i$ and $X_j$ is a function $C(j-i)$ of $|j-i|$, the variance of $S_N = \sum_{i=1}^N X_i$ is written:

$$\mathrm{Var}(S_N) = \sum_{q=1-N}^{N-1} C(q)(N-1-|q|). \tag{8.80}$$

If the correlations are integrable, that is, $\sum_{q=-\infty}^{+\infty} |C(q)| < \infty$, the variance leading order as $N \to \infty$ is easily estimated (with $t = N\tau$ where $\tau$ is the actual time step):

$$\mathrm{Var}(S_N) \sim N \sum_{q=-\infty}^{+\infty} C(q) = 2t\, D_{\mathrm{eff}} \quad \text{with} \quad D_{\mathrm{eff}} = \frac{1}{2\tau} \sum_{q=-\infty}^{+\infty} C(q). \tag{8.81}$$

This case is encountered if $\xi < \infty$, or if $\xi = \infty$ and correlations exhibit an integrable power-law decay $C(q) \sim |q|^{-\gamma}$ with $\gamma > 1$. One consistently recovers the expression $D = \mathrm{Var}(X)/2\tau$ in the case of i.i.d. variables. This shows that integrable correlations do not destroy the normal behavior, only turning the diffusion coefficient into a renormalized one. The sum $\sum_{q=-\infty}^{+\infty} C(q)$ can be interpreted as the square step-length of an *asymptotically equivalent ideal random walk*, that is, a random walk with independent steps and satisfying the same diffusion law. Considering the random walk with a coarser resolution $k$, the integration of short-range correlations into the effective size of $k$-steps is highly reminiscent of the block-spin RG and indeed provides the basis of a renormalization transformation.

The definition of RG operator, $\mathcal{R}p(x) = \sqrt{2}\,(p \otimes p)(x\sqrt{2})$, used in the central limit theorem derivation relies explicitly on the statistical independence of variables in the sequences under investigation. Otherwise, there would be no relation between the distribution of $\sum_{i=1}^N X_i$ and $\mathcal{R}p$ and no way to relate the fixed points and asymptotic behavior of the RG flow with the statistical law describing the collective behavior of the random variables $\{X_i\}$, through the size-dependence of their sum. Accordingly, in the case of correlated random variables, as already illustrated in Section 8.5.5, a different operator $\mathcal{R}$ should be devised. It should account for the integration of correlations within a subsequence $(X_i)_{i=1\ldots k}$ of length $k$ (that is, correlations of range smaller than $k$) into effective parameters of the renormalized distribution for the aggregated variable $\sum_{i=1}^{k} X_i$. When iterated to infinity, this procedure integrates recursively the influence of correlations at all scales, and hence gives access to their overall contributions to the collective statistical behavior, namely the scaling properties of $\sum_{i=1}^{N} X_i$ with respect to $N$.

Possible asymptotic behaviors for the sum of correlated random variables (assumed to be centered, with no loss of generality) are of four types.

(1) In the case of integrable correlations, the probability distribution function of $N^{-1/2}(\sum_{i=1}^{N} X_i)$ converges to a Gaussian distribution with a *renormalized variance* $\sigma_{\text{eff}}^2 \neq \sigma^2$ fully encapsulating the net large-scale influence of correlations ($\sigma_{\text{eff}}^2 > \sigma^2$ in the case of positive correlations).
(2) In the case of stronger correlations, an *anomalous rescaling* is required to obtain a well-defined asymptotic behavior: $N^{-\nu} \sum_{i=1}^{N} X_i$ converges to a Gaussian distribution for a unique appropriate value of the rescaling exponent $\nu$ ($\nu > 1/2$ in the case of positive correlations).
(3) A still more drastic discrepancy might occur, where even a modified rescaling is not enough to recover a well-defined asymptotic behavior, and a novel statistical law emerges, stating the convergence in law of $N^{-\nu}(\sum_{i=1}^{N} X_i)$ toward a certain *non-Gaussian distribution*.
(4) The fourth case covers all the remaining situations, where no well-defined non-trivial behavior emerges (convergence to 0 of the rescaled sum $N^{-\nu} \sum_{i=1}^{N} X_i$ if $\nu > \nu_c$, divergence if $\nu < \nu_c$, and no regular behavior as $N \to \infty$ for $\nu = \nu_c$).

In the case of correlated variables, the sum $S(t) = \sum_{i=1}^{t} X_i$ is an aggregated variable and its probability distribution function will be an emergent feature encapsulating the influence of correlations between all variables $(X_s)_{1 \le s \le t}$, that is, of range smaller than $t$. Keeping the condition of finite variance and the Gaussian character for the RG fixed point $p^*(x, t)$, as defined in (8.66), we obtain the family of *fractal Brownian motions* (associated with the names of Hurst and Mandelbrot) (Lesne 1998).

### *8.5.7 From lattice random walks to the Wiener process using the renormalization group*

A marginal but nevertheless highly fruitful application of the RG in this probabilistic context is to bridge random walks and continuous stochastic processes. This illustrates, in the special case of diffusion processes, the power of the RG to unify discrete and continuous models within the same universality class and, more generally, to bridge discrete and continuous models accounting for the same large-scale phenomena. The RG offers a rigorous way to pass from discrete models to their continuous limits and, conversely, to validate discretization of continuous models.

Trying to bridge lattice random walks with a continuous description brings out a difficulty: as the step size $a$ goes to 0, one has to decrease accordingly the duration $\tau$, but choosing by which amount is not easy, since the walker velocity is ill defined (it depends on the observation scale). The proper joint rescaling can be guessed from previously obtained knowledge about the system; it can also be obtained in a systematic way using RG methods. Let us implement this second approach in the case of normal diffusion (see Section 8.5.3).

In the context of random walks, renormalization expresses the consequence of a joint rescaling of space (by a factor of $K$) and time (by a factor of $k$) at the level of the transition probabilities $P_{a,\tau}(x, y, t)$ (density of the probability of jumping from $x$ to $y$ in time $t$) ruling the random walk (Lesne 1998):

$$[R_{K,k} P_{a,\tau}](x, y, t) \equiv K^d P_{a,\tau}(Kx, Ky, kt) \quad \text{in dimension } d, \tag{8.82}$$

where $x$, $y$ are restricted to the lattice $(a\mathbb{Z})^d$ and the time to $\tau\mathbb{N}$. It should be noted that the variables involved in $R_{K,k} P_{a,\tau}$ run over a finer lattice, $x, y \in [(aK)\mathbb{Z}]^d$, with a shorter time step, $t \in (\tau/k)\mathbb{N}$. The mean-square displacement is expressed as $\mathcal{D}(P, t) \equiv [\sum_x \sum_y |x - y|^2 P(x, y, t)]$, hence is transformed according to:

$$\mathcal{D}(R_{K,k} P_{a,\tau}, t) = \frac{1}{K^2} \mathcal{D}(P_{a,\tau}, kt). \tag{8.83}$$

Taking $K = k^\nu$ (thus recovering the operator $\mathcal{R}_{\nu,k}$ used throughout this chapter), the limit $\lim_{k\to\infty} \mathcal{R}_{\nu,k} P_{a,\tau}$, if it exists, will be a continuous transition probability $P^*(x, y, t)$ defined on $\mathbb{R}^d \times \mathbb{R}^d \times \mathbb{R}$. The procedure[11] for transition probabilities is then identical to that developed for probability distribution functions in Section 8.5.6: if $\lim_{k\to\infty} \mathcal{R}_{\nu,k} P_{a,\tau} = P^*$, then

$$\mathcal{D}(P^*, t) = \frac{1}{k^{2\nu}} \mathcal{D}(P^*, kt) \quad \text{hence} \quad \mathcal{D}(t) \sim t^{2\nu}. \tag{8.84}$$

This procedure can be implemented with $\nu = 1/2$ in the case of ideal random walks (independent and identically distributed steps). The self-similarity of these walks ensures the existence of a limit $P^*$, which is the transition probability of a *Wiener process* with diffusion coefficient $D = a^2/2\tau$, namely, in dimension $d$:

$$P_D^*(x, y, t) = \frac{1}{(4\pi d D t)^{-d/2}} \, \mathrm{e}^{-(x-y)^2/4dDt}. \tag{8.85}$$

This shows that *all ideal lattice random walks belong to the same universality class*, that of the Wiener process.

[11] The set $(R_{K,k})_{K\geq 0, k\geq 0}$ is a bona fide continuous renormalization group with a Lie-group structure: $R_{K_1,k_1} \circ R_{K_2,k_2} = R_{K_1K_2,k_1k_2}$. In consequence, it would have been sufficient to investigate the action of the two infinitesimal generators $(\partial R_{K,k}/\partial k)|_{K=1,k=1}$ and $(\partial R_{K,k}/\partial K)|_{K=1,k=1}$ spanning its Lie algebra (that is, its tangent vector space).

The link developed here between lattice random walks and continuous diffusion processes relies on the *self-similarity* of diffusion. This feature, already acknowledged by Perrin on experimental grounds, is a strong property, allowing us to bridge microscopic and macroscopic dynamics: as soon as a normal diffusive motion is observed, the diffusion law $R^2(t) \sim Dt$ involves one and the same diffusion coefficient $D$, whatever the observation scale is (except of course if some crossover occurs in the features of the underlying medium). In consequence, the best description of diffusive motion is the self-similar continuous Wiener process, that embeds any lattice random walk. We have shown here that, conversely, a lattice random walk supplemented with the self-similarity assumption (that is, invariance upon renormalization) allows us to recover the continuous process as an RG fixed-point, hence full knowledge at any scale.

In a more intricate setting where no simpler alternative approaches are available, this RG approach has been fruitfully applied to diffusion in disordered systems, the issue being to determine whether or not the disorder, accounted for as a noise term in the transition probabilities, modifies the normal diffusion law obtained in the unperturbed situation (Bricmont and Kupiainen 1991).

### 8.5.8 A statistical view of criticality and anomalous diffusion

A great interest of this probabilistic viewpoint is to reveal the origin of criticality and more generally anomalous behaviors as being a *"statistical catastrophe"* associated with the failure of the central limit theorem and even, possibly, of the weaker law of large numbers. These two theorems describe a "normal" situation, in which bounded and weakly correlated microscopic events build up a deterministic macroscopic behavior with moderate Gaussian fluctuations, whose relative amplitude behaves as $1/\sqrt{N}$ where $N$ is the size of the system (e.g. the number of particles). They appear as the cornerstone supporting the relation between statistical mechanics and thermodynamics and underlying mean-field approaches. As discussed in Sections 8.5.4 and 8.5.6, departure from this standard thermodynamic behavior arises in two typical instances.

(1) Elementary events $X$ (uncorrelated or weakly correlated) have arbitrarily large strength, i.e. an *infinite variance* $\langle X^2 \rangle = \infty$, hence direct macroscopic consequences. In this case, the partial average $\int_0^{x_0} x^2 P_X(x)\mathrm{d}x$ is dominated by $x_0$ (whereas it becomes independent of large enough $x_0$ if $\langle X^2 \rangle < \infty$). Similarly, the net result of an accumulation of individual events will be dominated by the largest (their rarity here being not low enough to balance their strength) rather than by those of mean size. In other words, *the notion of "typical event" is irrelevant.*
(2) Elementary events (still having a finite variance) are *strongly correlated*; then any minor change of one of these elements propagates to the whole, again with macroscopic

consequences. This is reflected in the divergence of the correlation length, and accordingly that of the correlation time, since fluctuations correlated at all scales will require a diverging time to relax to 0.

In the first case, one speaks of anomalous behavior. Criticality is currently more associated with the second instance. It thus reflects a qualitative change in the collective behavior due to the reorganization of the correlations between the elements that occurs through the emergence of a statistical singularity (thus strictly observed only in an infinite-size system). In both cases, but for different reasons, a single event might rule the whole behavior, i.e. *a microscopic event might have macroscopic consequences*, that obviously prevents any decoupling between microscopic and macroscopic scales. An RG approach is required to give account of the large-scale collective behavior. In the first case, the core of the RG procedure is an anomalous rescaling. In the second instance, the RG proceeds via an iterated integration of correlations and a rescaling of the lengths, thus decreasing the apparent correlation length and achieving a reduction in the relevant degrees of freedom of the system. In both cases, the RG tames or circumvents the singularity in an adaptive way: the transformation allowing this reduction itself brings information about the collective behavior and associated scaling laws.

This once more puts forward the common core of the RG: a *self-consistent* procedure to determine anomalous features, e.g. anomalous exponents that cannot be determined straightforwardly by a plain dimensional analysis, designed to reveal a non-trivial multiscale organization of the system, jointly in space, phase space and time. Only special values of the exponents involved in the joint rescaling and coarse graining (in other words, a consistent choice of joint rescaling factors) lead to a non-trivial limiting behavior. More generally, the RG points at the possible origins of anomalous large-scale behaviors and underlines their universality: they can be explained on *purely statistical grounds, irrespective of the specific physical features* (nature of the elements, mechanisms and detailed expression of the interactions) and involving only generic conditions about the dimension of the real space, the dimension of the phase space, or the interaction range.

Although a Wiener process trajectory (Brownian motion) exemplifies a fractal curve with no characteristic length, its criticality is less strong than for strongly correlated diffusion processes. Indeed, the normal diffusion law follows from a mere dimensional analysis of the macroscopic diffusion equation or, at the microscopic level, from a mere application of an averaging procedure. It corresponds to the *similarity of the first kind* according to the classification of Barenblatt (1987). In contrast, the exponents arising in the case of long-range correlations, for instance those associated with the so-called fractal Brownian motions, do not follow from dimensional analysis; by no means can they be guessed and their determination requires sophisticated RG recursive averaging. The first instance corresponds to what could be called *mean-field criticality*, also observed in a purely temporal context at a bifurcation point. The second case corresponds to the *similarity of the second kind* and fails to satisfy any mean-field approach; it can be termed *anomalous criticality*. It is also encountered at the transition to chaos (Section 8.3.3).

In conclusion, the RG formulated within a probabilistic framework offers a systematic tool to derive a wealth of *new limit theorems*, replacing the celebrated central limit theorem in critical or other anomalous situations where it fails to apply. The *central limit theorem appears as the mean-field, normal behavior* whereas criticality departs from its range of validity owing to long-range correlations arising between elementary units. Similarly, Brownian motion appears as the normal diffusion process, contrasting with anomalous diffusion processes originating in an anomalous distribution of the elementary events or their strong correlations.

## 8.6 Conclusions and perspectives

Let us summarize the RG features we have focused on in this chapter.

- We presented the RG historical approach in non-linear physics, for *curing singular perturbation series* (Section 8.2) in the same context, aim and spirit as multiscale approaches (presented in Chapter 7) but in a more systematic way;
- Then we gave a brief account of the standard RG devised to *capture criticality* and associated scaling laws (Section 8.3).
- We underlined that far beyond the scope of critical phenomena, the RG offers a generalized and systematic multiscale analysis to derive long-time and large-scale behaviors and provides a systematic means of *extracting large-scale features* and *reducing the number of degrees of freedom* (Section 8.4).
- We detailed how the RG provides a constructive and rigorous way to establish *new limit theorems and statistical laws*, beyond the central limit theorem (Section 8.5).

In conclusion, we shall first underline the invaluable contribution of the RG on the epistemological level, changing significantly our way of considering and devising models, hence deeply changing our theoretical approach to real systems. The RG is not only a powerful calculational tool for determining asymptotic behaviors and their scaling exponents. It is also associated with a deep frameshift in our way of analyzing and modeling real phenomena. Indeed, the RG does not analyze or compare solutions of a given model but *it acts at the level of a set of models*. Implementing an RG procedure amounts to studying a flow in a space of models, where the role of the time variable is played by the rescaling factor. In other words, it amounts to investigating quantitatively the *relationships between models at different scales*.

Since $\mathcal{R}$ acts upon the representation of the system (that is, its models) while preserving the system itself (that is, its physical reality), it should also preserve its symmetries. In order that the original model and its renormalized versions share the same symmety properties, the renormalization operator has to commute with all symmetry transformations preserving the system. Accordingly, symmetries appear as additional constraints, but also as a guideline in defining $\mathcal{R}$ and the relevant subspace of models in which to consider its action. The RG is nothing but a special symmetry group expressing the system scale invariance.

RG approaches benefit, in a constructive way, from the remarkable feature of scale invariant hierarchical structures and phenomena: *the fact that their emergent features are encapsulated in the inter-level relationship and in the equations expressing their scale invariance.* The RG provides a somehow "transverse" view, neither microscopic nor macroscopic but capturing the multiscale organization as a whole and turning its consistency into quantitative predictions. In consequence, the RG applies when there are simple relationships between the properties at different space, phase space and time scales. It allows us to capture recursively the whole multiscale organization of the system as regards its influence on large-scale, asymptotic observable behavior.

In multiscale approaches (Chapter 7), large-scale equations emerge under the form of a *solvability condition*, expressing a posteriori the consistency of the a priori decomposition in small-scale and large-scale contributions, i.e. ensuring that the actual characteristic scales coincide with the assumed ones. We have seen in the present chapter that in RG approaches, the large-scale descriptions arise, basically in a similar spirit, from *renormalization equations* showing that large-scale consequences of small-scale mechanisms are thoroughly accounted for in the effective parameters and other renormalized quantities, wherever the boundary between small scales and large scales lies. In this regard, RG methods appear as *systematic and constructive procedures* to derive macroscopic models from microscopic ones.

RG theory today offers several perspectives and novel directions to be explored further. For instance, it seems promising to investigate the physical interpretation of other RG flow features (other than fixed points), to generalize the renormalization group analysis using recent extensions of Lie group theory, to develop the RG on the basis of more general covariance properties, and to extend the RG into an assembly of local versions in the spirit of gauge theory extending global symmetries into space-dependent local versions.

## References

Barenblatt, G. I. (1987). *Dimensional Analysis*, New York: Gordon & Breach.
Bricmont, J. and Kupiainen, A. (1991). Renormalization group for diffusion in a random medium, *Phys. Rev. Lett.* **66**, 1689.
Brown, L. M. (1993). *Renormalization. From Lorentz to Landau (and Beyond)*, Berlin: Springer.
Cencini, M., Mazzino, A., Musacchio, S. and Vulpiani A. (2006). Large-scale influences on meso-scale modelling for scalar transport, *Physica D* **220**, 146.
Chen, L. Y., Goldenfeld N. and Oono, Y. (1994). Renormalization group theory for global asymptotic analysis, *Phys. Rev. Lett.* **73**, 1311.
Collet, P. and Eckmann, J. P. (1981). *Iterated Maps of the Interval as Dynamical Systems*, Basel: Birkhaüser.
Feigenbaum, M. J. (1978). Quantitative universality for a class of nonlinear transformations, *J. Stat. Phys.* **19**, 29.
Fisher, M. E. (1974). The renormalization-group and the theory of critical behavior, *Rev. Mod. Phys.* **46**, 597.
Fisher, M. E. (1998). Renormalization group theory: its basis and formulation in statistical physics, *Rev. Mod. Phys.* **70**, 653.
Gnedenko, B. V. and Kolmogorov, A. N. (1954). *Limit Distributions for the Sums of Independent Random Variables*, Cambridge, MA: Addison-Wesley.
Goldenfeld, N. (1992). *Lectures on Phase Transitions and the Renormalization-Group*, Reading, MA: Addison-Wesley.
Hatta, Y. and Kunihiro, T. (2002). Renormalization group method applied to kinetic equations: roles of initial values and time, *Ann. Phys.* **298**, 24.
Jona-Lasinio, G. (1975). The renormalization-group: a probabilistic view, *Nuovo Cimento B* **26**, 99.
Jona-Lasinio, G. (2001). Renormalization-group and probability theory, *Phys. Rep.* **352**, 439.
Jona-Lasinio, G. (2003). Self-similar random fields: from Kolmogorov to renormalization group, in *The Kolmogorov Legacy in Physics*, ed. Livi, R. and Vulpiani, A., p. 213, Berlin: Springer.
Kadanoff, L. P. (1966). Scaling laws for Ising models near $T_c$, *Physics* **2**, 263.
Kunihiro, T. and Tsumura, K. (2006). Application of the renormalization-group method to the reduction of transport equations, *J. Phys. A* **39**, 8089.
Lesne, A. (1998). *Renormalization Methods*, New York: Wiley.
Lévy, P. (1954). *Théorie de l'Addition des Variables Aléatoires*, Paris: Gauthiers-Villars, Paris; republished by J. Gabay, Paris, 1992.
Lindstedt, A. (1882). Ueber die integration einer fur die strorungstheorie ichstigen Differentialgleichung, *Astron. Nachr.* **103**, 211.
Mazzino, A., Musacchio, S. and Vulpiani, A. (2004). Multiple-scale analysis and renormalization for preasymptotic scalar transport, *Phys. Rev. E* **71**, 011113.
Nayfeh, A. H. (1973). *Perturbation Methods*, New York: Wiley.
Oono, Y. (2000). Renormalization and asymptotics, *Int. J. Mod. Phys. B* **14**, 1327.
Plischke, M. and Bergersen, B. (1994). *Equilibrium Statistical Physics*, Singapore: World Scientific.
Rayleigh, Lord (1917). On the reflection of light from a regularly stratified medium, *Proc. R. Soc. London, Ser. A.* **93**, 565.
Stanley, H. E. (1999). Scaling, universality, and renormalization: three pillars of modern critical phenomena, *Rev. Mod. Phys* **71**, S358.

Stueckelberg, E. C. G. and Petermann, A. (1953). La normalisation des constantes dans la théorie des quanta, *Helv. Phys. Acta* **26**, 499.
Wilson, K. G. (1971). Renormalization group and critical phenomena, *Phys. Rev. B* **4**, 3174.
Wilson, K. G. (1975). Renormalization-group methods, *Adv. Math.* **16**, 170.

# Index

WITHDRAWN
Carleton College Library